2006
INTERNATIONAL BUILDING CODE®

STUDY COMPANION

INTERNATIONAL
CODE COUNCIL®

2006 International Building Code

Study Companion

ISBN-13: 978-1-58001-524-0
ISBN-10: 1-58001-524-7

Publications Manager: Mary Lou Luif
Project Editor: Roger Mensink
Manager of Development: Doug Thornburg

First Printing: January 2007
Second Printing: April 2007
Third Printing: February 2008
Fourth Printing: February 2009
Fifth Printing: February 2012

Printed in the United States of America

TABLE OF CONTENTS

INTRODUCTION

This study companion provides practical learning assignments for independent study of the provisions of the 2006 *International Building Code®* (IBC®). The independent study format affords a method for the student to complete the study program in an unlimited amount of time. Progressing through the workbook, the learner can measure his or her level of knowledge by using the exercises and quizzes provided for each study session.

The workbook is also valuable for instructor-led programs. In jurisdictional training sessions, community college classes, vocational training programs and other structured educational offerings, the study guide and the IBC can be the basis for code instruction.

All study sessions begin with a general learning objective, the specific sections or chapters of the code under consideration and a list of questions summarizing the key points of study. Each session addresses selected topics from the IBC and includes code text, a commentary on the code provisions and illustrations representing the provisions under discussion. Quizzes are provided at the end of each study session. Before beginning the quizzes, the student should thoroughly review the referenced IBC provisions, particularly the key points.

The workbook is structured so that after every question the student has an opportunity to record his or her response and the corresponding code reference. The correct answers are indicated in the back of the workbook in the answer key.

This study companion was initially developed by Douglas W. Thornburg, AIA, C.B.O., for the 2000 *International Building Code.* In this publication, he has updated and revised the material based on the 2006 *International Building Code.* Mr. Thornburg is currently Technical Director of Product Development for the International Code Council®. In addition to authoring numerous educational texts and resource materials, he instructs seminars nationally on both the IBC and IRC®. Mr. Thornburg has over 25 years experience in the application and enforcement of building codes. Technical assistance on this publication was provided by Steve Van Note, C.B.O., ICC Senior Technical Staff.

Questions or comments concerning this workbook are encouraged. Please direct your comments to ICC at *studycompanion@iccsafe.org*.

2006 IBC Chapters 1, 34 and 35
Administration

OBJECTIVE: To obtain an understanding of the administrative provisions of the *International Building Code®* (IBC®), including the scope and purpose of the code, duties of the building official, issuance of permits, inspection procedures, special inspections, existing buildings and referenced standards.

REFERENCE: Chapters 1, 34 and 35, 2006 *International Building Code*

KEY POINTS:
- What is the purpose and scope of the *International Building Code*?
- What are the limitations for use of the *International Residential Code®* (IRC®)?
- When materials, methods of construction or other requirements are specified differently in separate provisions, which requirement governs?
- When there is a conflict between a general requirement and a specific requirement, which provision is applicable?
- When do the provisions of the appendices apply?
- What are the powers and duties of the building official in regard to the application and interpretation of the code? Right of entry?
- Under which conditions may the building official grant modifications to the code?
- How may alternative materials, designs and methods of construction be approved?
- When is a permit required? What types of work are exempted from permits?
- Is work exempted from a permit required to comply with the provisions of the code?
- What is the process outlined for obtaining a permit?
- Which documents must be submitted as part of the permit application process? What information is required on the construction documents?
- Which conditions or circumstances would bring the validity of a permit into question?
- When does a permit expire? What must occur when a permit expires prior to completion of a building?
- For what is a registered design professional responsible? When is such an individual required?
- What submittal documents must be provided with each permit application? Under what conditions are submittal documents not required?

KEY POINTS:
(Cont'd)

- What information is required on the construction documents? What special requirements apply to fire protection system shop drawings, means of egress layouts and exterior wall envelopes?
- What specific information must be included on a site plan?
- What is the process for the review and approval of construction documents? How are phased approvals to be addressed?
- What is the role of a design professional in responsible charge? When is such an individual to be utilized?
- What is a deferred submittal? What is the process for deferring the submittal of construction documents?
- What is the time limit of a permit for a temporary structure? To what criteria must a temporary structure comply?
- How are permit fees to be established? How is work started prior to permit issuance addressed?
- What types of inspections are specifically required by the code? When are inspections required?
- Are third-party inspection agencies permitted to perform the required inspections?
- Who is responsible for notifying the building official that work is ready for an inspection? Who must provide access for the inspection to take place?
- How shall an inspection be recorded? What if the inspection is failed?
- When is a certificate of occupancy required? For what reasons is revocation permitted?
- When may a temporary certificate of occupancy be issued?
- What is the purpose of a board of appeals? Who shall serve on the board?
- Which limitations are placed upon the authority of the board of appeals?
- When should a stop work order be issued?
- How must an existing building be viewed when additions or alterations are made?
- How are nonstructural repairs to an existing building to be handled? Glass replacement?
- What should happen when there is a change in use or occupancy of an existing building?
- How should moved structures be addressed?
- What accessibility issues are encountered when an existing building is altered or when its use changes?
- How can compliance alternatives be used to evaluate an existing building's degree of safety?
- What are referenced standards? How are they used in the IBC?

Code Text: *The provisions of the* International Building Code *shall apply to the construction, alteration, movement, enlargement, replacement, repair, equipment, use and occupancy, location, maintenance, removal and demolition of every building or structure or any appurtenances connected or attached to such buildings or structures.* See exception for dwellings and townhouses regulated by the *International Building Code.*

Discussion and Commentary: The *International Building Code* is intended to regulate the broad spectrum of construction activities associated with buildings and structures. General provisions allow for a comprehensive overview of regulations; however, where more specific circumstances exist, any applicable specific requirements will take precedence.

Fundamental purposes of the provisions of the *International Building Code:*

- Safety of building occupants
- Stop panic
- Safety of fire fighters and emergency responders
- Safety and protection of others' property
- Safety and protection of own property

The appendices of the *International Building Code* address a diverse number of issues that may be of value to a jurisdiction in developing a set of construction regulations. It should be noted that the provisions contained in the appendices do not apply unless specifically adopted.

Code Text: *Detached one- and two-family dwellings and multiple single-family dwellings (townhouses) not more than three stories above grade plane in height with a separate means of egress and their accessory structures shall comply with the* International Residential Code.

Discussion and Commentary: Many residential structures are exempt from the requirements of the *International Building Code* and are regulated instead by a separate and distinct document, the *International Residential Code*. The IRC® contains prescriptive requirements for the construction of detached single-family dwellings, detached duplexes, townhouses and all structures accessory to such buildings. Limited to three stories in height with individual egress facilities, these residential buildings are fully regulated by the IRC for building, plumbing, mechanical, electrical and energy provisions.

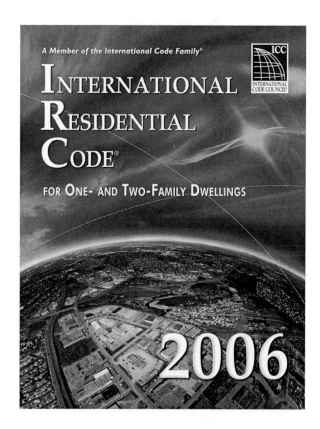

In addition to the general requirements, a townhouse is also limited by definition in the IRC. To fall under the scope of the IRC, each townhouse must be a single unit from the foundation to the roof, with at least two sides open to the exterior.

Topic: Application
Reference: IBC 101.2.1

Category: Administration
Subject: Appendix Chapters

Code Text: *Provisions in the appendices shall not apply unless specifically adopted.*

Discussion and Commentary: The appendix chapters of the IBC address subjects that have been deemed inappropriate as mandatory portions of the code. Rather, the appendices are optional, with each jurisdiction adopting all, some or none of the appendix chapters, depending on its needs for enforcement in any given area. There are various reasons why certain issues are found in the appendices. Often, the provisions are limited in application or interest. Some appendix chapters are merely extensions of requirements set forth in the body of the code. Others address issues that are often thought of as outside of the scope of a traditional building code. Whatever the reason, the appendix chapters are not applicable unless specifically adopted.

IBC Appendix Chapters

Appendix A	Employee Qualifications
Appendix B	Board of Appeals
Appendix C	Group U — Agricultural Buildings
Appendix D	Fire Districts
Appendix E	Supplementary Accessibility Requirements
Appendix F	Rodentproofing
Appendix G	Flood-Resistant Construction
Appendix H	Signs
Appendix I	Patio Covers
Appendix J	Grading
Appendix K	International Code Council Electrical Code Administrative Provisions

Appendix chapters, where not adopted as a portion of a jurisdiction's building code, may still be of value in application of the code. Provisions in the appendices might provide some degree of assistance in evaluating proposed alternative designs, methods or materials of construction.

Topic: Creation, Appointment and Deputies **Category:** Administration
Reference: IBC 103   **Subject:** Department of Building Safety

Code Text: *The Department of Building Safety is hereby created and the official in charge thereof shall be known as the building official. The building official shall be appointed by the chief appointing authority of the jurisdiction. In accordance with the prescribed procedures of this jurisdiction and with the concurrence of the appointing authority, the building official shall have the authority to appoint a deputy building official, the related technical officers, inspectors, plan examiners and other employees. Such employees shall have powers as delegated by the building official.*

Discussion and Commentary: The building official is an appointed officer of the jurisdiction and charged with the administrative responsibilities of the department of building safety. It is not uncommon for the jurisdiction to use a different position title to identify the building official, such as Chief Building Inspector, Superintendent of Central Inspection or Director of Code Enforcement. Regardless of the jurisdictional title, the code recognizes the individual in charge as the building official.

Inspectors, plan reviewers and other technical staff members are typically given some degree of authority to act for the building official in the decision-making process, including the making of appropriate interpretations of various provisions of the code.

Topic: General Requirements

Category: Administration

Reference: IBC 104.1

Subject: Duties and Powers of Building Official

Code Text: *The building official is hereby authorized and directed to enforce the provisions of the IBC. The building official shall have the authority to render interpretations of the IBC and to adopt policies and procedures in order to clarify the application of its provisions. Such interpretations, policies and procedures shall be in compliance with the intent and purpose of the IBC. Such policies and procedures shall not have the effect of waiving requirements specifically provided for in the IBC.*

Discussion and Commentary: It is important that the building official be knowledgeable to the point of being able to rule on those issues that are not directly addressed or that are unclear in the code. The basis for such a determination, which often takes some research to discover, is the intent and purpose of the *International Building Code*.

(Jurisdiction)

Department of Building Safety

Photo

_____
Name of individual

_____
Job function

The individual identified on the badge is a duly authorized employee of (the Jurisdiction) and is a designated representative of the Department of Building Safety.

Valid
through   _____   _____
                    Date   Building Official

Although the IBC gives broad authority to the building official in interpreting the code, this authority also comes with great responsibility. The building official must restrict all decisions to the intent and purpose of the code, with the waiving of any requirements being strictly prohibited.

Code Text: *The building official, member of the board of appeals or employee charged with the enforcement of the IBC, while acting for the jurisdiction in good faith and without malice in the discharge of the duties required by the IBC or other pertinent law or ordinance, shall not thereby be rendered liable personally and is hereby relieved from personal liability for any damage accruing to persons or property as a result of any act or by reason of an act or omission in the discharge of official duties.*

Discussion and Commentary: The protection afforded by the code regarding employee liability is limited to only those acts that occur in good faith. Absolute immunity from all tort liability is not provided where an employee acts maliciously. An employee is not relieved from personal liability where malice can be shown, and it is probable that the jurisdiction will not provide for the employee's defense.

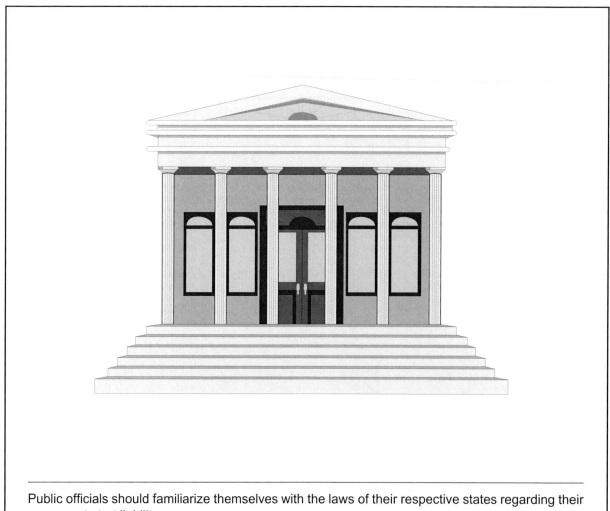

Public officials should familiarize themselves with the laws of their respective states regarding their exposure to tort liability.

Code Text: *The provisions of the IBC are not intended to prevent the installation of any material or to prohibit any design or method of construction not specifically prescribed by the IBC, provided that any such alternative has been approved. An alternative material, design or method of construction shall be approved where the building official finds that the proposed design is satisfactory and complies with the intent of the provisions of the IBC.*

Discussion and Commentary: The building official is granted broad authority in the acceptance of alternative materials, designs and methods of construction. Arguably the most important provision of the IBC, the intent is to implement the adoption of new technologies. Furthermore, it gives the code even more of a performance character. The provisions encourage state-of-the-art concepts in construction, design and materials, as long as they meet the performance level intended by the IBC.

Building official may approve alternative materials, design and methods of construction, if:

- Proposed alternative is satisfactory
- Proposed alternative complies with intent of code
- Material, method or work is equivalent in:
  - (1) Quality
  - (2) Strength
  - (3) Effectiveness
  - (4) Fire Resistance
  - (5) Durability
  - (6) Safety

Advisable for building official to:
- Require sufficient evidence or proof
- Record any action granting approval
- Enter information into the files

The building official should ensure that any necessary substantiating data or other evidence that shows the alternative to be equivalent in performance is submitted. Moreover, where tests are performed, reports of such tests must be retained by the building official.

Topic: Research Reports

Category: Administration

Reference: IBC 104.11.1

Subject: Duties and Powers of Building Official

Code Text: *Supporting data, where necessary to assist in the approval of materials or assemblies not specifically provided for in the IBC, shall consist of valid research reports from approved sources.*

Discussion and Commentary: The most familiar research reports are probably the "evaluation reports," identified as ICC Evaluation Service Reports and maintained by ICC Evaluation Service, Inc. ICC Evaluation Service Reports are developed based upon acceptance criteria for products that are alternatives to what is specified in the code and that are structural in nature and/or affect life safety. These reports are made available to building regulators, contractors, specifiers, architects, engineers and anyone else with an interest in the building industry or construction. The building official can also recognize the evaluation of products, assemblies and systems by other third-party agencies having the necessary credentials.

Although ICC Evaluation Service Reports are generally recognized nationally as valid reports developed by an approved source, the building official is the final authority on the acceptance of any research report for the purpose of accepting an alternate material, method or design.

Topic: Permits Required and Exempted	**Category:** Administration
Reference: IBC 105.1, 105.2	**Subject:** Permits

Code Text: *Any owner or authorized agent who intends to construct, enlarge, alter, repair, move, demolish, or change the occupancy of a building or structure . . . shall first make application to the building official and obtain the required permit.* See 13 exceptions where a building permit is not required. *Exemptions from permit requirements of* the IBC *shall not be deemed to grant authorization for any work to be done in any manner in violation of the provisions of* the IBC *or any other laws or ordinances of this jurisdiction.*

Discussion and Commentary: Except in those few cases specifically listed, such as small accessory structures and finish work, all construction-related work requires a permit and is subject to subsequent inspections.

Work exempt from permit:

- One-story detached accessory buildings limited to 120 square feet
- Fences not over 6 feet in height
- Oil derricks
- Retaining walls limited to 4 feet in height, unless supporting a surcharge or impounding Class I, II or III-A liquids
- Water tanks supported directly on grade, limited to capacity of 5,000 gallons and a ratio of height to diameter not exceeding 2 to 1
- Sidewalks and driveways limited to 30 inches above grade, not over any basement or story below, and not part of an accessible route
- Painting, papering, carpeting, cabinets, counter tops and similar finish work
- Temporary motion picture, television and theater stage sets and scenery
- Prefabricated swimming pools accessory to a Group R-3 occupancy when capacity is limited to 5,000 gallons, depth limited to 24 inches and installed entirely above ground
- Shade cloth structures used for nursery or agricultural purposes
- Swings and other playground equipment accessory to detached one- and two-family dwellings
- Window awnings supported by an exterior wall in Groups R-3 and U, where the maximum projection is 54 inches
- Movable fixtures, racks, cases, counters and partitions limited to 5 feet 9 inches in height

---

Whether or not a building permit is required by the code, it is intended that all work be done in accordance with the code requirements. The owner is responsible for all construction being done properly and safely.

Topic: Payment, Schedule and Valuations **Category:** Administration
Reference: IBC 108 **Subject:** Fees

Code Text: *A permit shall not be valid until the fees prescribed by law have been paid. On buildings, structures, electrical, gas, mechanical and plumbing systems or alterations requiring a permit, a fee for each permit shall be paid as required, in accordance with the schedule as established by the applicable governing authority. The applicant for a permit shall provide an estimated permit value at the time of application. Permit valuations shall include total value of the work, including materials and labor, for which a permit is being issued, such as electrical, gas, mechanical, plumbing equipment and other permanent systems.*

Discussion and Commentary: Fees are typically established at a level that will provide enough funds to adequately pay for the costs of operating the various building department functions, including administration, plan review and inspection. It is common that the fee schedule be based on the projected construction cost (valuation) of the work. Although there are several different methods for determining the appropriate valuation, it is important that a realistic and consistent approach be taken so that permit fees are applied fairly and accurately.

Fee schedules are not established by the code. Rather, the local jurisdiction is required to establish the appropriate fees to fund the operation of the building safety department.

Code Text: *Construction or work for which a permit is required shall be subject to inspection by the building official and such construction or work shall remain accessible and exposed for inspection purposes until approved. Approval as a result of an inspection shall not be construed to be an approval of a violation of the provisions of the IBC or of other ordinances of the jurisdiction.*

Discussion and Commentary: The inspection function is possibly the most critical activity in the entire code enforcement process. At the varied stages of construction, an inspector often performs the final check of the building for safety-related compliance. If necessary, the building official may require a preliminary inspection of the site, building or structure to gain information that may be of assistance in the issuance of a permit.

Required inspections (where applicable):

- Footing and foundation
- Concrete slab or under-floor
- Lowest floor elevation
- Frame
- Lath or gypsum board
- Fire-resistant penetrations
- Energy efficiency
- Others as required by the building official
- Special inspections
- Final

It may be necessary to have materials removed in order to provide inspection access or observation for a portion of the building. The responsibility rests with the permit applicant to make the work available for inspection, with no expenses to be borne by the jurisdiction.

Code Text: *Work shall not be done beyond the point indicated in each successive inspection without first obtaining the approval of the building official. The building official, upon notification, shall make the requested inspections and shall either indicate the portion of the construction that is satisfactory as completed, or notify the permit holder or his or her agent wherein the same fails to comply with the IBC. Any portions that do not comply shall be corrected and such portions shall not be covered or concealed until authorized by the building official.*

Discussion and Commentary: It is important that each successive inspection be approved prior to continuing further work. This practice helps to control the concealment of any work that must be inspected, resulting in the unnecessary removal of materials that might block access.

Any request for inspection is the responsibility of the building permit holders or their duly authorized agent. They must contact the building official when the work is ready for inspection, as well as provide access to that work.

Topic: Use and Occupancy

Reference: IBC 110.1

Category: Administration

Subject: Certificate of Occupancy

Code Text: *No building or structure shall be used or occupied, and no change in the existing occupancy classification of a building or structure or portion thereof shall be made until the building official has issued a certificate of occupancy therefor as provided herein. Issuance of a certificate of occupancy shall not be construed as an approval of a violation of the provisions of the IBC or of other ordinances of the jurisdiction.*

Discussion and Commentary: The certificate of occupancy is the tool with which the building official can regulate and control the uses and occupancies of the various buildings and structures within the jurisdiction. The code makes it unlawful to use or occupy a building unless a certificate of occupancy has been issued for that specific use.

The building official is permitted to suspend or revoke a certificate of occupancy for any of the following reasons: 1) when the certificate is issued in error, 2) when incorrect information is supplied, or 3) when the building is in violation of the code.

Code Text: *Additions or alterations of any building or structure shall conform with the requirements of* the IBC *for new construction. Additions or alterations shall not be made to an existing building or structure which will cause the existing building or structure to be in violation of any provisions of* the IBC. *Portions of the structure not altered and not affected by the alteration are not required to comply with the code requirements for a new structure.*

Discussion and Commentary: Any construction that takes place on a building will necessitate compliance with the building code in effect at the time of the work. However, only those portions being remodeled or added need comply. Existing portions of the building where no work is being performed are exempt from the current code requirements.

Conformance with the code in effect is also mandated for a change in use or occupancy. When approved by the building official, full conformance is not required where it is determined that the new use is less of a life-safety and fire-safety hazard than the previous use.

Code Text: *The provisions of Section 3410 are intended to maintain or increase the current degree of public safety, health and general welfare in existing buildings while permitting repair, alteration, addition and a change of occupancy without requiring full compliance with Chapters 2 through 33, or Sections 3401.3, and 3403 through 3407, except where compliance with other provisions of the IBC is specifically required in Section 3410. For repairs, alterations, additions and changes of occupancy to existing buildings that are evaluated in accordance with Section 3410, compliance with Section 3410 shall be accepted by the building official.*

Discussion and Commentary: A thorough investigation and evaluation of an existing building can be the basis for determining the level of compliance when an existing building is altered or repaired, or when the building or portion thereof undergoes a change in use. Based on the results of the evaluation process, it can be determined if the existing building, along with any modifications, is deemed to be in compliance with the IBC.

TABLE 3410.7
SUMMARY SHEET — BUILDING CODE

Existing occupancy _____ Proposed occupancy _____

Year building was constructed _____ Number of stories _____ Height in feet _____

Type of construction _____ Area per floor _____

Percentage of open perimeter _____% Percentage of height reduction _____%

Completely suppressed: Yes ____ No ____ Corridor wall rating _____

Compartmentation: Yes ____ No ____ Required door closers: Yes ____ No ____

Fire-resistance rating of vertical opening enclosures _____

Type of HVAC system _____, serving number of floors _____

Automatic fire detection: Yes ___ No ___, type and location _____

Fire alarm system: Yes ___ No ___, type _____

Smoke control: Yes ___ No ___, type _____

Adequate exit routes: Yes ___ No ___ Dead ends: _____ Yes ____ No ____

Maximum exit access travel distance _____ Elevator controls: Yes ____ No ____

Means of egress emergency lighting: Yes ___ No ___ Mixed occupancies: Yes ____ No ____

SAFETY PARAMETERS	FIRE SAFETY (FS)	MEANS OF EGRESS (ME)	GENERAL SAFETY (GS)
3410.6.1 Building Height 3410.6.2 Building Area 3410.6.3 Compartmentation			
3410.6.4 Tenant and Dwelling Unit Separations 3410.6.5 Corridor Walls 3410.6.6 Vertical Openings			
3410.6.7 HVAC Systems 3410.6.8 Automatic Fire Detection 3410.6.9 Fire Alarm System			
3410.6.10 Smoke control 3410.6.11 Means of Egress 3410.6.12 Dead ends	* * * * * * * * * * * *		
3410.6.13 Maximum Exit Access Travel Distance 3410.6.14 Elevator Control 3410.6.15 Means of Egress Emergency Lighting	* * * * * * * *		
3410.6.16 Mixed Occupancies 3410.6.17 Automatic Sprinklers 3410.6.18 Incidental Use		* * * * ÷ 2 =	
Building score — total value			

* * * *No applicable value to be inserted.

In evaluating an existing building undergoing a change in use, alteration or repair, the process addresses three issues: fire safety, means of egress and general safety. Each category must be deemed satisfactory in order for the existing building to be considered compliant.

Topic: Application

Reference: IBC Chapter 35

Category: Referenced Standards

Subject: Standards

Code Text: *This chapter lists the standards that are referenced in various sections of the IBC. The standards are listed herein by the promulgation agency of the standard, the standard identification, the effective date and title, and the section or sections of the IBC that reference the standard. The application of the referenced standards shall be as specified in Section 102.4.*

Discussion and Commentary: Limited in their scope, standards define more precisely the general provisions set forth in the code. The *International Building Code* references several hundred different standards, each addressing a specific aspect of building design or construction. In general terms, the standards referenced by the IBC are primarily materials, testing, installation or engineering standards.

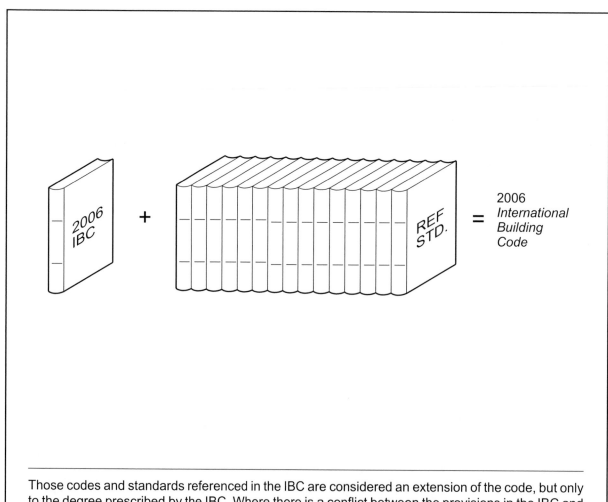

Those codes and standards referenced in the IBC are considered an extension of the code, but only to the degree prescribed by the IBC. Where there is a conflict between the provisions in the IBC and any referenced code or standard, the provisions of the IBC apply.

Quiz

Study Session 1
IBC Chapters 1, 34 and 35

1. The *International Residential Code* is applicable to townhouses a maximum of _____ above grade plane in height and provided with separate means of egress.

    a. 35 feet
    b. 40 feet
    c. three stories
    d. four stories

    Reference_____

2. Provisions of the appendix do not apply unless _____.

    a. specified in the code
    b. applicable to unique conditions
    c. specifically adopted
    d. relevant to fire or life safety

    Reference_____

3. If there is a conflict in the code between a general requirement and a specific requirement, the _____ requirement shall apply.

    a. general
    b. specific
    c. least restrictive
    d. most restrictive

    Reference_____

4. The _____ is considered by the code as the term to describe the individual in charge of the department of building safety.

    a. building official
        b. code official

    c. code administrator
        d. chief building inspector

Reference_____

5. The building official has the authority to _____ the provisions of the code.

    a. ignore
        b. waive

    c. modify
        d. interpret

Reference_____

6. Used materials may be utilized under which of the following conditions?

    a. They meet the requirements for new materials.

    b. They are limited to 10 percent of the total materials.

    c. Used materials may never be used in new construction.

    d. A representative sampling is tested for compliance.

Reference_____

7. The building official has the authority to grant modifications to the code _____.

    a. for only those issues not affecting life safety or fire safety

    b. for individual cases where the strict letter of the code is impractical

    c. where the intent and purpose of the code cannot be met

    d. related only to administrative functions

Reference_____

8. In order for an alternative material, design or method of construction to be considered acceptable, it must be equivalent to the code based on all but which of the following criteria?

    a. durability
        b. economics

    c. strength
        d. fire resistance

Reference_____

9. Tests performed by _____ may be required by the building official where there is insufficient evidence of code compliance.

    a. the owner   b. the contractor

    c. an approved agency   d. a design professional

Reference_____

10. A permit is not required for the construction of a one-story detached accessory structure when it has a maximum floor area of _____ square feet.

    a. 100   b. 120

    c. 150   d. 200

Reference_____

11. Movable fixtures, cases, counters and partitions are exempt from a building permit where they have a maximum height of _____.

    a. 5 feet, 0 inches   b. 5 feet, 6 inches

    c. 5 feet, 9 inches   d. 6 feet, 0 inches

Reference_____

12. Unless an extension is authorized, a permit becomes invalid when work on the site does not commence within _____ after permit issuance.

    a. 90 days   b. 180 days

    c. one year   d. two years

Reference_____

13. The building permit, or a copy of the permit, shall be kept _____ until completion of the project.

    a. at the job site

    b. by the permit applicant

    c. by the contractor

    d. by the design professional in responsible charge

Reference_____

14. When a building permit is issued, the construction documents shall be approved as _____.

    a. "Approved for Construction"

    b. "Conditional Approval"

    c. "Accepted as Reviewed"

    d. "Reviewed for Code Compliance"

Reference_____

15. Unless otherwise mandated by state or local laws, the approved construction documents shall be retained by the building official for a minimum of _____ from the date of completion of the permitted work.

    a. 90 days   b. 180 days

    c. 1 year   d. 2 years

Reference_____

16. Which one of the following inspections is not specifically identified by the *International Building Code* as a required inspection?

    a. footing inspection

    b. frame inspection

    c. soil classification

    d. energy efficiency inspection

Reference_____

17. Whose duty is it to provide access to work in need of inspection?

    a. the permit holder   b. the owner

    c. the contractor   d. the owner's agent

Reference_____

18. The certificate of occupancy shall contain all of the following information except:

    a. the name of the owner

    b. the name of the building official

    c. the type of construction

    d. the maximum height and area

Reference_____

19. A temporary certificate of occupancy is valid for what maximum period of time?

    a. 30 days   b. 60 days

    c. 180 days   d. a period set by the building official

Reference_____

20. The board of appeals is not authorized to rule on an appeal based on a claim that _____.

    a. the provisions of the code do not fully apply

    b. a code requirement should be waived

    c. the rules have been incorrectly interpreted

    d. a better form of construction is provided

Reference_____

21. Unless in compliance with the code for new structures, alterations to an existing structure are permitted to increase the force in any structural element by a maximum of _____percent.

    a. 0 (no increase is permitted)

    b. 5

    c. 10

    d. 25

Reference_____

22. Where permitted for an existing building, a fire escape must be designed to support a minimum live load of _____ pounds per square foot.

    a. 40   b. 50

    c. 100   d. 125

Reference_____

23. Where alterations occur to an existing building, an access ramp with a slope of 1:8 is permitted where necessitated by space limitations, provided the maximum rise of the ramp is _____ inches.

    a. 3   b. 6

    c. 12   d. 30

Reference_____

24. Which of the following categories is not specifically addressed in the evaluation of existing buildings for compliance alternatives?

    a.   fire safety   b.   means of egress

    c.   general safety   d.   structural safety

Reference_____

25. ACI 318-05 is a referenced standard addressing _____.

    a.   structural concrete   b.   wood construction

    c.   structural steel buildings   d.   gypsum board

Reference_____

26. A permit is required for a prefabricated above-ground swimming pool that is a minimum of _____ in depth and has a capacity of more than _____.

    a.   18 inches; 4,000 gallons   b.   24 inches; 5,000 gallons

    c.   30 inches; 5,000 gallons   d.   36 inches; 6,000 gallons

Reference_____

27. The installation of window awnings is exempt from a permit, provided the awnings each project a maximum of _____ inches from the exterior wall.

    a.   30   b.   36

    c.   48   d.   54

Reference_____

28. A permit may be suspended or revoked for all of the following reasons, except _____.

    a.   where it is issued in error

    b.   on the basis of incomplete information

    c.   where issued in violation of a jurisdictional ordinance

    d.   where other permits by the contractor have been voided

Reference_____

29. Unless extended by the building official, what is the maximum time period granted for a permit issued on a temporary structure?

   a. 90 days   b. 180 days

   c. 1 year   d. 2 years

   Reference_____

30. The provisions for compliance alternatives in the repair, alteration, addition and change of occupancy in existing buildings are applicable to all of the following occupancy classifications, except _____.

   a. Group A   b. Group E

   c. Group I   d. Group R

   Reference_____

31. A permit is not required for the installation of a self-contained refrigeration system, provided it contains a maximum of _____ pounds of refrigerant and is actuated by a maximum _____-horsepower motor.

   a. 5, 1   b. 10, 1

   c. 5, $1^1/_2$ d. 10, $1^1/_2$

   Reference _____

32. The means of egress layout required as a part of the construction documents for which of the following occupancies must include the number of occupants to be accommodated?

   a. F-1   b. I-1

   c. R-2   d. R-3

   Reference _____

33. The final permit valuation shall be determined by the _____.

   a. owner   b. building official

   c. design professional   d. general contractor

   Reference _____

34. Who is responsible for assuring that the work is accessible and exposed for inspection purposes?

    a. owner
    b. contractor

    c. permit applicant
    d. design professional

Reference _____

35. A stop work order shall be in writing and given to any of the following individuals except the _____.

    a. owner of the property involved

    b. owner's agent

    c. permit holder

    d. person doing the work

Reference _____

2006 IBC Chapter 3 and Section 508
Use and Occupancy Classification

OBJECTIVE: To gain an understanding of how an occupancy is classified based on its intended use and how a building with mixed uses is addressed.

REFERENCE: Chapter 3 and Section 508, 2006 *International Building Code*

KEY POINTS:
- What are the 10 general occupancy groups?
- How is a space that is intended to be occupied at different times for different purposes to be addressed?
- How is an occupancy that is not specifically described to be classified?
- Which types of activities are considered assembly uses? What is their general classification?
- How are small assembly uses classified where accessory to a different occupancy?
- What is the classification for restaurants and cafes? Theaters? Places of religious worship, conference rooms and libraries? Arenas? Grandstands?
- What is the primary use classified as Group B?
- Group E occupancies describe educational uses for individuals of what age group?
- Which types of day care are considered Group E occupancies?
- Manufacturing operations fall into what occupancy group? How do the two divisions of factory-use differ from each other?
- What type of operations or materials cause a use to be considered Group H?
- How does the amount of hazardous materials affect the occupancy classification?
- Which occupancies address physical hazards? Health hazards? Semiconductor fabrication facilities?
- Which characteristics are typical of a Group I occupancy?
- In which institutional occupancies are the occupants considered incapable of self-preservation?
- For which types of institutional uses may the *International Residential Code* be utilized?
- What general type of building is considered a Group M occupancy?
- How are residential occupancies classified?
- What is the key difference between a Group R-1 and Group R-2 occupancy?

- What is a congregate living facility? How should such a facility be classified?
- When is a residential use permitted to be constructed under the provisions of the *International Residential Code*?
- What do storage occupancy classifications have in common with those of manufacturing uses?
- How is a vehicle repair garage classified? An aircraft repair hangar?
- What is the classification of an enclosed parking garage? An open parking garage?
- What is a utility occupancy? How does its classification differ from that of other occupancies?
- What is an incidental use area? How must such an area be separated from the remainder of the building? When is sprinkler protection required?
- What is the option to utilizing Table 508.2 for the separation or protection of incidental use areas?
- What is an accessory occupancy? What benefit is derived from such a designation?
- Which other options are available for addressing multiple occupancies within a building?
- What is the concept of the nonseparated occupancy provisions? What conditions apply to buildings with nonseparated occupancies?
- What is the basis for separated occupancies? How are the minimum required fire-resistive separations determined?
- How is the fire-resistance rating for an occupancy separation determined? How does the presence of an automatic sprinkler system affect the required rating?

Code Text: *Structures or portions of structures shall be classified with respect to occupancy in one or more of the groups listed. Where a structure is proposed for a purpose which is not specifically provided for in* the IBC, *such structure shall be classified in the group which the occupancy most nearly resembles, according to the fire safety and relative hazard involved.*

Discussion and Commentary: The perils contemplated by the occupancy groupings are divided into two general categories: those related to people and those related to content. People-related hazards include the number and density of the occupants, their age and mobility, and their awareness of surrounding conditions. Content-related hazards include the storage and use of hazardous materials, as well as the presence of large quantities of combustible materials.

Assembly **B**usiness

Educational **F**actory

Hazardous **I**nstitutional

Mercantile **R**esidential

Storage **U**tility

Proper occupancy classification is critical in making appropriate code determinations throughout a project. In the classification process, the building official must use judgment in the determination of the potential hazards of an affected occupancy.

Code Text: *Assembly Group A occupancy includes, among others, the use of a building or structure, or a portion thereof, for the gathering of persons for purposes such as civic, social or religious functions, recreation, food or drink consumption or awaiting transportation.* See exceptions for assembly buildings and assembly spaces with an occupant load of less than 50.

Discussion and Commentary: The conditions related to a typical Group A occupancy suggest a moderate hazard use. This use often includes sizable numbers of people who are generally mobile and aware of the surrounding conditions. The extremely high occupant density level often present in an assembly occupancy is what distinguishes Group A from other occupancies. Where the occupant load of an assembly building does not exceed 50, a Group B classification is more appropriate owing to the lesser hazard.

Group A-1

Motion picture theaters
Theaters
Symphony and
  concert halls

Group A-2

Banquet halls
Night clubs
Restaurants
Taverns

Group A-3

Amusement arcades
Art galleries
Bowling alleys
Places of worship
Community halls
Conference rooms
Exhibition halls
Lecture halls
Libraries
Museums
Passenger stations

Group A-4

Arenas
Skating rinks
Swimming pools
Tennis courts

Group A-5

Amusement park
  structures
Bleachers
Grandstands
Stadiums

Unique conditions are represented by the classifications of Groups A-1, A-2, A-4 and A-5. However, the category Group A-3 includes a variety of broad and diverse assembly uses. It is not uncommon to find high combustible loading in Group A-3 occupancies.

Code Text: *Business Group B occupancy includes, among others, the use of a building or structure, or a portion thereof, for office, professional or service-type transactions, including storage of records and accounts.*

Discussion and Commentary: Business occupancies typically have a low to moderate fire load, a moderate density level, and occupants who are usually mobile and have a general awareness of the surrounding conditions. As such, business occupancies are grouped into a classification based upon a relatively moderate fire hazard level. Group B occupancies are not restricted by occupant load, as the number of people in a business use, such as an office, can range from one person to thousands of people.

Group B

Animal hospitals, kennels and ponds
Banks
Barber and beauty shops
Car wash
Civil administration
Clinic-outpatient
Educational occupancies above the 12th grade
Laboratories; testing and research
Motor vehicle showrooms
Post offices
Print shops
Professional services
Radio and television stations
Training and skill development

As is the case for many of the occupancy groups, a review of the building's intended uses is necessary to determine the amount of hazardous materials that may be stored, handled or used. If the amounts exceed a specified quantity, then a Group H classification will be in order.

Code Text: *Educational Group occupancy includes, among others, the use of a building or structure, or a portion thereof, by six or more persons at any one time for educational purposes through the 12th grade. The use of a building or structure, or portion thereof, for educational, supervision or personal care services for more than five children older than $2^1/_2$ years of age, shall be classified as a Group E occupancy.*

Discussion and Commentary: Educational occupancies address classroom uses for students of high school age and younger. Education facilities limited to use by older students, such as college classrooms, are classified as Group B occupancies; however, a Group A classification should be considered for lecture halls and similar large occupant load spaces.

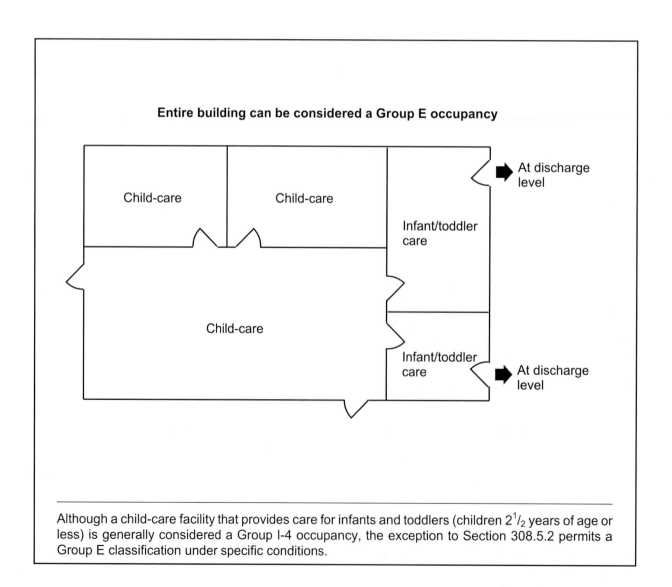

Although a child-care facility that provides care for infants and toddlers (children $2^1/_2$ years of age or less) is generally considered a Group I-4 occupancy, the exception to Section 308.5.2 permits a Group E classification under specific conditions.

Code Text: *Factory Industrial Group F occupancy includes, among others, the use of a building or structure, or a portion thereof, for assembling, disassembling, fabricating, finishing, manufacturing, packaging, repair or processing operations that are not classified as a Group H hazardous or Group S storage occupancy.*

Discussion and Commentary: Although the potential hazard and fire severity varies among the many uses categorized as Group F occupancies, the uses still share elements in common. The occupants are adults who are awake and who generally have enough familiarity with the premises to be able to exit the building with reasonable efficiency. The presence of combustible materials in the industrial process causes a classification of Group F-1, which is by far the most common factory use.

Group F-1

Aircraft
Appliances
Automobiles
Bakeries
Business machines
Carpets and rugs
Clothing
Electric generation
Plants
Electronics
Food processing
Furniture
Laundries
Millwork
Paper mills or products
Plastic products
Printing or publishing
Refuse incineration
Textiles
Woodworking

Group F-2

Brick and masonry
Ceramic products
Foundries
Glass products
Gypsum
Ice
Metal products

Classification as a Group F-2 occupancy is strictly limited because of the restrictions placed on such uses. The fabrication or manufacture of noncombustible materials, as well as the finishing, packaging or processing operations, cannot involve a significant fire hazard.

Code Text: *High-hazard Group H occupancy includes, among others, the use of a building or structure, or a portion thereof, that involves the manufacturing, processing, generation or storage of materials that constitute a physical or health hazard in quantities in excess of those allowed in control areas constructed and located as required in Section 414.*

Discussion and Commentary: There is only one fundamental type of Group H occupancy—that which is designated based solely on excessive quantities of hazardous materials contained therein. The quantities of hazardous materials that necessitate a Group H classification vary, based on the type, quantity, condition (use or storage) and environment of the materials. Where the use does not exceed the maximum allowable quantities set forth in the code, a classification other than Group H is appropriate.

Where Hazardous Materials and Processes are Involved

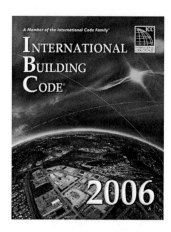

References for Detailed Provisions

Although the *International Building Code* is limited to general construction regulations and occupancy-specific requirements, the *International Fire Code*® (IFC®) sets forth special detailed provisions relating to hazardous materials and the specific conditions of their storage, use and handling.

Topic: Exceptions to Group H   **Category:** Occupancy Classification
Reference: IBC 307.1, Exceptions   **Subject:** Group H Occupancies

Code Text: *The following shall not be classified in Group H, but shall be classified in the occupancy which they most nearly resemble.* See a listing of 15 conditions under which a Group H occupancy is not warranted. *Hazardous materials in any quantity shall conform to the requirements of the IBC, including Section 414, and the* International Fire Code.

Discussion and Commentary: Although some degree of hazardous materials is found in most buildings, the occupancy is designated as Group H only where the quantities are excessive and the hazards are not adequately addressed. The most common condition for a non-H classification is where the amount of hazardous materials contained in the building does not exceed the maximum allowable quantities shown in Table 307.1(1) for physical hazards and Table 307.1(2) for health hazards. Footnotes to both tables can be used to increase the permitted quantities.

[F] TABLE 307.1(1)—continued
MAXIMUM ALLOWABLE QUANTITY PER CONTROL AREA OF HAZARDOUS MATERIALS POSING A PHYSICAL HAZARD[a, j, m, n, p]

MATERIAL	CLASS	GROUP WHEN THE MAXIMUM ALLOWABLE QUANTITY IS EXCEEDED	STORAGE[b]			USE-CLOSED SYSTEMS[b]			USE-OPEN SYSTEMS[b]	
			Solid pounds (cubic feet)	Liquid gallons (pounds)	Gas (cubic feet at NTP)	Solid pounds (cubic feet)	Liquid gallons (pounds)	Gas (cubic feet at NTP)	Solid pounds (cubic feet)	Liquid gallons (pounds)
Pyrophoric material	N/A	H-2	4[c, g]	(4)[c, g]	50[c, g]	1[g]	(1)[g]	10[c, g]	0	0
Unstable (reactive)	4	H-1	1[c, g]	(1)[c, g]	10[d, g]	0.25[g]	(0.25)[g]	2[c, g]	0.25[g]	(0.25)[g]
	3	H-1 or H-2	5[d, e]	(5)[d, e]	50[d, e]	1[d]	(1)	10[d, e]	1[d]	(1)[d]
	2	H-3	50[d, e]	(50)[d, e]	250[d, e]	50[d]	(50)[d]	250[d, e]	10[d]	(10)[d]
	1	N/A	NL	NL	N/L	NL	N/L	NL	NL	NL
Water reactive	3	H-2	5[d, e]	(5)[d, e]	N/A	5[d]	(5)[d]	N/A	1[d]	(1)[d]
	2	H-3	50[d, e]	(50)[d, e]	N/A	50[d]	(50)[d]	N/A	10[d]	(10)[d]
	1	N/A	NL	NL	N/A	NL	NL	N/A	NL	NL

For SI: 1 cubic foot = 0.023 m³, 1 pound = 0.454 kg, 1 gallon = 3.785 L.

NL = Not Limited; N/A = Not Applicable; UD = Unclassified Detonable

a. For use of control areas, see Section 414.2.
b. The aggregate quantity in use and storage shall not exceed the quantity listed for storage.
c. The quantities of alcoholic beverages in retail and wholesale sales occupancies shall not be limited providing the liquids are packaged in individual containers not exceeding 1.3 gallons. In retail and wholesale sales occupancies, the quantities of medicines, foodstuffs, consumer or industrial products, and cosmetics containing not more than 50 percent by volume of water-miscible liquids with the remainder of the solutions not being flammable, shall not be limited, provided that such materials are packaged in individual containers not exceeding 1.3 gallons.
d. Maximum allowable quantities shall be increased 100 percent in buildings equipped throughout with an automatic sprinkler system in accordance with Section 903.3.1.1. Where Note e also applies, the increase for both notes shall be applied accumulatively.
e. Maximum allowable quantities shall be increased 100 percent when stored in approved storage cabinets, day boxes, gas cabinets, exhausted enclosures or safety cans. Where Note d also applies, the increase for both notes shall be applied accumulatively.
f. The permitted quantities shall not be limited in a building equipped throughout with an automatic sprinkler system in accordance with Section 903.3.1.1.
g. Permitted only in buildings equipped throughout with an automatic sprinkler system in accordance with Section 903.3.1.1.
h. Containing not more than the maximum allowable quantity per control area of Class IA, IB or IC flammable liquids.
i. Inside a building, the maximum capacity of a combustible liquid storage system that is connected to a fuel-oil piping system shall be 660 gallons provided such system complies with the International Fire Code.
j. Quantities in parenthesis indicate quantity units in parenthesis at the head of each column.
k. A maximum quantity of 200 pounds of solid or 20 gallons of liquid Class 3 oxidizers is allowed when such materials are necessary for maintenance purposes, operation or sanitation of equipment. Storage containers and the manner of storage shall be approved.
l. Net weight of the pyrotechnic composition of the fireworks. Where the net weight of the pyrotechnic composition of the fireworks is not known, 25 percent of the gross weight of the fireworks, including packaging, shall be used.
m. For gallons of liquids, divide the amount in pounds by 10 in accordance with Section 2703.1.2 of the International Fire Code.
n. For storage and display quantities in Group M and storage quantities in Group S occupancies complying with Section 414.2.5, see Tables 414.2.5(1) and 414.2.5(2).
o. Densely packed baled cotton that complies with the packing requirements of ISO 8115 shall not be included in this material class.
p. The following shall not be included in determining the maximum allowable quantities:
    1. Liquid or gaseous fuel in fuel tanks on vehicles.
    2. Liquid or gaseous fuel in fuel tanks on motorized equipment operated in accordance with this code.
    3. Gaseous fuels in piping systems and fixed appliances regulated by the International Fuel Gas Code.
    4. Liquid fuels in piping systems and fixed appliances regulated by the International Mechanical Code.

Where one of the 15 exceptions is applied, the classification is based on the general use. For example, a warehouse containing quantities below the maximum allowable would simply be classified as Group S-1. A manufacturing facility would be classified as a Group F-1 occupancy.

Code Text: *Institutional Group I occupancy includes among others, the use of a building or structure, or a portion thereof, in which people are cared for or live in a supervised environment, having physical limitations because of health or age are harbored for medical treatment or other care or treatment, or in which people are detained for penal or correctional purposes or in which the liberty of the occupants is restricted.*

Discussion and Commentary: The institutional uses classified as Group I occupancies are of three broad types. The first is a facility in which care is provided for the very young, sick or injured. The second category includes those facilities in which the personal liberties of the inmates or residents are restricted. Thirdly, supervised care facilities are regulated. Though the hazard due to combustible contents is quite low, the occupants' lack of mobility limits their egress ability.

Group I-1

Residential board and care facilities
Assisted living facilities
Halfway houses
Group homes
Congregate care facilities
Alcohol rehabilitation facilities
Alcohol and drug centers
Convalescent facilities

Group I-2

Hospitals
Nursing homes
Mental hospitals
Detoxification facilities
Infant care (24-hour basis)

Group I-3

Prisons
Jails
Reformatories
Detention centers
Correctional centers
Prerelease centers

Group I-4

Custodial care facilities
(less than 24 hours)

Where the number of children, patients or residents in institutional uses is five or less, the hazards are similar in nature to a residential use. In most cases, an institutional facility with such a low occupant load would be considered a Group R-3 occupancy.

Code Text: *Mercantile Group M occupancy includes, among others, buildings and structures or a portion thereof, for the display and sale of merchandise, and involves stocks of goods, wares or merchandise incidental to such purposes and accessible to the public.*

Discussion and Commentary: A Group M occupancy is a retail or wholesale facility, or a store. An entire building can be classified as a Group M occupancy, such as a department store, or a portion of a building can be considered a mercantile use, such as the sales room in a manufacturing facility. A service station, including a canopy over the pump islands, is also classified as a Group M occupancy. In limited instances, a sales operation is designated as a Group B occupancy, as in the case of automobile showrooms.

Group M

Department stores
Drug stores
Markets
Motor fuel-dispensing facilities
Retail or wholesale stores
Sales rooms

When classifying the occupancy of a storage area accessory to the sales area in a retail store, it is appropriate to apply the provisions that address the specific hazards of the use. In most situations, it is appropriate to classify the incidental storage area as a Group S-1 occupancy.

Code Text: *Residential Group R occupancy includes, among others, the use of a building or structure, or a portion thereof, for sleeping purposes when not classified as an Institutional Group I or when not regulated by the* International Residential Code.

Discussion and Commentary: Residential occupancies are characterized by: 1) their use by people for living and sleeping purposes, 2) a relatively low potential fire severity, and 3) the worst fire record of all structure types. Because occupants of these types of buildings spend up to one-third of each day sleeping, there is a high potential of a fire to rage out of control before the occupants awaken. After awakening, the residents will typically be disoriented for a short period of time, further decreasing the opportunity for immediate egress.

Group R-1

Boarding houses (transient)
Hotels
Motels

Group R-2

Apartment houses
Boarding houses (not transient)
Convents
Dormitories
Fraternities
Sororities
Monasteries
Vacation timeshare properties

Group R-3

One- and two-family dwellings
(unless regulated by IRC)
Adult-care facilities (5 or fewer)
Child-care facilities (5 or fewer)
Congregate living facilities
($\leq$ 16 occupants)

Group R-4

Residential care facilities
(6-16 occupants)
Assisted living facilities
(6-16 occupants)

Detached one- and two-family dwellings, as well as townhouses, are not regulated by the *International Building Code* when limited to the conditions of the exception to Section 101.2. They are to be designed and constructed in accordance with the *International Residential Code*.

Code Text: *Storage Group S occupancy includes among others, the use of a building or structure, or a portion thereof, for storage that is not classified as a hazardous occupancy.*

Discussion and Commentary: Where a warehouse or other storage facility does not contain significant amounts of hazardous commodities (as determined by Section 307), it should be considered a Group S occupancy. A facility used for the storage of combustible goods is classified as Group S-1, whereas a Group S-2 occupancy shall be used only for the storage of noncombustible materials. If it is reasonable to believe that a storage building will house combustible goods for any significant period of time, it would be appropriate to consider the structure a Group S-1 occupancy, designed and constructed accordingly. Motor-vehicle-related uses are also included in the Group S category, with repair garages classified as Group S-1 and parking garages (both open and enclosed) as Group S-2 occupancies.

Group S-1	Group S-2
Aerosols, Level 2 and Level 3	Aircraft hangar
Aircraft repair hangar	Asbestos
Bags; cloth, burlap, paper	Cement in bags
Belting; canvas, leather	Chalk and crayons
Books	Dairy products
Paper in rolls	Dry cell batteries
Cardboard and cardboard boxes	Electric motors
Clothing	Food products
Furniture	Fresh fruits and vegetables
Grains	Frozen foods
Lumber	Glass
Motor vehicle repair garages	Gypsum board
Tires, bulk storage of	Meats
Tobacco, cigars, cigarettes	Metals
Upholstery and mattresses	Open parking garages
	Enclosed parking garages

Although the goods being stored in a Group S-2 occupancy must be noncombustible, the code permits a limited amount of combustibles in the packaging or support materials. Wood pallets, paper cartons, paper wrappings, plastic trim and film wrapping are permitted for such purposes.

Code Text: *Buildings and structures of an accessory character and miscellaneous structures not classified in any specific occupancy shall be constructed, equipped and maintained to conform to the requirements of the IBC commensurate with the fire and life hazard incidental to their occupancy.*

Discussion and Commentary: Those structures not ordinarily occupied by people are typically classified as Group U occupancies. The fire load in these structures varies considerably but is usually not excessive. Because these types of uses are not normally occupied, the concern for fire severity is not very great, and as a group they constitute a low hazard. Several of the structures regulated as Group U occupancies are never occupied, such as fences, towers and tanks.

Group U

Agricultural buildings
Barns
Carports
Fences more than 6 feet in height
Greenhouses
Livestock shelters
Private garages
Retaining walls
Sheds
Stables
Tanks
Towers

Private garages and carports classified as Group U occupancies are generally limited to 1,000 square feet and one story in height. However, such structures are permitted to be 3,000 square feet where no repair work is done and no fuel is dispensed.

Topic: Occupancy Classification   **Category:** Mixed Use and Occupancy
Reference: IBC 508.2.1, 508.2.2   **Subject:** Incidental Use Areas

Code Text: *An incidental use area shall be classified in accordance with the occupancy of that portion of the building in which it is located or the building shall be classified as a mixed occupancy and shall comply with Section 508.3. Incidental use areas shall be separated or protected, or both, in accordance with Table 508.2.*

Discussion and Commentary: It is common to find uses that are typical of the general occupancy classification of the building, yet which create a hazard different from the other hazards found in the occupancy. An example would be a chemistry laboratory classroom in a high school building. The code addresses such conditions by requiring incidental areas to be separated from the remainder of the building with fire-resistance-rated construction, or to be protected with a fire-extinguishing system. As an option, the provisions for a mixed-occupancy building could apply.

TABLE 508.2
INCIDENTAL USE AREAS

ROOM OR AREA	SEPARATION AND/OR PROTECTION
Furnace room where any piece of equipment is over 400,000 Btu per hour input	1 hour or provide automatic fire-extinguishing system
Rooms with boilers where the largest piece of equipment is over 15 psi and 10 horsepower	1 hour or provide automatic fire-extinguishing system
Refrigerant machinery rooms	1 hour or provide automatic sprinkler system
Parking garage (Section 406.2)	2 hours; or 1 hour and provide automatic fire-extinguishing system
Hydrogen cut-off rooms, not classified as Group H	1-hour in Group B, F, M, S and U occupancies. 2-hour in Group A, E, I and R occupancies.
Incinerator rooms	2 hours and automatic sprinkler system
Paint shops, not classified as Group H, located in occupancies other than Group F	2 hours; or 1 hour and provide automatic fire-extinguishing system
Laboratories and vocational shops, not classified as Group H, located in Group E or I-2 occupancies	1 hour or provide automatic fire-extinguishing system
Laundry rooms over 100 square feet	1 hour or provide automatic fire-extinguishing system
Storage rooms over 100 square feet	1 hour or provide automatic fire-extinguishing system
Group I-3 cells equipped with padded surfaces	1 hour
Group I-2 waste and linen collection rooms	1 hour
Waste and linen collection rooms over 100 square feet	1 hour or provide automatic fire-extinguishing system
Stationary storage battery systems having a liquid capacity of more than 100 gallons used for facility standby power, emergency power or uninterrupted power supplies	1-hour in Group B, F, M, S and U occupancies. 2-hour in Group A, E, I and R occupancies.

For SI: 1 square foot = 0.0929 m^2, 1 pound per square inch = 6.9 kPa,
1 British thermal unit per hour = 0.293 watts, 1 horsepower = 746 watts,
1 gallon = 3.785 L.

(Continued)

The separation and protection of incinerator rooms is unique in that both methods are required. Although a fire-extinguishing system is required within the incinerator room, it is still necessary to provide a fire-resistance-rated separation utilizing minimum 2-hour fire barriers.

Code Text: *Where Table 508.2 requires a fire-resistance-rated separation, the incidental use area shall be separated from the remainder of the building by a fire barrier constructed in accordance with Section 706 or a horizontal assembly constructed in accordance with Section 711, or both. Where Table 508.2 permits an automatic fire-extinguishing system without a fire barrier, the incidental use area shall be separated from the remainder of the building by construction capable of resisting the passage of smoke. Where an automatic fire-extinguishing system or an automatic sprinkler system is provided in accordance with Table 508.2, only the incidental use areas need be equipped with such a system.*

Discussion and Commentary: In utilizing Table 508.2, it is common that two options are available for addressing rooms or areas considered incidental use areas. A fire barrier may often be used to isolate the specific hazard from the remainder of the building. As an alternative, a sprinkler system or other fire-extinguishing system may be used to limit any fire in the incidental use area to that space only. By incorporating smoke containment construction, little if any smoke created would be transferred to other portions of the building.

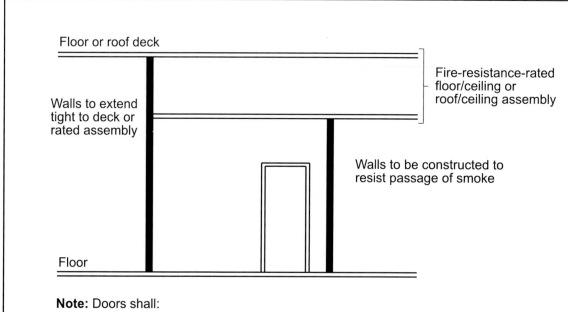

Note: Doors shall:
- be self-closing or automatic-closing upon detection of smoke
- have no air transfer openings
- have no excessive undercuts

Construction to resist the passage of smoke

Where a fire-extinguishing system is used to provide protection for an incidental use area, the area must still be separated from the remainder of the building. The separation must be constructed such that smoke will be contained. Enclosure doors also are regulated.

Code Text: *Each portion of a building shall be individually classified in accordance with Section 302.1. Where a building contains more than one occupancy group, the building or portion thereof shall comply with Sections 508.3.1 (Accessory Occupancies), 508.3.2 (Nonseparated Occupancies), 508.3.3 (Separated Occupancies), or a combination of these sections.* See exceptions for 1) occupancies separated in accordance with Section 509 (Special Provisions), and 2) Group H-1, H-2 and H-3 occupancies required by Table 415.3.2 to be located in a separate and detached building.

Discussion and Commentary: It is not uncommon for two or more distinct occupancy classifications to occur in the same building. Where such conditions exist, the code requires that such multiple occupancies be either 1) isolated from each other using fire-resistive separation elements (fire barriers and/or horizontal assemblies), or 2) imposed with special provisions that eliminate the need for such fire separations.

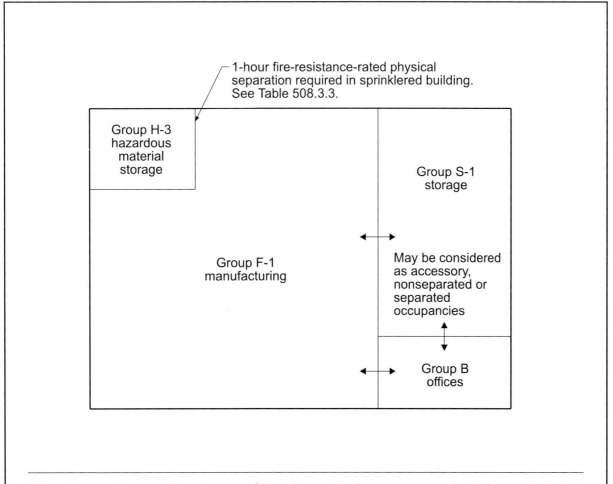

Although compliance with only one of the three mixed-occupancy methods is required, it is acceptable to utilize two or even all three methods within the same building.

Code Text: *Accessory occupancies are those occupancies subsidiary to the main occupancy of the building or portion thereof. Aggregate accessory occupancies shall not occupy more than 10 percent of the area of the story in which they are located and shall not exceed the tabular values in Table 503, without height and area increases in accordance with Sections 504 and 506 for such accessory occupancies.* See exceptions for 1) accessory assembly occupancies less than 750 square feet in floor area, 2) assembly occupancies that are accessory to Group E occupancies, and 3) accessory religious education rooms and religious auditoriums with occupant loads of less than 100.

Discussion and Commentary: The mixed-occupancy method of "Accessory Occupancies" is one of the three design options that the code provides when dealing with mixed-occupancy buildings. This approach is only applicable where one or more of the occupancies is quite small in relationship to the major occupancy in the building. The aggregate floor area of all accessory occupancies is limited to 10 percent of the floor area of the story in which the accessory occupancies are located. In addition, the aggregate floor area of the accessory occupancies cannot exceed the allowable floor area taken directly from Table 503.

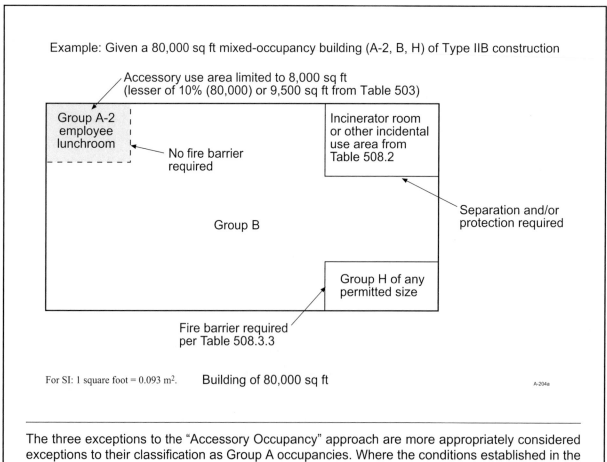

Example: Given a 80,000 sq ft mixed-occupancy building (A-2, B, H) of Type IIB construction

Accessory use area limited to 8,000 sq ft
(lesser of 10% (80,000) or 9,500 sq ft from Table 503)

Group A-2 employee lunchroom

No fire barrier required

Incinerator room or other incidental use area from Table 508.2

Separation and/or protection required

Group B

Group H of any permitted size

Fire barrier required per Table 508.3.3

For SI: 1 square foot = 0.093 m².   Building of 80,000 sq ft   A-204a

The three exceptions to the "Accessory Occupancy" approach are more appropriately considered exceptions to their classification as Group A occupancies. Where the conditions established in the exceptions exist, the specific uses listed are not to be classified as Group A.

Code Text: *Accessory occupancies shall be individually classified in accordance with Section 302.1. Code requirements shall apply to each portion of the building based on the occupancy classification of that accessory space, except that the most restrictive applicable provisions of Section 403 and Chapter 9 shall apply to the entire building or portion thereof.*

Discussion and Commentary: The occupancy classification of a use that is regulated under the provisions for accessory occupancies is based solely upon the specific use of that area. Although the size of the occupancy may be quite small in comparison with the remainder of the building, the accessory occupancy has its own unique hazards that must be adequately addressed.

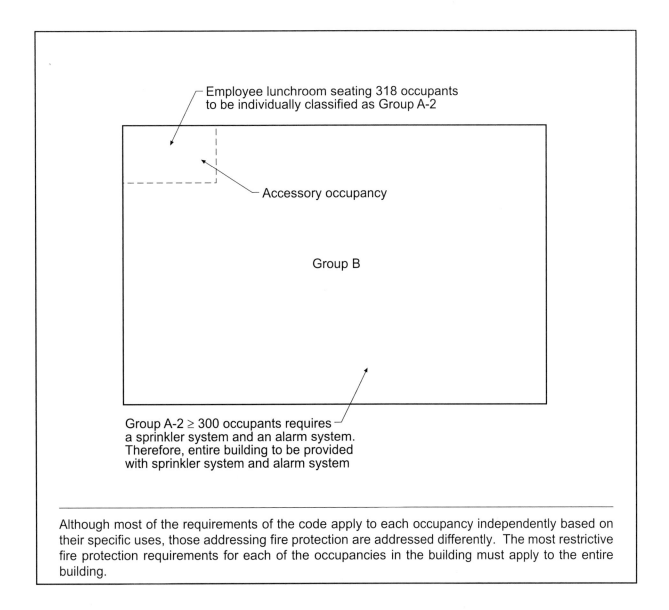

Employee lunchroom seating 318 occupants to be individually classified as Group A-2

Accessory occupancy

Group B

Group A-2 ≥ 300 occupants requires a sprinkler system and an alarm system. Therefore, entire building to be provided with sprinkler system and alarm system

Although most of the requirements of the code apply to each occupancy independently based on their specific uses, those addressing fire protection are addressed differently. The most restrictive fire protection requirements for each of the occupancies in the building must apply to the entire building.

Code Text: *The allowable area and height of the building shall be based on the allowable area and height for the main occupancy in accordance with Section 503.1. No separation is required between accessory occupancies or the main occupancy.* See exception for Group H-2, H-3, H-4 and H-5 occupancies.

Discussion and Commentary: Where the methodology of "Accessory Occupancies" is utilized, the allowable height and area of the accessory occupancies, as well as that of the major occupancy, is based solely on the building's major occupancy. There is no mandate to apply the more restrictive height and area provisions of each of the occupancies involved, as required under the "Nonseparated Occupancies" method, nor to go through calculations based on the unity formula as required for "Separated Occupancies."

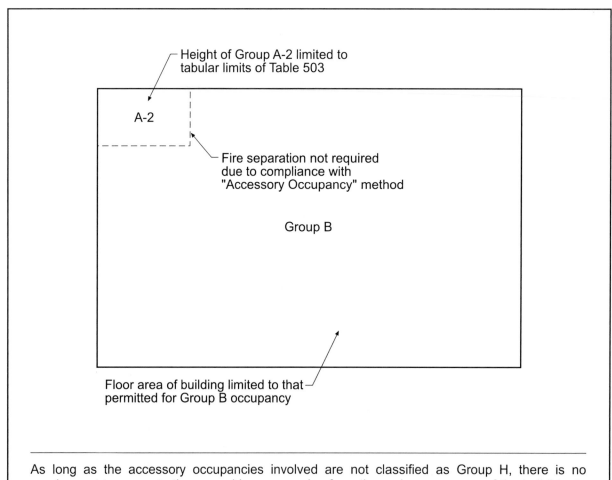

As long as the accessory occupancies involved are not classified as Group H, there is no requirement to separate the assembly occupancies from the major occupancy of the building. In addition, where two or more accessory occupancies are present, they do not need to be separated from each other.

Code Text: *Nonseparated occupancies shall be individually classified in accordance with Section 302.1. Code requirements shall apply to each portion of the building based on the occupancy classification of that space except that the most restrictive applicable provisions of Section 403 and Chapter 9 shall apply to the entire building or portion thereof.*

Discussion and Commentary: The allowance for "Nonseparated Occupancies," one of the alternatives to the physical separation of different occupancies, is based on the most limiting requirements for building size and fire-protection features, such as sprinklers, standpipes and alarm systems. Where occupancies are regulated by the nonseparated occupancy provisions, a physical separation is permitted, but it is not required to be fire-resistance rated.

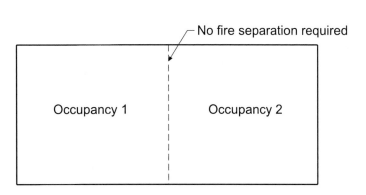

- Type of construction limited by:
  - Lesser height limit of Occupancy 1 or 2
  - Lesser floor area limit of Occupancy 1 or 2

- Most restrictive fire-protection system requirements of Occupancy 1 and 2 (also high-rise provisions where applicable)

Where the provisions for nonseparated occupancies are utilized in a high-rise building as defined by Section 403, the special high-rise provisions are applicable to the entire building, including those portions that are not located in the high-rise portion.

Topic: Height, Area and Separation   **Category:** Mixed Use and Occupancy
Reference: IBC 508.3.2.2, 508.3.2.3   **Subject:** Nonseparated Occupancies

Code Text: *The allowable area and height of the building or portion thereof shall be based on the most restrictive allowances for the occupancy groups under consideration for the type of construction of the building in accordance with Section 503.1. No separation is required between occupancies.* See exception for Group H-2, H-3, H-4 and H-5 occupancies.

Discussion and Commentary: Where the option for nonseparated occupancies is utilized to address a mixed-occupancy building, it is necessary to determine the maximum building size for each of the occupancies that are not appropriately separated. The maximum allowable height and area for each occupancy would be based upon the building's type of construction. The most restrictive height and area of those nonseparated occupancies would then be the limiting size for the combination of such occupancies.

Given: A Type VB building contains both Group B and Group E occupancies.

Determine: The height and area limitations if the occupancies are not separated under the provisions of Section 508.3.2.

OCCUPANCY	ALLOWABLE HEIGHT[1]	ALLOWABLE AREA[1]
Group B[2]	2 stories	9,000 square feet
Group E[2]	1 story	9,500 square feet

[1] Based on Table 503 assuming no permitted increases.

[2] Most restrictive fire protection requirements of Chapter 9 also applicable to entire building.

∴ Thus, for nonseparated occupancies, the maximum building size would be 1 story and 9,000 square feet.

Nonseparated Occupancies

The use of the nonseparated occupancies method is not applicable to high-hazard occupancies. Those areas or spaces classified as Group H occupancies must be isolated from other occupancies within the building by fire barriers and/or horizontal assemblies in accordance with Table 508.3.3 for occupancy separations.

Code Text: *Separated occupancies shall be individually classified in accordance with Section 302.1. Each fire area shall comply with this code based on the occupancy classification of that portion of the building.*

Discussion and Commentary: Under the provisions for "Separated Occupancies," each of the distinct uses is to be individually classified as to occupancy. This approach is consistent with that for accessory occupancies and nonseparated occupancies. The concept of separated occupancies provides for a fire-resistance-rated separation in order to isolate the hazards associated with a specific occupancy from other portions of the building.

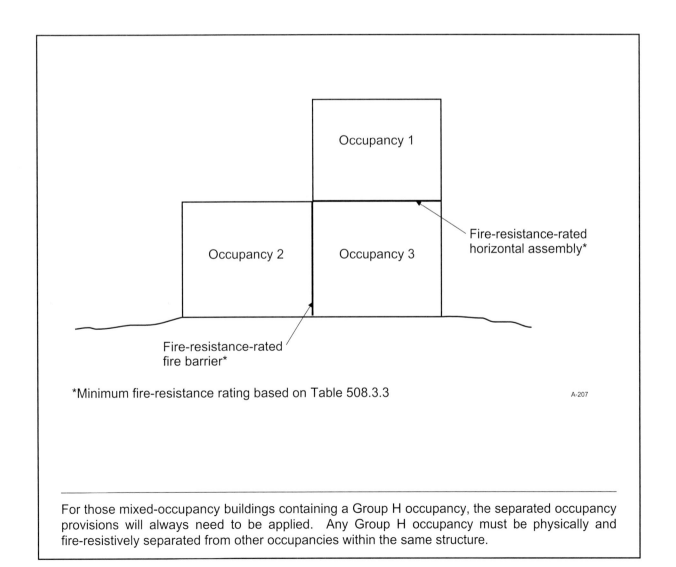

Occupancy 1

Occupancy 2

Occupancy 3

Fire-resistance-rated horizontal assembly*

Fire-resistance-rated fire barrier*

*Minimum fire-resistance rating based on Table 508.3.3

A-207

For those mixed-occupancy buildings containing a Group H occupancy, the separated occupancy provisions will always need to be applied. Any Group H occupancy must be physically and fire-resistively separated from other occupancies within the same structure.

Topic: Allowable Height and Area   **Category:** Mixed Use and Occupancy
Reference: IBC 508.3.3.2, 508.3.3.3   **Subject:** Separated Occupancies

Code Text: *In each story, the building area shall be such that the sum of the ratios of the actual floor area of each occupancy divided by the allowable area of each occupancy shall not exceed one. Each occupancy shall comply with the height limitations based on the type of construction of the building in accordance with Section 503.1.*

Discussion and Commentary: The approach to separated occupancies mandates that the ratios of the actual and allowable floor areas be calculated in order to determine compliance. Often known as the "unity formula," this calculation recognizes the relationship between the permitted sizes of the various occupancies involved. The unity formula is only applicable where the separated occupancy method is utilized and does not apply to accessory occupancies or nonseparated occupancies.

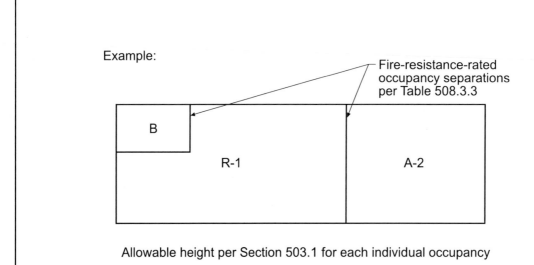

Example:

Fire-resistance-rated occupancy separations per Table 508.3.3

B

R-1

A-2

Allowable height per Section 503.1 for each individual occupancy

$$\frac{\text{Actual area A-2}}{\text{Allowable area A-2}} + \frac{\text{Actual area B}}{\text{Allowable area B}} + \frac{\text{Actual area R-1}}{\text{Allowable area R-1}} \leq 1.0$$

The height limitations for separated occupancies are based upon the general provisions of Section 503.1. The height limit, in both feet and stories, is to be measured from the grade plane, and the measurement must include all intervening fire areas.

Code Text: *Individual occupancies shall be separated from adjacent occupancies in accordance with Table 508.3.3. Required separations shall be fire barriers constructed in accordance with Section 706 or horizontal assemblies constructed in accordance with Section 711, or both, so as to completely separate adjacent occupancies.*

Discussion and Commentary: A matrix, Table 508.3.3, has been established to identify any required fire-resistance-rated separation between various occupancies. The table is based on the perceived degree of dissimilarity between the occupancies involved. Where Table 508.3.3 requires a level of fire-resistance between the adjoining occupancies, fire barriers and/or horizontal separations are to be used. The intended result is that the hazards associated with one occupancy be completely isolated from those present in the remainder of the building.

TABLE 508.3.3
REQUIRED SEPARATION OF OCCUPANCIES (HOURS)

OCCUPANCY	A[e], E		I		R[d]		F-2, S-2[c,d], U[d]		B[b], F-1, M[b], S-1		H-1		H-2		H-3, H-4, H-5	
	S	NS	S	NS	S	NS	S	NS	S	NS	S	NS	S	NS	S	NS
A[e], E[e]	N	N	1	2	1	2	N	1	1	2	NP	NP	3	4	2	3[a]
I	—	—	N	N	1	NP	1	2	1	2	NP	NP	3	NP	2	NP
R[d]	—	—	—	—	N	N	1	2	1	2	NP	NP	3	NP	2	NP
F-2, S-2[c,d], U[d]	—	—	—	—	—	—	N	N	1	2	NP	NP	3	4	2	3[a]
B[b], F-1, M[b], S-1	—	—	—	—	—	—	—	—	N	N	NP	NP	2	3	1	2[a]
H-1	—	—	—	—	—	—	—	—	—	—	N	NP	NP	NP	NP	NP
H-2	—	—	—	—	—	—	—	—	—	—	—	—	N	NP	1	NP
H-3, H-4, H-5	—	—	—	—	—	—	—	—	—	—	—	—	—	—	N	NP

For SI: 1 square foot = 0.0929 m².
S = Buildings equipped throughout with an automatic sprinkler system installed in accordance with Section 903.3.1.1.
NS = Buildings not equipped throughout with an automatic sprinkler system installed in accordance with Section 903.3.1.1.
N = No separation requirement.
NP = Not permitted.
a. For Group H-5 occupancies, see Section 903.2.4.2.
b. Occupancy separation need not be provided for storage areas within Groups B and M if the:
    1. Area is less than 10 percent of the floor area;
    2. Area is equipped with an automatic fire-extinguishing system and is less than 3,000 square feet; or
    3. Area is less than 1,000 square feet.
c. Areas used only for private or pleasure vehicles shall be allowed to reduce separation by 1 hour.
d. See Section 406.1.4.
e. Commercial kitchens need not be separated from the restaurant seating areas that they serve.

Although the title of Section 508.3.3 implies that the method described requires a physical and/or fire-resistive separation, that is not always the case. A number of the occupancies do not require such a separation due to the lack of hazard dissimilarity between the uses that occur.

Quiz

Study Session 2
IBC Chapter 3 and Section 508

1. An institutional occupancy is typically considered Group _____.

    a. A   b. B

    c. I   d. R

    Reference_____

2. A Group _____ occupancy is the general classification for miscellaneous and utility structures.

    a. A   b. M

    c. S   d. U

    Reference_____

3. Unless classified as a mixed occupancy, an "incidental use area" is to be classified _____.

    a. in accordance with the occupancy where it is located

    b. based on its relative fire and life-safety hazard

    c. as a Group B occupancy

    d. as an accessory occupancy

    Reference_____

4. Unless protected by an automatic fire-extinguishing system, what is the minimum fire separation between a 350-square-foot laundry room and the manufacturing building it is incidental to?

    a. no fire separation is required

    b. 1-hour fire partition

    c. 1-hour fire barrier and/or horizontal assembly

    d. smoke barrier

    Reference_____

5. What minimum level of protection is required for an incinerator room located in a manufacturing facility?

    a. 1-hour fire partition only

    b. automatic fire-extinguishing system only

    c. 1-hour fire barrier and an automatic sprinkler system

    d. separation of two hours and an automatic sprinkler system

    Reference_____

6. What minimum level of protection is required for a chemistry laboratory/classroom in a high school?

    a. 1-hour fire barrier/horizontal assembly only

    b. automatic fire-extinguishing system only

    c. both a 1-hour fire barrier/horizontal assembly and an automatic fire-extinguishing system

    d. either a 1-hour fire barrier/horizontal assembly or an automatic fire-extinguishing system

    Reference_____

7. Where an automatic fire-extinguishing system without a fire barrier is utilized for the protection of incidental use areas, what minimum level of separation is required?

    a. 1-hour fire partition

    b. 1-hour fire barrier/horizontal assembly

    c. 1-hour smoke barrier walls and horizontal assemblies

    d. construction capable of resisting the passage of smoke

    Reference_____

8. Any accessory occupancy area classified as a Group _____ occupancy must always be separated by a fire barrier horizontal assembly from other occupancies in the building.

    a. A   b. E

    c. H   d. I

Reference_____

9. Where the provisions for nonseparated occupancies are used for a mixed-occupancy building, the most restrictive _____ requirements shall apply to the nonseparated uses.

    a. fire protection system   b. means of egress

    c. occupancy classification   d. interior finish

Reference_____

10. Where the provisions for separated occupancies are used for a nonsprinklered mixed-occupancy building, the minimum separation between a Group A-2 and Group B occupancy shall be a _____.

    a. 1-hour fire partition

    b. 1-hour fire barrier/horizontal assembly

    c. 2-hour fire barrier/horizontal assembly

    d. 2-hour fire wall

Reference_____

11. Where the provisions for separated occupancies are used for a sprinklered mixed-occupancy building, the minimum separation between a Group H-2 and Group F-1 occupancy shall be a _____.

    a. 1-hour fire partition

    b. 1-hour fire barrier/horizontal assembly

    c. 2-hour fire barrier/horizontal assembly

    d. 2-hour fire wall

Reference_____

12. An accessory assembly area need not be considered a separate occupancy where the floor area is a maximum of _____ square feet.

   a. 120
   b. 400
   c. 750
   d. 1,000

   Reference_____

13. A lunchroom in a Group E middle school shall be separated at what minimum level from the remainder of the school building?

   a. no separation is required

   b. 1-hour fire barrier

   c. 2-hour fire barrier

   d. 3-hour fire barrier

   Reference_____

14. Which of the following uses is typically considered a Group A-4 occupancy?

   a. restaurant with a dance floor
   b. school library

   c. outdoor football stadium
   d. indoor hockey arena

   Reference_____

15. Which of the following uses is not considered a Group B occupancy?

   a. convenience store
   b. motor vehicle showroom

   c. car wash
   d. college classroom building

   Reference_____

16. A manufacturing facility involved in the manufacture of _____ is considered a Group F-2 occupancy.

   a. soaps and detergents
   b. ceramic products

   c. automobiles
   d. agricultural machinery

   Reference_____

17. A Class IIIA combustible liquid has a closed cup flash point at or above _____ and below _____.

   a. 73°F, 100°F   b. 100°F, 140°F

   c. 140°F, 200°F   d. 200°F, 212°F

   Reference_____

18. Buildings containing materials that present a deflagration hazard are typically considered _____ occupancies.

   a. Group H-1   b. Group H-2

   c. Group H-3   d. Group H-5

   Reference_____

19. A child-care facility providing care on a 24-hour basis to six or more infants/toddlers ($2^1/_2$ years of age or less) is classified as a Group _____ occupancy.

   a. E   b. I-1

   c. I-2   d. R-4

   Reference_____

20. Prior to any permitted increases, the maximum allowable quantity per control area of a Class IB flammable liquid permitted in a storage condition in a one-story Group F-1 occupancy is _____ gallons.

   a. 15   b. 30

   c. 60   d. 120

   Reference_____

21. A facility used for supervised residential care and housing more than 16 persons is classified as a Group _____ occupancy.

   a. I-1   b. I-4

   c. R-3   d. R-4

   Reference_____

22. Which of the following uses is not considered a Group M occupancy?

    a. wholesale store   b. sales room

    c. motor vehicle showroom   d. motor vehicle service station

    Reference_____

23. A sorority house is typically considered a Group _____ occupancy.

    a. R-1   b. R-2

    c. R-3   d. R-4

    Reference_____

24. Which of the following uses is not considered a residential care/assisted living facility?

    a. halfway house   b. drug abuse center

    c. convalescent facility   d. detoxification facility

    Reference_____

25. Which of the following uses is not considered a Group S-2 occupancy?

    a. open parking garage   b. enclosed parking garage

    c. dry cell battery storage   d. stable

    Reference_____

26. Where an accessory occupancy is not required to be separated from the major use by a fire barrier, it is limited to a maximum floor area of _____ of the area of the story in which it is located.

    a. 10 percent   b. 15 percent

    c. 25 percent   d. 33 percent

    Reference_____

27. Where the provisions for separated occupancies are utilized for a fully sprinklered building housing both a Group A-2 occupancy and a Group R-1 occupancy, the minimum required separation between the two occupancies shall be a _____.

    a. 1-hour fire partition

    b. 1-hour fire barrier/horizontal assembly

    c. 2-hour fire barrier/horizontal assembly

    d. 2-hour fire wall

    Reference_____

28. A public library is typically classified as a _____ occupancy.

    a. Group A-3   b. Group A-4

    c. Group B   d. Group M

    Reference_____

29. A manufacturing facility utilizing highly toxic materials exceeding the maximum allowable quantities set forth in Table 307.1(2) is considered a _____ occupancy.

    a. Group F-1   b. Group F-2

    c. Group H-3   d. Group H-4

    Reference_____

30. Aerosol storage buildings shall be classified as Group _____ occupancies when constructed in accordance with the *International Fire Code*.

    a. H-2   b. H-3

    c. S-1   d. S-2

    Reference_____

31. What is the largest occupant load permitted in an assembly building for it to be classified as a Group B occupancy?

    a. 15   b. 30

    c. 49   d. 99

    Reference _____

32. Where a training and skill development use occurs in other than a school or academic program, it is classified as a(n) Group _____ occupancy.

    a. A-3   b. B

    c. E   d. M

Reference _____

33. A congregate living facility with an occupant load of 12 persons is to be classified as a Group _____ occupancy.

    a. R-1   b. R-2

    c. R-3   d. R-4

Reference _____

34. By definition, transient residential dwelling unit or sleeping unit has a maximum occupancy period of _____ days.

    a. 14   b. 30

    c. 90   d. 180

Reference _____

35. Under the provisions for separated occupancies, what is the minimum hourly fire-resistance-rated separation required between a Group F-1 occupancy and a Group S-1 occupancy where located in a nonsprinklered building?

    a. 0 hours (no separation requirement)

    b. 1-hour

    c. 2-hour

    d. 3-hour

Reference _____

Study Session

3

2006 IBC Chapter 6
Types of Construction

OBJECTIVE: To gain an understanding of how a building is classified as a specific type of construction, based on the construction materials and the various building elements' resistance to fire.

REFERENCE: Chapter 6, 2006 *International Building Code*

KEY POINTS:
- What do the various types of construction indicate?
- How are the required fire-resistance ratings of building elements determined?
- Why are exterior walls regulated by additional criteria?
- Why are exterior walls protected differently based on fire separation distance?
- At what minimum distance is the protection of exterior walls unnecessary?
- Which types of materials are required to be used as building elements of a Type I or II building?
- How do the two different categories of Type I construction differ in fire protection? Type II construction?
- Which types of materials are required for use in the exterior walls of a Type III structure? In the interior building elements?
- What is another name for Type IV construction?
- How shall exterior walls be constructed? Interior building elements?
- What are the minimum construction details for columns used in a building of Type IV construction?
- In Type IV buildings, what is the minimum size of heavy-timber members used in the floor and roof framing? Floors? Roofs? Partitions?
- Where the minimum dimensions for Type IV solid sawn members are prescribed, how are the equivalent sizes established for glued laminated members?
- Type V buildings may be constructed of which building materials?
- How does a Type VA building differ from a Type VB building?
- In noncombustible Type I and II buildings, where may fire-retardant-treated wood be used?
- Which specific allowances are provided for combustible materials in Type I and Type II buildings?

KEY POINTS:
(Cont'd)

- What are the limitations for the use of fire-retardant-treated wood in the roof construction of noncombustible buildings? In nonbearing partitions? In nonbearing exterior walls?
- Which building elements are considered structural frame elements for the determination of fire resistance? Secondary members?
- When are bracing members considered part of the structural frame?
- Under which conditions may the required fire resistance of roof supports be reduced?
- At what height may the required fire resistance of roof construction be eliminated? In which occupancies is the elimination not applicable?
- For which building elements are heavy-timber members and 1-hour fire-resistance-rated construction interchangeable?
- How can a sprinkler system affect a building's type of construction classification?
- How are interior nonbearing walls regulated for fire resistance based on construction type? Exterior nonbearing walls?

Code Text: *Buildings and structures erected or to be erected, altered or extended in height or area shall be classified in one of the five construction types defined in Sections 602.2 through 602.5.*

Discussion and Commentary: There are two major groupings based on the construction materials: noncombustible construction (Types I and II) and noncombustible or combustible construction (Types III, IV and V). These groupings are divided into two more categories: protected, where the major structural elements are provided with some degree of fire resistance, and unprotected, where no fire protection of the building elements is typically mandated. Protected construction is further distinguished in Type I buildings where the required protection for many structural elements exceeds a 1-hour fire-resistance rating.

Noncombustible	Exterior and interior (bearing or nonbearing) walls, floors, roofs and structural elements are to be of noncombustible materials	I	A	B
		II	A	B
Noncombustible or combustible	Exterior walls are to be of noncombustible materials	III	A	B
		IV	HT	
		V	A	B

A-208

It is the intent of the *International Building Code* that each building be classified as a single type of construction. The construction materials and the degree to which such materials are protected determine the classification based on the criteria of Table 601 and Chapter 6.

Code Text: *The building elements shall have a fire-resistance rating not less than that specified in Table 601.*

Discussion and Commentary: The building elements regulated by Table 601 for types of construction include structural frame members, such as columns, girders and trusses; bearing walls, both interior and exterior; floor construction, including supporting beams and joists; and roof construction, consisting of supporting beams, joists, rafters and other members. The required fire-resistance rating for each of these elements is based on the specific type of construction assigned to the building. The required fire-resistance rating can be as high as a 3-hour or as little as a 0-hour.

TABLE 601
FIRE-RESISTANCE RATING REQUIREMENTS FOR BUILDING ELEMENTS (hours)

BUILDING ELEMENT	TYPE I		TYPE II		TYPE III		TYPE IV	TYPE V	
	A	B	A[e]	B	A[e]	B	HT	A[e]	B
Structural frame[a]	3[b]	2[b]	1	0	1	0	HT	1	0
Bearing walls Exterior[g] Interior	3 3[b]	2 2[b]	1 1	0 0	2 1	2 0	2 1/HT	1 1	0 0
Nonbearing walls and partitions Exterior	See Table 602								
Nonbearing walls and partitions Interior[f]	0	0	0	0	0	0	See Section 602.4.6	0	0
Floor construction Including supporting beams and joists	2	2	1	0	1	0	HT	1	0
Roof construction Including supporting beams and joists	$1^1/_2$[c]	1[c, d]	1[c, d]	0[d]	1[d]	0[d]	HT	1[c, d]	0

For SI: 1 foot = 304.8 mm.

a. The structural frame shall be considered to be the columns and the girders, beams, trusses and spandrels having direct connections to the columns and bracing members designed to carry gravity loads. The members of floor or roof panels which have no connection to the columns shall be considered secondary members and not a part of the structural frame.

b. Roof supports: Fire-resistance ratings of structural frame and bearing walls are permitted to be reduced by 1 hour where supporting a roof only.

c. Except in Group F-1, H, M and S-1 occupancies, fire protection of structural members shall not be required, including protection of roof framing and decking where every part of the roof construction is 20 feet or more above any floor immediately below. Fire-retardant-treated wood members shall be allowed to be used for such unprotected members.

d. In all occupancies, heavy timber shall be allowed where a 1-hour or less fire-resistance rating is required.

e. An approved automatic sprinkler system in accordance with Section 903.3.1.1 shall be allowed to be substituted for 1-hour fire-resistance-rated construction, provided such system is not otherwise required by other provisions of the code or used for an allowable area increase in accordance with Section 506.3 or an allowable height increase in accordance with Section 504.2. The 1-hour substitution for the fire resistance of exterior walls shall not be permitted.

f. Not less than the fire-resistance rating required by other sections of this code.

g. Not less than the fire-resistance rating based on fire separation distance (see Table 602).

Where a structure is separated by one or more fire walls, the code treats those individual compartments created by the fire walls as separate buildings. Thus, each separate compartment would be considered a distinct building for the purpose of classification by type of construction.

Code Text: *Exterior walls shall have a fire-resistance rating not less than that specified in Table 602.*

Discussion and Commentary: The rationale behind exterior wall protection is that an owner has no control over what occurs on adjacent property. The property line concept provides a convenient means of protecting one building from another insofar as radiant heat could potentially be transmitted from one building to another during a fire. The requirements are based on "fire separation distance," which must be considered for all exterior walls. Where such walls are also bearing walls, the provisions of Table 601 also apply, governed by the more restrictive of the hourly ratings. Additional provisions for exterior walls are found in Section 704.

Table 602 Regulates Exterior Walls Only

- Table 602 used in conjunction with Table 601 for fire resistance of exterior bearing walls
- Only Table 602 used for nonbearing exterior walls
- Based primarily on occupancy type
- Highest required rating for exterior wall is 3 hours
- Final threshold at $\geq 30'$
- Additional provisions for exterior walls and openings in Section 704

TABLE 602
FIRE-RESISTANCE RATING REQUIREMENTS FOR EXTERIOR WALLS BASED ON FIRE SEPARATION DISTANCE[a, e]

FIRE SEPARATION DISTANCE = X (feet)	TYPE OF CONSTRUCTION	OCCUPANCY GROUP H	OCCUPANCY GROUP F-1, M, S-1	OCCUPANCY GROUP A, B, E, F-2, I, R, S-2, U[b]
X < 5[c]	All	3	2	1
5 ≤ X < 10	IA	3	2	1
	Others	2	1	1
10 ≥ X < 30	IA, IB	2	1	1[d]
	IIB, VB	1	0	0
	Others	1	1	1[d]
X ≥ 30	All	0	0	0

For SI: 1 foot = 304.8 mm.
a. Load-bearing exterior walls shall also comply with the fire-resistance rating requirements of Table 601.
b. For special requirements for Group U occupancies see Section 406.1.2.
c. See Section 705.1.1 for party walls.
d. Open parking garages complying with Section 406 shall not be required to have a fire-resistance rating.
e. The fire-resistance rating of an exterior wall is determined based upon the fire separation distance of the exterior wall and the story in which the wall is located.

The "Fire separation distance" is defined in Section 702.1 as the distance measured from the building face to the closest interior lot line, to the centerline of a street, alley or public way, or to an imaginary line between two buildings on the property.

Category: Type of Construction

Subject: Exterior Walls

Code Text: Fire separation distance *is the distance measured from the building face to one of the following: 1) the closest interior lot line; 2) to the centerline of a street, alley or public way; or 3) to an imaginary line between two buildings on the property. The distance shall be measured at right angles from the face of the wall.*

Discussion and Commentary: The atmospheric separation provided between a building and an adjoining structure provides resistance to fire spread due to radiant heat transfer. Many provisions throughout the IBC, such as those regulating projections and parapets, are based upon the degree of separation provided. A measurement at a right angle from the building face addresses heat transfer from the building of fire incident toward other structures and properties.

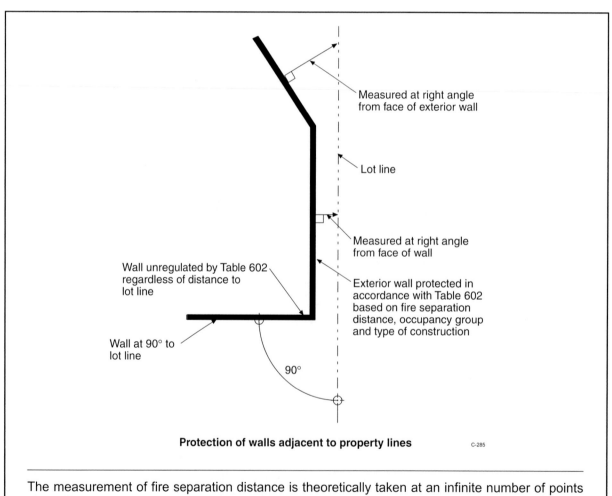

Protection of walls adjacent to property lines

C-285

The measurement of fire separation distance is theoretically taken at an infinite number of points along the exterior wall. In reality, zones are created adjacent to the building under consideration, with higher degrees of regulation mandated for those zones closest to the building.

Code Text: The minimum fire-resistance rating for exterior bearing walls shall be *not less than the fire-resistance rating based on fire separation distance.* In addition to the fire-resistance rating requirements for exterior walls based on fire separation distance, *load-bearing exterior walls shall also comply with the fire-resistance rating requirements of Table 601.*

Discussion and Commentary: When analyzing an exterior wall for its required level of fire resistance, it is necessary to use both Tables 601 and 602. Table 601 addresses the potential need for structural stability of exterior bearing walls under fire conditions. Exterior nonbearing walls are not regulated by this table. The concern of radiant heat transfer from an adjoining burning building results in Table 602 regulating the exterior wall rating based upon its fire separation distance (typically the distance from the building face to the lot line). This concern exists for both bearing and nonbearing exterior walls.

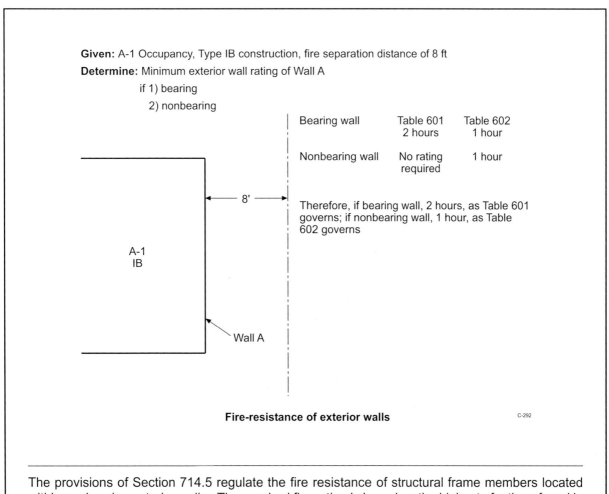

Given: A-1 Occupancy, Type IB construction, fire separation distance of 8 ft

Determine: Minimum exterior wall rating of Wall A

    if 1) bearing

      2) nonbearing

	Table 601	Table 602
Bearing wall	2 hours	1 hour
Nonbearing wall	No rating required	1 hour

Therefore, if bearing wall, 2 hours, as Table 601 governs; if nonbearing wall, 1 hour, as Table 602 governs

A-1
IB

8'

Wall A

Fire-resistance of exterior walls

C-292

The provisions of Section 714.5 regulate the fire resistance of structural frame members located within nonbearing exterior walls. The required fire rating is based on the highest of ratings found in Table 601 (structural frame, exterior bearing wall) and Table 602 (fire separation distance).

Code Text: *Types I and II construction are those types of construction in which the building elements listed in Table 601 are of noncombustible materials, except as permitted in Section 603 and elsewhere in the IBC.*

Discussion and Commentary: Type I buildings are noncombustible, and the building elements are also provided with a mandated degree of fire resistance. This type of construction requires the highest level of fire protection specified in the code. Type II buildings are also of noncombustible construction; however, the level of fire resistance is usually less than that required for Type I structures. Buildings of Type II construction may have a limited degree of fire resistance (Type IIA) or no fire resistance whatsoever (Type IIB). There are limited allowances for the use of fire-retardant-treated wood in nonbearing partitions, nonbearing exterior walls and roof construction.

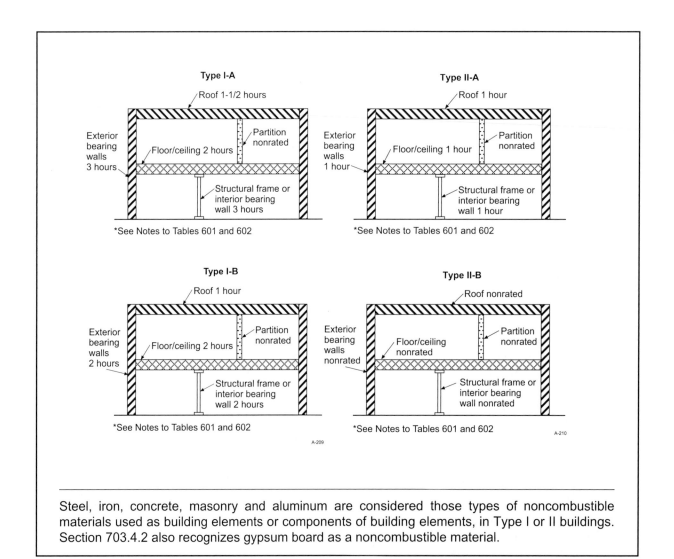

Steel, iron, concrete, masonry and aluminum are considered those types of noncombustible materials used as building elements or components of building elements, in Type I or II buildings. Section 703.4.2 also recognizes gypsum board as a noncombustible material.

Code Text: *Type III construction is that type of construction in which the exterior walls are of noncombustible materials and the interior building elements are of any material permitted by the code. Fire-retardant-treated wood framing complying with Section 2303.2 shall be permitted within exterior wall assemblies of a 2-hour rating or less.*

Discussion and Commentary: Type III buildings are considered combustible buildings and are either protected or unprotected. This building type was developed out of the necessity to prevent conflagrations in heavily built-up areas where buildings were erected side-by-side in congested downtown business districts. To limit the spread of fire from building to building, exterior walls were required to be of both noncombustible and fire-resistant construction.

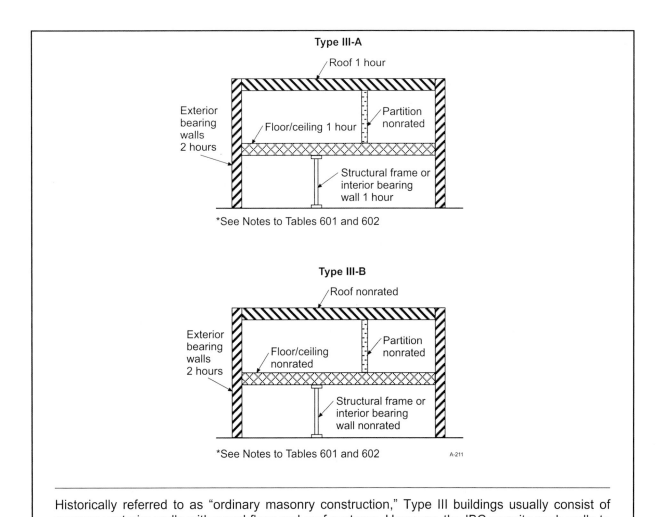

Historically referred to as "ordinary masonry construction," Type III buildings usually consist of masonry exterior walls with wood floor and roof systems. However, the IBC permits such walls to contain fire-retardant-treated wood as an element of the exterior wall construction.

Code Text: *Type IV construction (Heavy Timber, HT) is that type of construction in which the exterior walls are of noncombustible materials and the interior building elements are of solid or laminated wood without concealed spaces. The details of Type IV construction shall comply with the provisions of Section 602.4. Fire-retardant-treated wood framing complying with Section 2303.2 shall be permitted within exterior wall assemblies with a 2-hour rating or less.*

Discussion and Commentary: Referred to as "heavy-timber," buildings of Type IV construction are essentially Type III buildings with an interior of timber members. To conform to Type IV construction, building members must be of substantial thickness. Given the characteristics of massive wood members, there is little chance for sudden structural collapse during or after a fire.

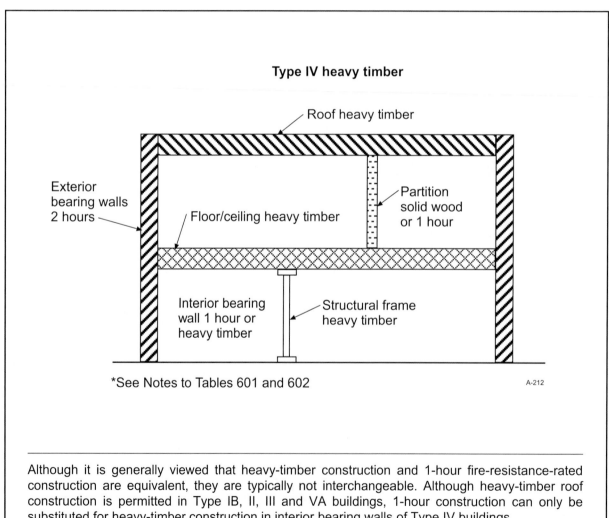

Although it is generally viewed that heavy-timber construction and 1-hour fire-resistance-rated construction are equivalent, they are typically not interchangeable. Although heavy-timber roof construction is permitted in Type IB, II, III and VA buildings, 1-hour construction can only be substituted for heavy-timber construction in interior bearing walls of Type IV buildings.

Code Text: *Minimum solid-sawn nominal dimensions are required for structures built using Type IV construction (HT). For glued-laminated members the equivalent net finished width and depths corresponding to the minimum nominal width and depths of solid sawn lumber are required as specified in Table 602.4.*

Discussion and Commentary: Solid-sawn wood members and glued-laminated timbers are manufactured with different methods and procedures: therefore; they do not have the same dimensions. However, they both have the same inherent fire-resistive capability that has been long recognized in the code. A comparison of the varied widths and depths indicates how the reduced width of glued-laminated members is offset by an increase in the depth, providing for similar cross-sectional areas.

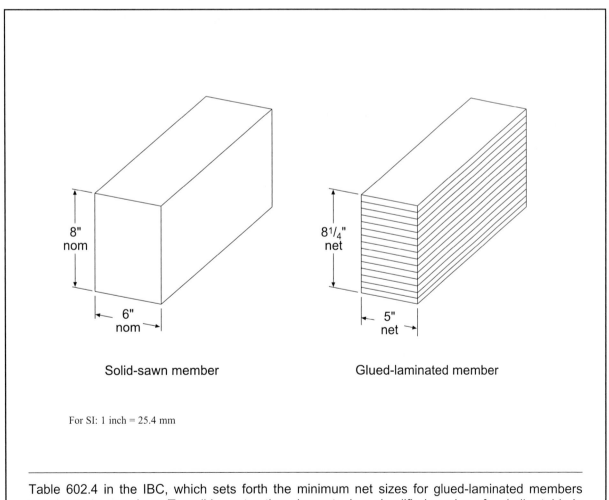

Solid-sawn member

Glued-laminated member

For SI: 1 inch = 25.4 mm

Table 602.4 in the IBC, which sets forth the minimum net sizes for glued-laminated members necessary to comply as Type IV construction elements, is a simplified version of a similar table in American Institute of Timber Construction (AITC) Standard 113.

Code Text: *Type V construction is that type of construction in which the structural elements, exterior walls and interior walls are of any materials permitted by the IBC.*

Discussion and Commentary: Type V buildings are essentially construction systems that will not fit into any of the other higher types of construction specified by the IBC. Although the construction normally considered Type V is the conventional light-frame wood building, any combination of approved materials can be considered Type V construction. Section 602.1.1 indicates that a building is not required to conform to the details of a type of construction higher than the type that meets the minimum requirements based on occupancy, even though certain features of such a building actually conform to a higher construction type.

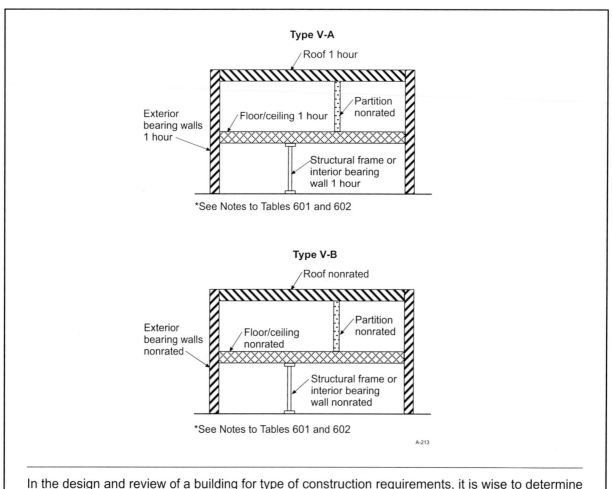

In the design and review of a building for type of construction requirements, it is wise to determine first if the structure can be built as a Type VB building, based on occupancy, location on property, height and floor area. If so, any other building type is also permitted.

Code Text: *Combustible materials shall be permitted in buildings of Type I and II construction in the following applications:* (22 applications listed).

Discussion and Commentary: Materials used in the construction of buildings classified as either Type I or Type II are intended to be noncombustible, thereby not increasing the potential fire loading (fuel contribution). There are, however, a number of applications where the presence of combustible building materials is desirable in otherwise noncombustible structures. Such materials are typically permitted where they are adequately protected, limited in use or amount, or installed in accordance with the *International Fire Code*, *International Mechanical Code*®, or other provisions of the *International Building Code*.

Combustible materials permitted in buildings of Type I and
Type II construction in the following applications:

- Fire-retardant-treated wood in:
  - Roof construction of most buildings.
  - Nonbearing partitions with fire-resistance rating ≤ 2 hours.
  - Nonbearing exterior walls requiring no fire rating.
- Thermal and acoustical insulation with limited flame spread.
- Foam plastics per Chapter 26.
- A, B or C roof coverings.
- Interior floor finish, trim, millwork such as, doors, frames, etc.
- Stages and platforms per Section 410.
- Blocking for handrails, fixtures, windows and door frames, etc.
- Light-transmitting plastics per Chapter 26.
- Nailing or furring strips per Section 803.4.
- Heavy timber for specific components.
- Additional applications as specified.

In Type I and II construction, the use of fire-retardant-treated wood is permitted in roof construction of all Type II buildings and those buildings of Type I construction that do not exceed two stories or have a top-story height of at least 20 feet.

Code Text: *Combustible materials shall be permitted in buildings of Type I or II construction in accordance with Sections 603.1.1 through 603.1.3. The use of nonmetallic ducts shall be permitted when installed in accordance with the limitations of the* International Mechanical Code. *The use of combustible piping materials shall be permitted when installed in accordance with the limitations of the* International Mechanical Code *and the* International Plumbing Code®. *The use of electrical wiring methods with combustible insulation, tubing, raceways and related components shall be permitted when installed in accordance with the limitations of the Electrical Code.*

Discussion and Commentary: The IMC contains requirements for nonmetallic ducts that address the issues of flammability, flame spread and smoke development. The IPC regulates the use of combustible piping materials, such as plastic, and also addresses those same characteristics applicable to nonmetallic ducts. Similar regulations apply to combustible wiring materials regulated under the Electrical Code. These provisions in Chapter 6 clarify that such combustible materials are acceptable for installation in buildings of Type I and II construction, provided they meet the limitations set forth in the appropriate code.

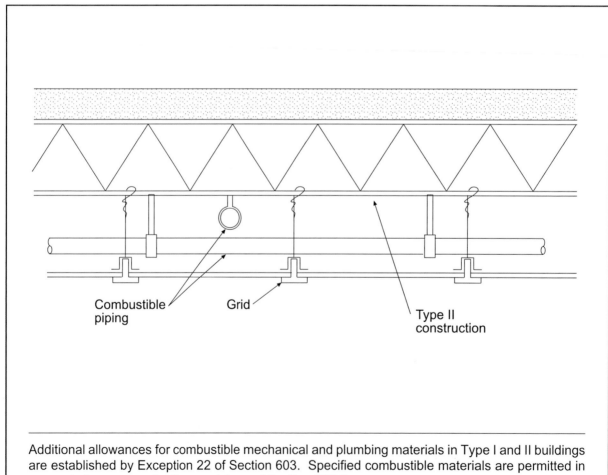

Combustible piping   Grid   Type II construction

Additional allowances for combustible mechanical and plumbing materials in Type I and II buildings are established by Exception 22 of Section 603. Specified combustible materials are permitted in concealed spaces under the provisions of Section 717.5.

Topic: Structural Frame	**Category:** Types of Construction
Reference: IBC Table 601, Note a	**Subject:** Building Elements

Code Text: *The structural frame shall be considered to be the columns and the girders, beams, trusses and spandrels having direct connections to the columns and bracing members designed to carry gravity loads. The members of floor or roof panels which have no connection to the columns shall be considered secondary members and not a part of the structural frame.*

Discussion and Commentary: To maintain stability of the building as a whole, the major structural elements are regulated for endurance when subjected to a fire. In addition to the columns, beams and girders, both interior bearing walls and exterior bearing walls are regulated to a level of fire resistance equal to or greater than that of other structural elements. Secondary members, such as floor joists, roof joists or rafters, are protected within the rated floor-ceiling or roof-ceiling assemblies.

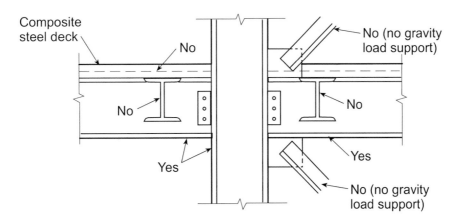

Structural Frame Considered to be:

- **Columns**
- **Girders**
- **Beams**
- **Trusses**
- **Spandrels**

Having direct connection to the columns and bracing members designed to carry gravity loads

Composite steel deck

Components of structural frame

A-017

Lateral force bracing is not considered part of the structural frame where it serves no other purpose than to resist the lateral loads. For example, lateral load bracing within exterior nonbearing walls or interior partitions would be protected by the wall or partition construction.

Code Text: *Except in Group F-1, H, M and S-1 occupancies, fire protection of structural members shall not be required, including protection of roof framing and decking where every part of the roof construction is 20 feet or more above any floor immediately below. Fire-retardant-treated wood members shall be allowed to be used for such unprotected members.*

Discussion and Commentary: Where there is limited potential for a fire to be of a severe nature at the roof structure due to its height above the floor below, an elimination of the required fire-resistance rating of the roof construction is permitted. Elimination of the required fire resistance is not allowed where combustible or hazardous materials are located adjacent to the roof.

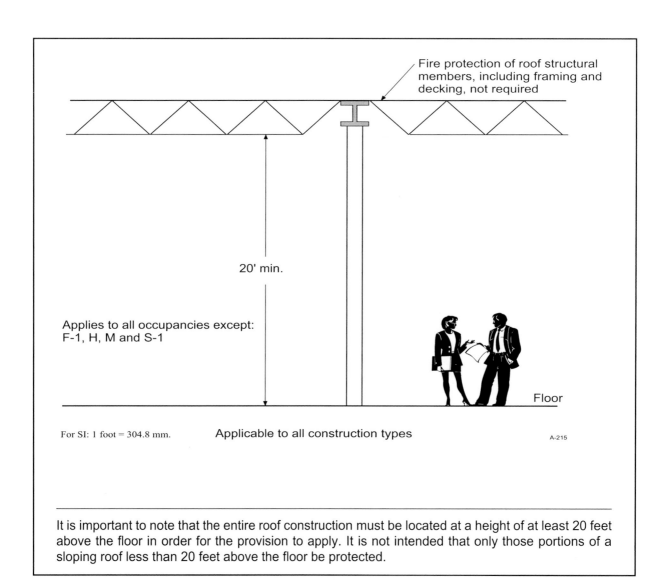

Fire protection of roof structural members, including framing and decking, not required

20' min.

Applies to all occupancies except: F-1, H, M and S-1

Floor

For SI: 1 foot = 304.8 mm.

Applicable to all construction types

A-215

It is important to note that the entire roof construction must be located at a height of at least 20 feet above the floor in order for the provision to apply. It is not intended that only those portions of a sloping roof less than 20 feet above the floor be protected.

Topic: Fire-Resistant Substitution   **Category:** Types of Construction
Reference: IBC Table 601, Note e   **Subject:** Building Elements

Code Text: *An approved automatic sprinkler system in accordance with Section 903.3.1.1 shall be allowed to be substituted for 1-hour fire-resistance-rated construction, provided such system is not otherwise required by other provisions of the code or used for an allowable area increase in accordance with Section 506.3 or an allowable height increase in accordance with Section 504.2. The 1-hour substitution for the fire resistance of exterior walls shall not be permitted.*

Discussion and Commentary: In buildings of Type IIA, IIIA or VA construction, the code allows for a reduction in the "built-in" fire protection, due to the presence of a sprinkler system that would not otherwise be required. This reduction does not apply to exterior walls, nor does it apply when a sprinkler system is used for an increase in allowable height or allowable area.

Type II-B Construction	+	Approved Automatic Sprinkler System*	=	Type II-A Construction
Type III-B Construction	+	Approved Automatic Sprinkler System*	=	Type III-A Construction
Type V-B Construction	+	Approved Automatic Sprinkler System*	=	Type V-A Construction

*Unless otherwise required, or used for allowable height and area increase

---

The application of this note to Table 601 may be extremely limited based on the language of Section 901.2, which states that *any fire protection system for which an exception or reduction to the provisions of the code has been granted shall be considered to be a required system.*

Quiz

Study Session 3
IBC Chapter 6

1. What types of construction are considered "noncombustible"?

    a. I, II   b. I, II, III, IV

    c. III, IV   d. III, IV, V

    Reference_____

2. Type III buildings shall have _____ exterior walls and interior elements _____.

    a. fire-resistant, of noncombustible materials

    b. noncombustible, of noncombustible materials

    c. noncombustible, of any material permitted by the code

    d. combustible, of any material permitted by the code

    Reference_____

3. A Type IV building is also described as _____ construction.

    a. ordinary   b. heavy timber

    c. combustible   d. provincial

    Reference_____

4. In buildings of Type III and IV construction, under what condition is fire-retardant-treated wood framing permitted within exterior wall assemblies?

    a. the wall has a 2-hour rating or less

    b. the fire separation distance exceeds 10 feet

    c. wood columns of heavy-timber sizes are used

    d. the wall is a nonbearing element

    Reference_____

5. Where supporting floor loads, wood columns of Type IV construction shall be of what minimum nominal size?

    a. 5 inches by 5 inches   b. 6 inches by 6 inches

    c. 6 inches by 8 inches   d. 8 inches by 8 inches

    Reference_____

6. Where used in floor framing, wood beams of Type IV construction shall be of what minimum nominal size?

    a. 4 inches by 8 inches   b. 4 inches by 10 inches

    c. 6 inches by 10 inches   d. 8 inches by 10 inches

    Reference_____

7. Roofs shall be without concealed spaces in buildings of Type _____ construction.

    a. I   b. III

    c. IV   d. V

    Reference_____

8. Which of the following materials is permitted in a building of Type VB construction?

    a. wood   b. steel

    c. masonry   d. all of the above

    Reference _____

9. In a building of Type IB construction, what is the minimum required fire-resistance rating of the floor construction?

    a. 3 hours   b. 2 hours

    c. 1 hour   d. 0 hours (no rating required)

Reference_____

10. Which one of the following members is not considered to be a part of the structural frame?

    a. columns

    b. girders

    c. bracing members carrying gravity loads

    d. floor joists

Reference_____

11. In a building of Type IIB construction, what is the minimum required fire-resistance rating of the roof construction?

    a. 0 hours (no rating required)   b. 1 hour

    c. $1^1/_2$ hours   d. 2 hours

Reference_____

12. In a building of Type IIIA construction, what is the minimum required fire-resistance rating of the floor construction?

    a. 0 hours (no rating required)   b. 1 hour

    c. $1^1/_2$ hours   d. 2 hours

Reference_____

13. In a one-story Type IA building, what is the minimum fire-resistance rating for the interior bearing walls supporting the roof only?

    a. 3 hours   b. 2 hours

    c. $1^1/_2$ hours   d. 1 hour

Reference_____

14. In Type I and II buildings limited to two stories in height, what building element is permitted to be constructed of fire-retardant-treated wood?

   a. structural frame

   b. bearing walls

   c. floor construction

   d. roof construction

   Reference_____

15. In a Type IA building housing a Group A-4 occupancy, fire protection of the roof structural members, framing and decking is not required where every portion of the roof construction is a minimum of _____ feet above the floor below.

   a. 18

   b. 20

   c. 25

   d. 35

   Reference_____

16. In a Type IIB building housing a Group I-2 occupancy, what is the minimum rating of an exterior nonbearing wall located with a fire separation distance of 8 feet?

   a. 3 hours

   b. 2 hours

   c. 1 hour

   d. 0 hours (no rating required)

   Reference_____

17. In a Type IIA building housing a Group R-2 occupancy, what is the minimum rating of an exterior nonbearing wall located with a fire separation distance of 3 feet?

   a. 3 hours

   b. 2 hours

   c. 1 hour

   d. 0 hours (no rating required)

   Reference_____

18. In a Type IIIA building housing a Group A-1 occupancy, what is the minimum rating of an exterior bearing wall located with a fire separation distance of 10 feet?

   a. 3 hours

   b. 2 hours

   c. 1 hour

   d. 0 hours (no rating required)

   Reference_____

19. In a Type VB building housing a Group B occupancy, what is the minimum rating of an exterior bearing wall located with a fire separation distance of 10 feet?

　　a. 3 hours　　　　　　　　　b. 2 hours

　　c. 1 hour　　　　　　　　　d. 0 hours (no rating required)

Reference_____

20. Considering all occupancies and construction types, for which minimum fire separation distance is no fire rating required for a nonbearing exterior wall?

　　a. 20 feet　　　　　　　　　b. 30 feet

　　c. 40 feet　　　　　　　　　d. 60 feet

Reference_____

21. In a building of Type I or II construction, nonbearing partitions having a maximum fire-resistance rating of _____ are permitted to be constructed of fire-retardant-treated wood.

　　a. 0 hours (no rating required)　　b. $\frac{1}{2}$ hour

　　c. 1 hour　　　　　　　　　d. 2 hours

Reference_____

22. A detached garage classified as a Group U occupancy shall not be required to have a fire-resistance rating where the fire separation distance is a minimum of _____ feet.

　　a. 0, no rating is required　　　b. 3

　　c. 5　　　　　　　　　　　d. 10

Reference_____

23. In a building of Type I or II construction, nonbearing exterior walls having a maximum fire-resistance rating of _____ are permitted to be constructed of fire-retardant-treated wood.

　　a. 0 hours (no rating required)　　b. $\frac{1}{2}$ hour

　　c. 1 hour　　　　　　　　　d. 2 hours

Reference_____

24. Show windows may be of combustible construction in Type I and II construction where located a maximum of _____ above grade.

    a. 15 feet   b. 35 feet

    c. one story   d. three stories

Reference_____

25. Which type of roof covering is not permitted on an office building of Type IIB construction?

    a. Class A   b. Class B

    c. Class C   d. nonclassified

Reference_____

26. Which of the following methods of construction is not permitted for partitions in Type IV structures?

    a. two layers of $^1/_2$-inch fire-retardant-treated structural wood panels

    b. 4 inches of laminated solid-wood construction

    c. two layers of 1-inch matched boards

    d. one-hour fire-resistance-rated construction

Reference_____

27. What is the minimum fire-resistive rating required for interior metal stud partitions in Type IIB construction?

    a. 0, no rating is required   b. 1 hour

    c. 2 hours   d. 3 hours

Reference_____

28. What is the minimum required vertical distance between the upper floor and the roof to allow the use of fire-retardant-treated wood in the roof construction of an eight-story Type IB building?

    a. 20 feet

    b. 25 feet

    c. there is no minimum distance required

    d. FRT wood is never permitted in such a case

Reference_____

29. For an office building of Type IB construction, what is the minimum required fire-resistance rating for an exterior bearing wall located on an interior lot line?

    a. 0, no rating is required   b. 1 hour

    c. 2 hours   d. 3 hours

Reference_____

30. In a building of Type IIA construction, heavy timber members may be used in lieu of one-hour fire-resistance-rated construction for which building element?

    a. structural frame members   b. interior bearing walls

    c. floor construction   d. roof construction

Reference_____

31. In a building of Type IV construction, 1-hour combustible construction is permitted in lieu of heavy-timber construction for which of the following building elements?

    a. floor construction   b. exterior bearing walls

    c. interior bearing walls   d. roof construction

Reference _____

32. Where an exterior bearing wall of a Group M occupancy of Type IIB construction is located 3 feet from an interior lot line, the wall must have a minimum fire-resistance rating of _____ hour(s).

    a. 0 (no rating required)   b. 1

    c. 2   d. 3

Reference _____

33. A nonbearing exterior wall of a Type IB open parking garage complying with Section 406 shall have a minimum fire-resistance rating of _____ hour(s) where the fire separation distance is 10 feet.

    a. 0 (no rating required)   b. 1

    c. 2   d. 3

Reference _____

34. A glued-laminated beam, where utilized in a Type IV building requiring a 6-inch by 10-inch solid-sawn member of nominal size, shall have a minimum net finished size of _____.

    a. $5^1/_4$ inches by $9^1/_4$ inches

    b.   6 inches by 10 inches

    c. $6^3/_4$ inches by $10^1/_2$ inches

    d.   5 inches by $10^1/_2$ inches

Reference _____

35. Under general conditions, thermal and acoustical insulation other than foam plastic is permitted to be installed in buildings of Type I or II construction, provided the insulation has a maximum flame spread index of _____.

    a.   25   b.   75

    c.   200   d.   450

Reference _____

2006 IBC Chapter 5
General Building Heights and Areas

OBJECTIVE: To gain an understanding of how a building is classified and regulated based on its floor area, height and number of stories.

REFERENCE: Chapter 5, 2006 *International Building Code*

KEY POINTS:
- How and why must buildings be identified by their address?
- How is the maximum basic floor area of a building determined? The basic height in feet and number of stories?
- How is the tabular allowable building area determined?
- How is a basement viewed in the calculation of maximum allowable floor area?
- How must multiple buildings located on the same lot be handled?
- What types of special industrial occupancies are exempt from the height and area limitations of Table 503?
- Under which conditions may the basic allowable height of a building be increased? How much of an increase is permitted?
- Which special provisions address the construction of towers, spires, steeples and other roof structures?
- What is a mezzanine?
- Under which conditions may a floor level be considered a mezzanine?
- Under what conditions is a mezzanine permitted to be enclosed or in some manner separated from the room in which it is located?
- Which conditions provide for an increase in the allowable floor areas specified in Table 503?
- What is the minimum width of a yard or public way that can provide for a floor area increase?
- How much of a building's perimeter must be considered "open" for the area increase to apply?
- What amount of area increase is permitted for sprinklered single-story buildings? For sprinklered multistory structures?
- How is the maximum total combined floor area for a multistory building determined?

KEY POINTS:
(Cont'd)

- Which occupancy groups are eligible for unlimited floor area in a one-story nonsprinklered building?
- Which criteria must be met for buildings of Groups A-4, B, F, M or S to be unlimited in floor area? Groups A-1 and A-2?
- What is the minimum width required for open space surrounding an unlimited area building?
- Under which conditions may that width be reduced?
- Under what limitations is a Group A-3 occupancy permitted to be of unlimited area?
- How may high-hazard occupancies be accommodated in an unlimited area building?
- Which limitations are placed on motion picture theaters of unlimited area?
- Which special provisions address the situation where an enclosed or open parking garage is located below another occupancy group? Below an open parking garage?
- What are the height limitations for apartment houses and other Group R-2 occupancies where the special provisions are met?
- How is an open parking garage regulated where located below Groups A, B, I, M or R?

Code Text: *Building area is the area included within surrounding exterior walls (or exterior walls and fire walls) exclusive of vent shafts and courts. Areas of the building not provided with surrounding walls shall be included in the building area if such areas are included within the horizontal projection of the roof or floor above.*

Discussion and Commentary: The building area must be determined in order to verify that it does not exceed the maximum allowable area as determined by Section 503.1. The building area is considered, in very general terms, the "footprint" of the building, excluding those unroofed areas and any projections that may extend beyond the exterior walls. Where complying mezzanines are located within a building, they are not assumed to contribute to the building area.

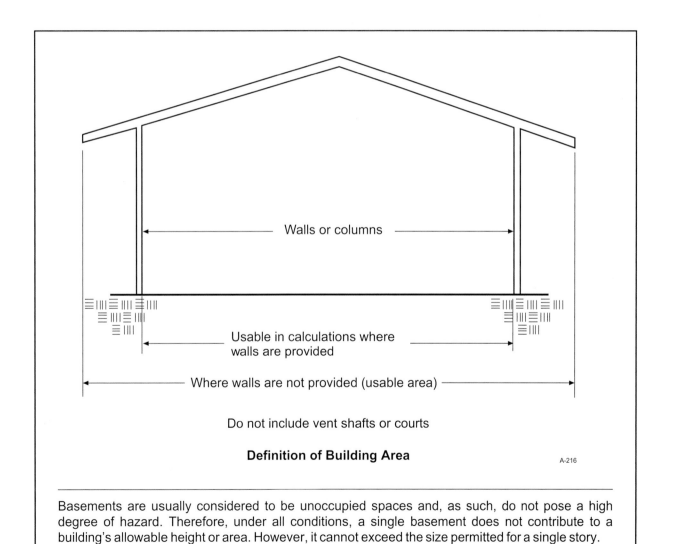

Walls or columns

Usable in calculations where walls are provided

Where walls are not provided (usable area)

Do not include vent shafts or courts

Definition of Building Area

A-216

Basements are usually considered to be unoccupied spaces and, as such, do not pose a high degree of hazard. Therefore, under all conditions, a single basement does not contribute to a building's allowable height or area. However, it cannot exceed the size permitted for a single story.

Code Text: *A "story above grade plane" is any story having its finished floor surface entirely above grade plane. A basement is that portion of a building that is partly or completely below grade plane. A basement shall be considered as a story above grade plane where the finished surface of the floor above the basement is 1) more than 6 feet above grade plane, or 2) more than 12 feet above the finished ground level at any point.*

Discussion and Commentary: A number of provisions in the IBC are applicable based on the location of the floor under consideration, relative to the exterior ground level. Therefore, it is necessary to define specifically the circumstances under which a floor level is considered a story above grade plane.

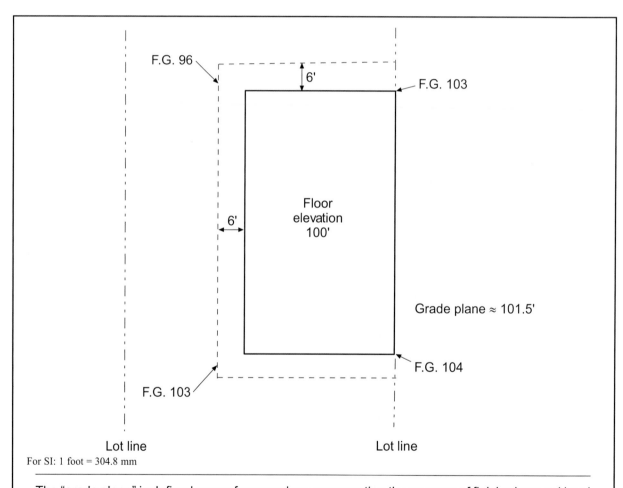

For SI: 1 foot = 304.8 mm

The "grade plane" is defined as a reference plane representing the average of finished ground level adjoining the building at exterior walls. It is measured at the lowest point between the building and the lot line, though never more than 6 feet from the building.

Code Text: *The height and area for buildings of different construction types shall be governed by the intended use of the building and shall not exceed the limits in Table 503 except as modified hereafter.*

Discussion and Commentary: Table 503 is the foremost code provision used in establishing equivalent risk (offsetting a building's inherent fire hazard—represented by occupancy group—with materials and construction features). Sections 504 and 506 give height and area increases to the limits of Table 503 for buildings with certain features. Table 503 has three components. The left column lists all of the occupancy classifications, the top row lists the various construction types, and each cell of the matrix contains the specific height (in stories above grade) and area (per floor) limitations for the occupancy group/type of construction combination in question.

TABLE 503
ALLOWABLE HEIGHT AND BUILDING AREAS[a]
Height limitations shown as stories and feet above grade plane.
Area limitations as determined by the definition of "Area, building," per story

		TYPE OF CONSTRUCTION								
		TYPE I		TYPE II		TYPE III		TYPE IV	TYPE V	
		A	B	A	B	A	B	HT	A	B
GROUP	HGT(feet) HGT(S)	UL	160	65	55	65	55	65	50	40
A-1	S	UL	5	3	2	3	2	3	2	1
	A	UL	UL	15,500	8,500	14,000	8,500	15,000	11,500	5,500
A-2	S	UL	11	3	2	3	2	3	2	1
	A	UL	UL	15,500	9,500	14,000	9,500	15,000	11,500	6,000
A-3	S	UL	11	3	2	3	2	3	2	1
	A	UL	UL	15,500	9,500	14,000	9,500	15,000	11,500	6,000
A-4	S	UL	11	3	2	3	2	3	2	1
	A	UL	UL	15,500	9,500	14,000	9,500	15,000	11,500	6,000
A-5	S	UL	UL	UL	UL	UL	UL	UL	UL	UL
	A	UL	UL	UL	UL	UL	UL	UL	UL	UL
B	S	UL	11	5	4	5	4	5	3	2
	A	UL	UL	37,500	23,000	28,500	19,000	36,000	18,000	9,000
E	S	UL	5	3	2	3	2	3	1	1
	A	UL	UL	26,500	14,500	23,500	14,500	25,500	18,500	9,500
F-1	S	UL	11	4	2	3	2	4	2	1
	A	UL	UL	25,000	15,500	19,000	12,000	33,500	14,000	8,500
F-2	S	UL	11	5	3	4	3	5	3	2
	A	UL	UL	37,500	23,000	28,500	18,000	50,500	21,000	13,000

(Continued)

The maximum height in feet above grade plane is shown along a top row and is based solely on construction type. Buildings must meet both the height in stories and the height in feet criteria to be considered in compliance with that type of construction.

Code Text: *Buildings and structures designed to house special industrial processes that require large areas and unusual heights to accommodate craneways or special machinery and equipment, including among others, rolling mills; structural metal fabrication shops and foundries; or the production and distribution of electric, gas or steam power, shall be exempt from the height and area limitations of Table 503.*

Discussion and Commentary: A limited number of buildings that house special industrial processes need extensive heights and/or areas for their operations. The activities that occur are generally of moderate to low hazard, and the buildings are not typically accessible to the public. Therefore, it has been deemed appropriate that no type of construction limitations should be placed on these unique structures.

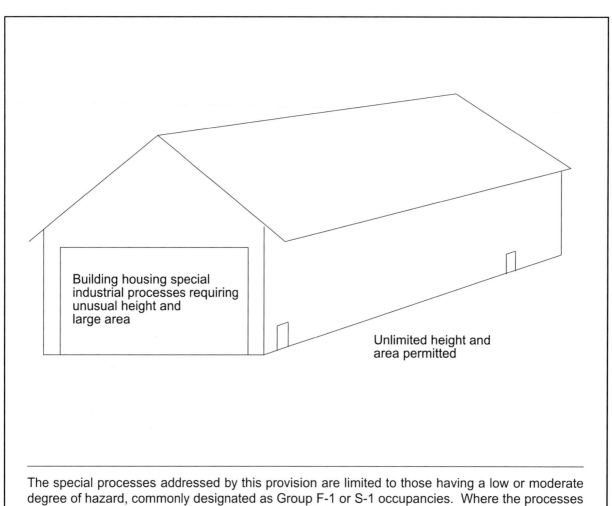

Building housing special industrial processes requiring unusual height and large area

Unlimited height and area permitted

The special processes addressed by this provision are limited to those having a low or moderate degree of hazard, commonly designated as Group F-1 or S-1 occupancies. Where the processes under consideration would necessitate a Group H classification because of the high hazards involved, the application of this provision is inappropriate.

Code Text: *Two or more buildings on the same lot shall be regulated as separate buildings or shall be considered as portions of one building if the height of each building and the aggregate area of buildings are within the limitations of Table 503 as modified by Sections 504 and 506. The provisions of the code applicable to the aggregate building shall be applicable to each building.*

Discussion and Commentary: In general, the provisions of Section 704.3 require an assumed imaginary line to be located between two buildings on the same site to regulate exterior wall and opening protection, as well as roof-covering requirements. This alternate method would provide protection equivalent to that of buildings on adjoining lots.

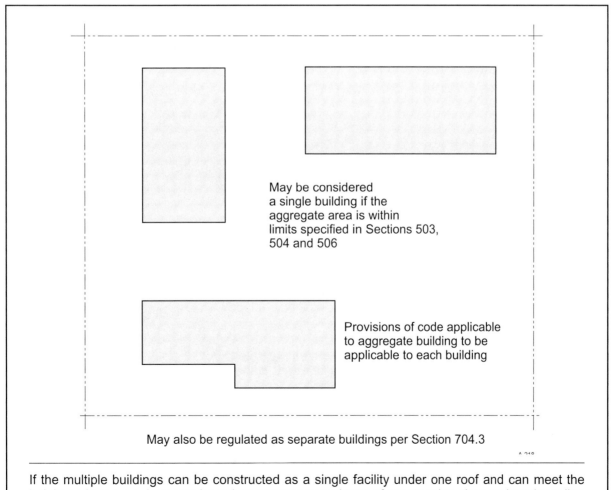

May be considered a single building if the aggregate area is within limits specified in Sections 503, 504 and 506

Provisions of code applicable to aggregate building to be applicable to each building

May also be regulated as separate buildings per Section 704.3

If the multiple buildings can be constructed as a single facility under one roof and can meet the height and area requirements based on occupancy and type of construction, then an imaginary line need not be assumed. The buildings will simply be considered a single structure.

Code Text: *Where a building is equipped throughout with an approved automatic sprinkler system installed in accordance with Section 903.3.1.1, the value specified in Table 503 for maximum height is increased by 20 feet (6096 mm) and the maximum number of stories is increased by one story. These increases are permitted in addition to the area increase in accordance with Sections 506.2 and 506.3.* See exceptions for specific Group I and H occupancies and for buildings utilizing the fire-resistance-rating substitution method.

Discussion and Commentary: The installation of an automatic sprinkler system improves the fire-safety aspects of a building to the degree that an increase in height is justified. The presence of a sprinkler system also can be used to provide the sizable area increase permitted in Section 506.3.

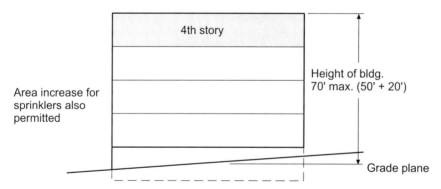

Maximum height and number of stories based upon occupancy and type of construction as set forth in Table 503.

Given: A Type VA office building is permitted to be 3 stories and 50 feet in height per Table 503.

If sprinkler system installed per Section 504.2, the story limit may be increased by one story, and the height can exceed the limit in Table 503 by 20 feet.

4th story

Area increase for sprinklers also permitted

Height of bldg.
70' max. (50' + 20')

Grade plane

Increase does not apply when sprinkler installed under following conditions:
1. Group I-2 fire areas of Type IIB, III, IV or V construction
2. Group H-1, H-2, H-3 or H-5 fire areas
3. Fire-resistance-rating substitution per Table 601, note e

A-219a

For SI: 1 foot = 304.8 mm

In a residential building protected with an NFPA 13R sprinkler system, the increase is permitted, provided the building does not exceed four stories or 60 feet. The provision does not intend for a residential sprinkler system to be installed in buildings over four stories in height.

Topic: Roof Structures

Category: Building Heights and Areas

Reference: IBC 504.3

Subject: Height Modifications

Code Text: *Towers, spires, steeples and other roof structures shall be constructed of materials consistent with the required type of construction of the building except where other construction is permitted by Section 1509.2.1. Such structures shall not be used for habitation or storage. The structures shall be unlimited in height if of noncombustible materials and shall not extend more than 20 feet (6096 mm) above the allowable height if of combustible materials.*

Discussion and Commentary: The types of structures addressed by this provision are intended to be unoccupied with no significant fire loading. It would seem logical that the height of such structures could be increased over that required for typical buildings. The only limitation occurs where the structure is of combustible materials, which would create a higher hazard.

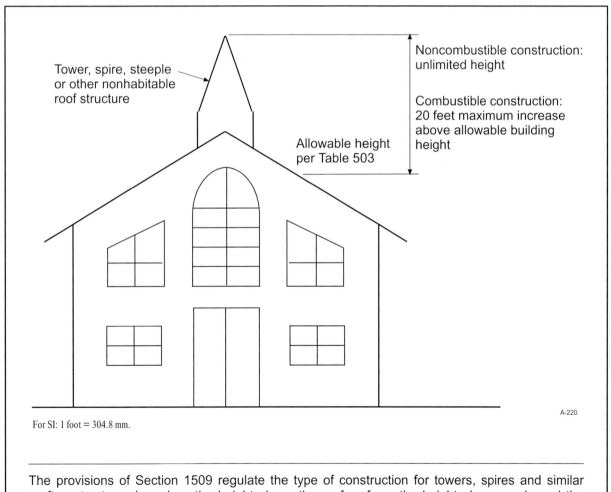

For SI: 1 foot = 304.8 mm.

A-220

The provisions of Section 1509 regulate the type of construction for towers, spires and similar rooftop structures based on the height above the roof surface, the height above grade and the largest cross-sectional dimension. Penthouses and equipment screening are also addressed.

Code Text: *A mezzanine is an intermediate level or levels between the floor and ceiling of any story and in accordance with Section 505. The aggregate area of a mezzanine or mezzanines within a room shall not exceed one-third of the floor area of that room or space in which they are located.* See exceptions which allow for increased mezzanine sizes in 1) special industrial occupancies of Type I or II construction, and 2) fully sprinklered Type I or II buildings provided with an approved emergency voice/alarm communication system.

Discussion and Commentary: Because of size limitation and openness (a mezzanine is open to the room in which it is located, with exceptions), an intermediate floor level within a room adds minimal hazard to the building and its occupants. The occupants of the mezzanine by means of sight, smell or hearing will be able to determine if there is some emergency or fire taking place either on the mezzanine or in the room in which the mezzanine is located.

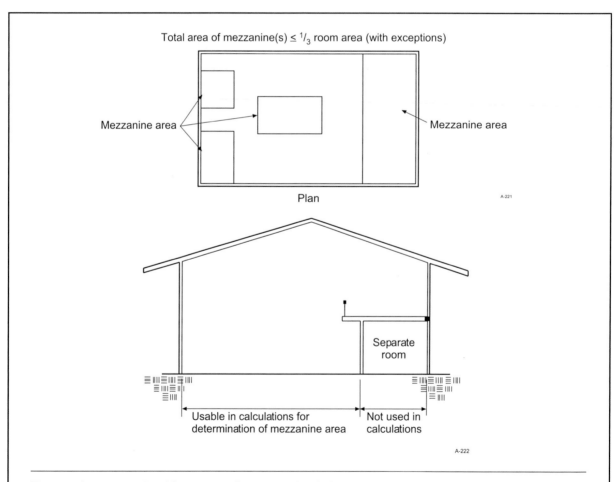

The maximum permitted floor area of a mezzanine is based on the floor area of the room in which it is located. Only those portions of the lower room that are unenclosed may be considered in the calculation of maximum mezzanine size.

Code Text: *A mezzanine or mezzanines in compliance with Section 505 shall be considered a portion of the story below. Such mezzanines shall not contribute to either the building area or number of stories as regulated by Section 503.1. The area of a mezzanine shall be included in determining the fire area defined in Section 702.*

Discussion and Commentary: There are two distinct benefits derived from the qualification of a floor level as a mezzanine. One, the mezzanine is not considered in the allowable number of stories, and two, for allowable area purposes, the mezzanine floor area does not increase the building area of the story in which it is located. However, in the determination of fire area size for sprinkler requirements, the floor area must be considered. The requirements for sprinkler systems are generally based on the fire load expected in an occupancy; thus, an increased floor area would increase the potential fire loading. The provisions of Section 505.3 also provide for a third benefit in regard to number of egress paths required.

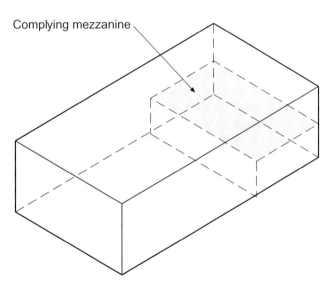

Complying mezzanine

Mezzanine:
- Does <u>not</u> contribute to floor area for maximum allowable area
- Does <u>not</u> contribute as an additional story
- Does contribute to floor area for fire area size determination

Example:
For 8,000 sq ft first floor as shown with 2,000 sq ft mezzanine, building area is 8,000 sq ft, building is one story in height, and fire area is 10,000 sq ft

For SI: 1 square foot = 0.93 m^2.

A-223

Although it is quite possible that an individual floor level within a building can meet all of the provisions of the IBC and qualify as a mezzanine, its actual designation is the choice of the designer. It may be more advantageous to treat the floor level simply as an additional story.

Code Text: *A mezzanine shall be open and unobstructed to the room in which such mezzanine is located except for walls not more than 42 inches high, columns and posts.* See exceptions addressing mezzanines 1) where the enclosed area has a maximum occupant load of 10, 2) having two or more means of egress with at least one providing direct access to an exit, 3) where the aggregate floor area of the enclosed space does not exceed 10 percent of the mezzanine area, 4) in industrial facilities, and 5) in fully sprinklered buildings where two or more means of egress are provided.

Discussion and Commentary: By definition, a mezzanine is intended to be open to the room or space below. This common environment allows individuals on either floor level to be aware of the conditions and hazards that may affect their safety. The IBC, through the application of one of five exceptions, permits the mezzanine to be enclosed when it has been determined that the enclosure creates little, if any concern.

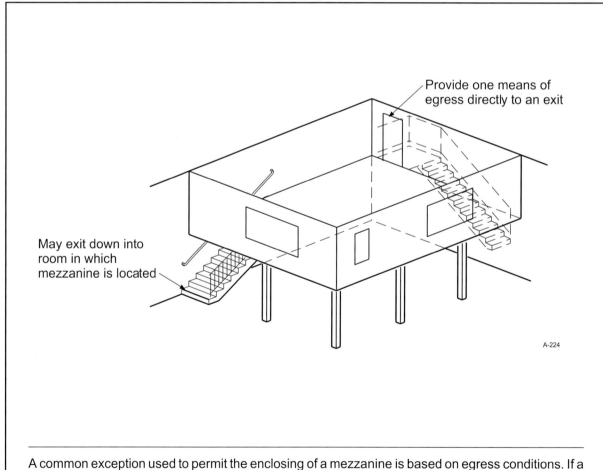

Provide one means of egress directly to an exit

May exit down into room in which mezzanine is located

A-224

A common exception used to permit the enclosing of a mezzanine is based on egress conditions. If a minimum of two means of egress are provided from the mezzanine level, with at least one such egress path leading directly to an exit, then the mezzanine is not required to be open to the room or space below.

Code Text: *The areas limited by Table 503 shall be permitted to be increased due to frontage (I_f) and automatic sprinkler system protection (I_s) in accordance with the following:*

$A_a = \{A_t = [A_t \times I_f] + [A_t \times I_s]\}$ *where:*

A_a *= Allowable area per story (square feet)*

A_t *= Tabular area per story in accordance with Table 503 (square feet)*

I_f *= Area increase factor due to frontage as calculated in accordance with Section 506.2*

I_s *= Area increase factor due to sprinkler protection as calculated in accordance with Section 506.3*

Discussion and Commentary: The tabular values established by Table 503 are merely the baseline when determining the maximum allowable heights and areas for buildings based on occupancy group and construction type. Increases to the tabular values are typically permitted where the building is surrounded by adequate open space and/or sprinklered throughout.

Example:

Given: A two-story type VB building housing a Group I-4 occupancy. The building is provided with an automatic sprinkler system throughout and located on the lot as shown.

Determine: The maximum allowable building area per floor.

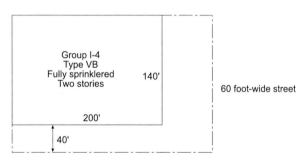

Solution:

Tabular area (Table 503) A_t = 9000 sq ft

Frontage increase (Section 506.2)

$I_f = [F/P - 0.25]\ W/30$
     $= [340/680 - 0.25]\ 30/30$ (where $W > 30$; value of 30 to be used)
     $= [0.50 - 0.25]$
     $= 0.25$

Sprinkler increase (Section 506.3)
     $I_s = 2\ (200\%)$

Total allowable area (Section 506.3)
     $A_a = A_t + [A_t + I_f] + [A_t + I_s]$
        $= 9000 + [9000(0.25)] + [9000(2)]$
        $= 9000 + 2250 + 18,000$
        $= 29,250$ sq ft per floor

For SI: 1 square foot = 0.093m^2, 1 foot = 304.8 mm

A single basement that is not considered a story above grade plane need not be considered in the determination of the total allowable building area, provided the basement does not exceed the area permitted for a one-story building.

Code Text: *Every building shall adjoin or have access to a public way to receive an area increase for frontage. Where a building has more than 25 percent of its perimeter on a public way or open space having a minimum width of 20 feet, the frontage increase shall be determined in accordance with the following:* $I_f = [F/P -0.25]$ W/30.

Discussion and Commentary: It is assumed that every building will adjoin a street, alley or yard on at least one side. Therefore, no frontage increase is given for the first 25 percent of a building's perimeter that is open. Credit is provided, however, where additional frontage is considered open (20 feet or more in width). The benefit of increased allowable building area is accrued based on better access for the fire department, as well as decreased exposure to adjoining properties.

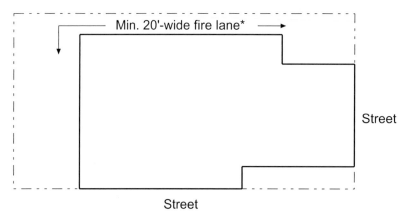

Entire perimeter considered for frontage increase

Min. 20'-wide fire lane*

Street

Street

Open space to be on same lot or dedicated for public use, and accessed from a street or approved fire lane

*Fire lane need only be provided to within 150 feet of exterior wall per Section 503.1.1 of the IFC.

A-226

For SI: 1 foot = 304.8 mm

Access must be provided from a street or an approved fire lane for any open space that is used for a frontage increase in allowable floor area. The *International Fire Code* mandates that a fire lane for fire apparatus be maintained with an unobstructed width of at least 20 feet.

Code Text: *Where a building is protected throughout with an approved automatic sprinkler system in accordance with Section 903.3.1.1, the area limitation in Table 503 is permitted to be increased by an additional 200 percent (I_s = 2) for buildings with more than one story above grade plane and an additional 300 percent (I_s = 3) for buildings with no more than one story above grade plane.* See exceptions for Group H-1, H-2 and H-3 occupancies, and for buildings utilizing the fire-resistance-rated substitution allowance.

Discussion and Commentary: Because of its excellent record of in-place fire suppression, an automatic sprinkler system provides for a sizable allowable area increase. The presence of the sprinkler system also allows a height increase as addressed in Section 504.2. However, sprinkler systems are considered an absolute necessity in occupancies associated with high hazard levels, such as Groups H-1, H-2 and H-3; thus, no size increases are permitted for such uses.

Examples

Given:	Group B occupancy single-story
	Type VB construction
	No open yards available
Find:	Total allowable area
	Basic allowable area = 9,000 sq ft (Table 503)
	Sprinkler increase (*Is*) = 27,000 sq ft (300%)
	Total allowable area = 36,000 sq ft
Given:	Same situation, however two stories in height
Find:	Total allowable area
	Basic allowable area = 9,000 sq ft (Table 503)
	Sprinkler increase (I_s) = 18,000 sq ft (200%)
	Total allowable area per floor = 27,000 sq ft

For SI: 1 foot = 304.8 mm

It is assumed that in many cases, fire department suppression activities will supplement an automatic sprinkler system. Because a multistory building presents more problems to the fire department than a single-story structure, a smaller increase in area is justified.

Code Text: *The maximum area of a building with more than one story above grade plane shall be determined by multiplying the allowable area of the first floor (A_a), as determined in Section 506.1, by the number of stories above grade plane as listed: 1) for buildings with two stories above grade plane, multiply by 2; 2) for buildings with three or more stories above grade plane, multiply by 3; and 3) no story shall exceed the allowable area per floor (A_a), as determined by Section 506.1 for the occupancy of that story.* See exceptions for unlimited area buildings and residential buildings sprinklered with an NFPA 13R system.

Discussion and Commentary: The tabular allowable building areas set forth in Table 503 are limited on a story-by-story basis. The presence of sufficient open space adjacent to a building provides for an increase above the tabular value. The protection afforded by an automatic sprinkler system justifies a significant allowable area increase. The IBC permits both increases to be used to provide the maximum building area permitted per story.

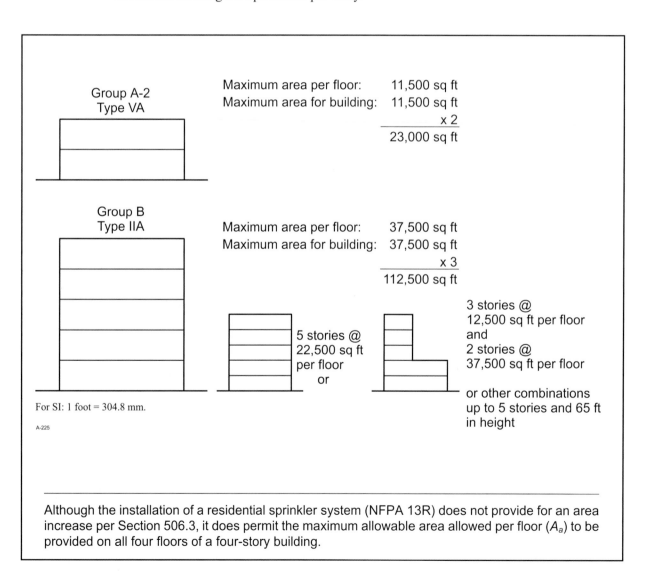

Group A-2
Type VA

Maximum area per floor: 11,500 sq ft
Maximum area for building: 11,500 sq ft
                                       x 2
                            23,000 sq ft

Group B
Type IIA

Maximum area per floor: 37,500 sq ft
Maximum area for building: 37,500 sq ft
                                       x 3
                           112,500 sq ft

5 stories @
22,500 sq ft
per floor
or

3 stories @
12,500 sq ft per floor
and
2 stories @
37,500 sq ft per floor

or other combinations
up to 5 stories and 65 ft
in height

For SI: 1 foot = 304.8 mm.

A-225

Although the installation of a residential sprinkler system (NFPA 13R) does not provide for an area increase per Section 506.3, it does permit the maximum allowable area allowed per floor (A_a) to be provided on all four floors of a four-story building.

Code Text: *The area of a one-story, Group B, F, M or S building or a one-story Group A-4 building, of other than Type V construction, shall not be limited when the building is provided with an automatic sprinkler system throughout in accordance with Section 903.3.1.1, and is surrounded and adjoined by public ways or yards not less than 60 feet (18 288 mm) in width. Provisions also apply to two-story buildings of such occupancies other than Group A-4.*

Discussion and Commentary: It is often beneficial to have very large, undivided floor areas for facilities such as arenas, office buildings, factories, retail centers and warehouses. The unlimited area provisions allow for an alternative to the higher types of construction that would normally be required. The installation of a sprinkler system and sufficient open space around the building reduce the potential fire severity to a reasonable level in these moderate-hazard occupancies.

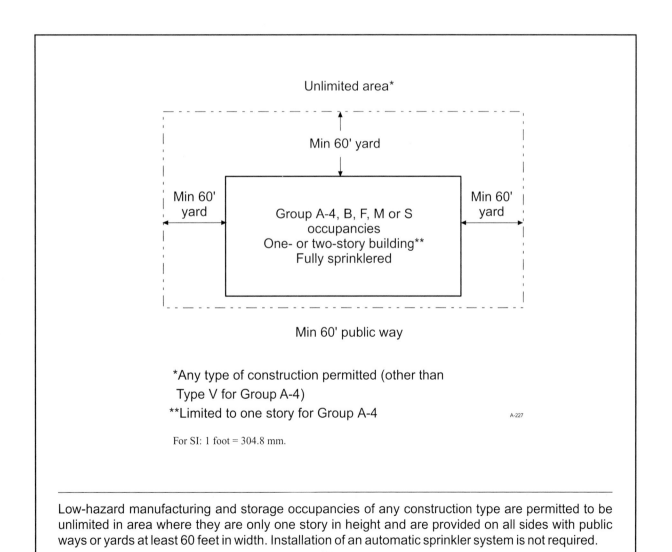

Unlimited area*

Min 60' yard

Min 60' yard

Min 60' yard

Group A-4, B, F, M or S occupancies
One- or two-story building**
Fully sprinklered

Min 60' public way

*Any type of construction permitted (other than Type V for Group A-4)
**Limited to one story for Group A-4

A-227

For SI: 1 foot = 304.8 mm.

Low-hazard manufacturing and storage occupancies of any construction type are permitted to be unlimited in area where they are only one story in height and are provided on all sides with public ways or yards at least 60 feet in width. Installation of an automatic sprinkler system is not required.

Topic: Reduced Open Space

Reference: IBC 507.5

Category: Building Heights and Areas

Subject: Unlimited Area Buildings

Code Text: *The permanent open space of 60 feet (18 288 mm) required in Sections 507.2, 507.3, 507.4, 507.6 and 507.10 shall be permitted to be reduced to not less than 40 feet (12 192 mm) provided the following requirements are met: 1) the reduced open space shall not be allowed for more than 75 percent of the perimeter of the building, 2) the exterior wall facing the reduced open space shall have a minimum fire-resistance rating of 3 hours, and 3) openings in the exterior wall facing the reduced open space shall have opening protectives with a fire-resistance rating of 3 hours.*

Discussion and Commentary: When it is necessary or desirable to reduce the open space around the perimeter of an unlimited area building, the code provides an alternative. An equivalent level of protection can be provided by increasing the level of exterior wall and opening protection.

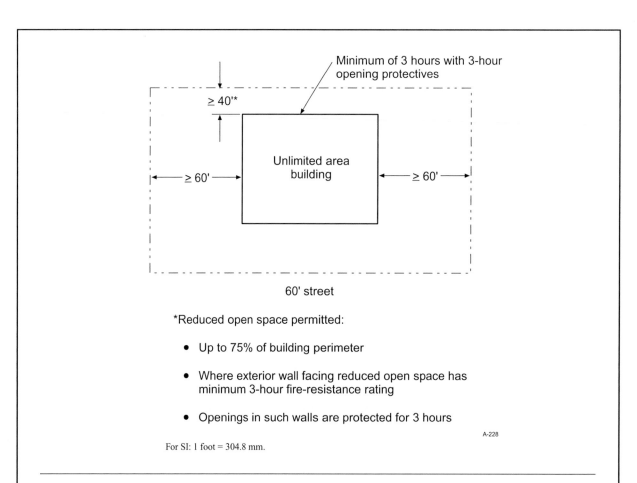

60' street

*Reduced open space permitted:

- Up to 75% of building perimeter

- Where exterior wall facing reduced open space has minimum 3-hour fire-resistance rating

- Openings in such walls are protected for 3 hours

A-228

For SI: 1 foot = 304.8 mm.

This provision is designed for warehouses, factories, retail stores, office buildings, Group A-3 uses and movie theaters where fire resistance at the exterior wall is easily accomplished. The reduction does not apply to other buildings permitted to be unlimited in area, such as educational uses and aircraft paint hangars.

Code Text: *The area of a one-story, Group A-3 building used as a place of religious worship, community hall, dance hall, exhibition hall, gymnasium, lecture hall, indoor swimming pool or tennis court of Type II construction shall not be limited when all of the following criteria are met:* See four conditions for allowance of unlimited area.

Discussion and Commentary: The Group A-3 occupancy classification includes the most diverse types of assembly uses assigned by the code. Traditionally, the allowable area of Group A occupancies is greatly limited as compared to most other occupancy groups. However, those assembly uses expected to have a relatively low fire load are permitted in unlimited area buildings subject to the special conditions prescribed by the code.

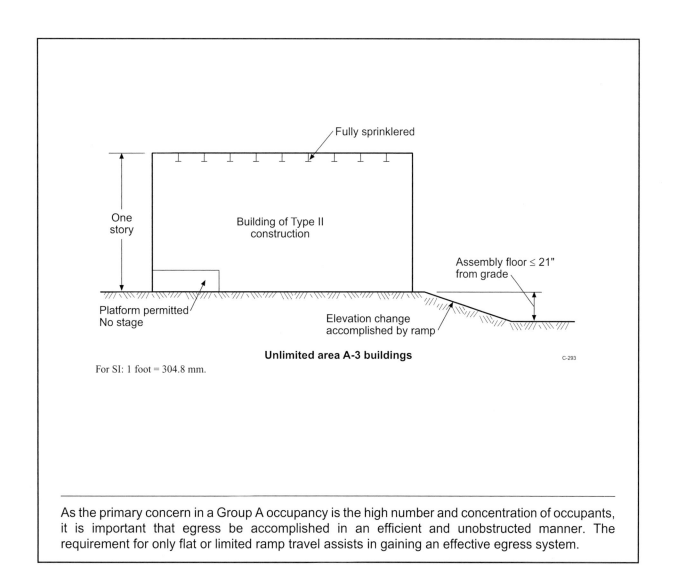

Unlimited area A-3 buildings

C-293

For SI: 1 foot = 304.8 mm.

As the primary concern in a Group A occupancy is the high number and concentration of occupants, it is important that egress be accomplished in an efficient and unobstructed manner. The requirement for only flat or limited ramp travel assists in gaining an effective egress system.

Code Text: *Group H-2, H-3 and H-4 occupancies shall be permitted in unlimited area buildings containing Group F and S occupancies, in accordance with Sections 507.3 and 507.4 and the limitations of Section 507.7. The aggregate floor area of the Group H occupancies located at the perimeter of the unlimited area building shall not exceed 10 percent of the area of the building nor the area limitations for the Group H occupancies as specified in Table 503 as modified by Section 506.2, based upon the percentage of the perimeter of each Group H fire area that fronts on a street or other unoccupied space. The aggregate floor area of Group H occupancies not located at the perimeter of the building shall not exceed 25 percent of the area limitations for the Group H occupancies as specified in Table 503.*

Discussion and Commentary: The aggregate allowable area of the permitted Group H occupancies in a factory or warehouse is dependent on the type of construction of the building and the location of the Group H occupancy in the buildings.

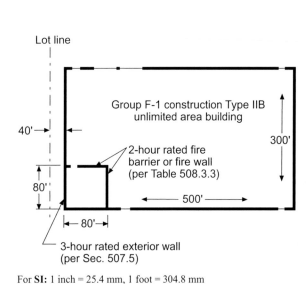

Lot line

Group F-1 construction Type IIB unlimited area building

2-hour rated fire barrier or fire wall (per Table 508.3.3)

40'

300'

80'

500'

80'

3-hour rated exterior wall (per Sec. 507.5)

For **SI:** 1 inch = 25.4 mm, 1 foot = 304.8 mm
1 square foot = 0.0929 m²

- Group H-2
- In Equation 5-2
  F = 160 ft
  P = 320 ft
  W = 30 ft
  (The maximum per Sec. 506.2.1)
- In Equation 5-2,

$$I_f = 100 \left[\frac{160}{320} -0.25 \right] \frac{30}{30}$$

$$I_f = 25\%$$

- Allowable area for H-2 = 7,000 + (0.25)(7,000)
  = 8,750 sq ft
- Check 10% of floor area criterium:
  (500)(300) = 150,000 sq ft
  150,000/10 = 15,000 sq ft
  15,000 > 8,750, ∴ 8,750 maximum allowable

- Actual area = 6,400, which is less than 8,750, therefore OK

Group H-2 at the corner of an unlimited area Group F or S building

C-294

More ready access to the Group H from the exterior of the building provides the fire department with an opportunity to respond more effectively to an incident. As such, the allowable floor area of the Group H can be far greater than where completely surrounded by the Group F or S use.

Topic: Open Parking Garages

Category: Building Heights and Areas

Reference: IBC 509.7

Subject: Special Provisions

Code Text: *Open parking garages constructed under Groups A, I, B, M and R shall not exceed the height and area limitations permitted under Section 406.3. The height and area of the portion of the building above the open parking garage shall not exceed the limitations in Section 503 for the upper occupancy. The height, in both feet and stories, of the portion of the building above the open parking garage shall be measured from grade plane and shall include both the open parking garage and the portion of the building above the parking garage.*

Discussion and Commentary: In the more common types of occupancies, it is desirable at times to provide tiers of parking below the major use of the building. In this special mixed-use condition, two different types of construction are permitted for determining the maximum allowable height and area. This special allowance is just one of several special provisions established in Section 509 that modify the general requirements of the code.

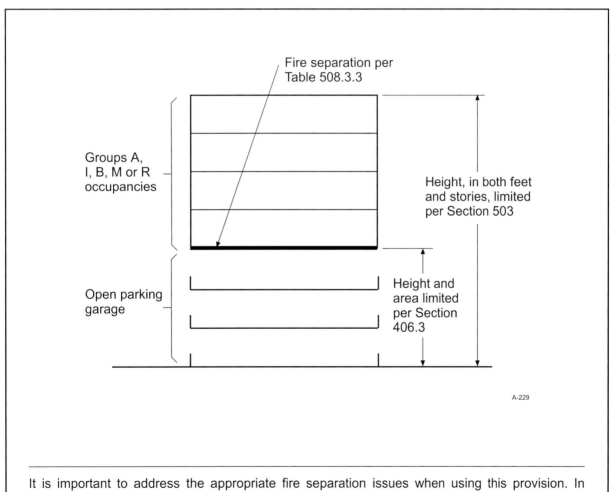

Fire separation per
Table 508.3.3

Groups A,
I, B, M or R
occupancies

Height, in both feet
and stories, limited
per Section 503

Open parking
garage

Height and
area limited
per Section
406.3

A-229

It is important to address the appropriate fire separation issues when using this provision. In addition, the structural members supporting the upper occupancy must be protected by the more restrictive fire-resistant assemblies of all of the occupancies involved.

Quiz

Study Session 4

IBC Chapter 5

1. Premises must be identified by numbers or addresses visible from the street, with a minimum character height of at least _____ inches.

   a. three   b. four

   c. six   d. eight

   Reference_____

2. Building height is measured to the _____.

   a. average height of the highest roof surface

   b. highest point of the highest roof surface

   c. average height of all of the roof surfaces

   d. highest point of the lowest roof surface

   Reference_____

3. The tabular allowable height and building area per floor for a Group I-2 occupancy of Type IB construction is _____ stories and _____ square feet.

   a. 3, 10,000   b. 2, 15,000

   c. 4, unlimited   d. unlimited, unlimited

   Reference_____

4. What is the maximum tabular allowable height, in feet, for a Group B occupancy of Type IIA construction?

a. 50

b. 55

c. 65

d. unlimited

Reference_____

5. A single basement does not need to be included in the total allowable area of a building, provided it does not exceed _____.

a. one-third the floor area permitted for any single story

b. the area permitted for a one-story building

c. twice the area permitted for a single story

d. the tabular area based on construction type and occupancy group

Reference_____

6. A basement is considered a story above grade plane where the finished surface of the floor above the basement is more than _____ feet above grade plane.

a. 3

b. 4

c. 5

d. 6

Reference _____

7. The maximum building area of a six-story building is limited to _____ times the allowable area permitted per story.

a. two

b. three

c. four

d. six

Reference_____

8. An allowable height increase of _____ and _____ is permitted in a building housing a Group B occupancy and protected throughout with an automatic sprinkler system.

a. 15 feet, one story

b. 20 feet, one story

c. 15 feet, two stories

d. 20 feet, two stories

Reference_____

9. In a Type IIB building, an increase in allowable height is not permitted for a fire area that contains which one of the following occupancy groups?

    a. Group A-1   b. Group H-4

    c. Group I-2   d. Group R-2

Reference_____

10. Combustible steeples are limited to a maximum height of _____ feet above the allowable building height.

    a. 15   b. 20

    c. 30   d. 40

Reference_____

11. The area of a mezzanine is not to be included in the determination of the _____.

    a. fire area   b. building area

    c. occupant load   d. plumbing fixture count

Reference_____

12. A minimum clear height of _____ is required above and below mezzanine floor construction.

    a. 6 feet, 8 inches   b. 7 feet

    c. 7 feet, 6 inches   d. 8 feet

Reference_____

13. In general, the aggregate area of mezzanines within a room is limited to _____ of the area of the room in which the mezzanines are located.

    a. 10 percent   b. 25 percent

    c. $33^1/_3$ percent   d. 50 percent

Reference_____

14. Portions of a mezzanine need not be open to the room in which the mezzanine is located, provided the enclosed space is limited in size to a maximum of _____ of the mezzanine area.

    a. 10 percent   b. 25 percent

    c. $33^1/_3$ percent   d. 50 percent

Reference_____

15. A mezzanine is not required to be open to the room in which it is located where the occupant load of the enclosed space does not exceed _____.

    a. 10   b. 20

    c. 30   d. 50

Reference_____

16. An allowable area increase for frontage is not permitted where a maximum of _____ of the building perimeter is sufficiently open.

    a. 10 percent   b. 25 percent

    c. $33^1/_3$ percent   d. 40 percent

Reference_____

17. In order to be considered as sufficiently open for an allowable area increase for frontage, the public way or open space must have a minimum width of _____ feet.

    a. 10   b. 20

    c. 25   d. 30

Reference_____

18. What is the maximum tabular height, in number of stories, for a Group R-1 occupancy of Type IIB construction?

    a. 3   b. 4

    c. 5   d. 6

Reference _____

19. An increase in allowable height is permitted for a Group _____ fire area in a building fully sprinklered in accordance with NFPA 13.

    a. H-2   b. H-3

    c. H-4   d. H-5

    Reference _____

20. The allowable building area may be increased by _____ in a multistory building protected throughout with an approved automatic sprinkler system installed in accordance with NFPA 13.

    a. 50 percent   b. 100 percent

    c. 200 percent   d. 300 percent

    Reference_____

21. In a fully-sprinklered single-story building, the allowable building area may be increased by _____.

    a. 50 percent   b. 100 percent

    c. 200 percent   d. 300 percent

    Reference_____

22. Which one of the following occupancy groups may be located in a nonsprinklered one-story building of Type IIB construction under the unlimited area building provisions?

    a. Group B   b. Group F-1

    c. Group S-2   d. Group U

    Reference_____

23. In order to reduce the required open space surrounding certain unlimited area buildings from 60 feet to 40 feet, walls facing the reduced open space shall have a minimum fire-resistance rating of _____.

    a. 45 minutes   b. 1 hour

    c. 2 hours   d. 3 hours

    Reference_____

24. A complying Group H-2 aircraft paint hangar may be unlimited in floor area when limited to one story, provided the hangar is surrounded by yards or public ways having a minimum width of _____.

    a. 40 feet

    b. 60 feet

    c. twice the height of the hangar

    d. one and one-half times the height of the hangar

Reference_____

25. Where utilizing the special provisions for height and area of Section 509, an enclosed parking garage is permitted beneath all but which of the following occupancy groups?

    a. Group A-3   b. Group F-2

    c. Group R-1   d. Group R-2

Reference_____

26. What is the maximum allowable height permitted for a noncombustible communications tower located on the roof of a Type IIB office building?

    a. 55 feet   b. 65 feet

    c. 75 feet   d. unlimited

Reference_____

27. In order for a Type IIIB office building to be considered for unlimited area, it must be limited to a maximum height of _____.

    a. one story   b. two stories

    c. 50 feet   d. 40 feet

Reference_____

28. A one-story Group A-3 gymnasium may be considered under the unlimited area provisions, provided the assembly floor is located a maximum of _____ above or below grade level.

    a. 21 inches   b. 24 inches

    c. 30 inches   d. 48 inches

Reference_____

29. An H-3 storage room located within an unlimited area manufacturing building of Type IIB construction is limited to _____ square feet where not located on the building's perimeter.

    a. 3,125   b. 3,500

    c. 12,500   d. 14,000

    Reference_____

30. A nine-story apartment building of Type IIA construction shall be located a minimum of _____ feet from any other building on the lot and from all lot lines.

    a. 10 feet   b. 30 feet

    c. 50 feet   d. 60 feet

    Reference_____

31. In buildings of Type I or II construction housing special industrial occupancies in accordance with Section 503.1.1, the aggregate floor area of mezzanines is limited to a maximum of _____ of the area of the room in which they are located.

    a. $^1/_4$ b. $^1/_3$

    c. $^1/_2$ d. $^2/_3$

    Reference _____

32. Where the aggregate area of all equipment platforms within a room is limited to a maximum of _____ of the area of the room in which they are located, the equipment platforms shall not be considered a portion of the floor below.

    a. $^1/_4$ b. $^1/_3$

    c. $^1/_2$ d. $^2/_3$

    Reference _____

33. The area of a complying covered mall building and any anchor stores is not limited where their maximum height is _____ stories.

    a. 2   b. 3

    c. 4   d. 5

    Reference _____

34. A Group E building of unlimited area shall be surrounded and adjoined by public ways and yards a minimum of _____ feet in width.

    a. 30   b. 40

    c. 60   d. 75

Reference _____

35. In a fully sprinklered building, an allowable area increase is not permitted for the floor area of a Group _____ occupancy.

    a. H-3   b. I-2

    c. I-3   d. R-2

Reference _____

Study Session

5

2006 IBC Sections 701 – 704
Fire-resistance-rated Construction I

OBJECTIVE: To gain an understanding of the fundamentals of fire-resistance-rated construction, the methods for the determination of fire resistance, and the regulation of exterior walls for fire-resistance rating and opening protection.

REFERENCE: Sections 701 through 704, 2006 *International Building Code*

KEY POINTS:
- Why are fire-resistance-rated materials and systems used in the construction of buildings?
- What is a fire-resistance rating? How is such a rating determined?
- What referenced standard is the basis for determining fire-resistance ratings?
- What is nonsymmetrical wall construction?
- How must interior walls and partitions of nonsymmetrical construction be tested? Exterior walls?
- Which alternative methods are available for determining the fire-resistance rating of different building elements?
- What is a prescriptive design of fire-resistance-rated building elements? What specific types of elements are addressed?
- Which types of materials can be evaluated for fire-resistance ratings through calculations?
- When is a material considered noncombustible? Is gypsum board considered a noncombustible material?
- How is a projection defined? What types of building elements are considered projections?
- What limits the extent of a projection?
- Which types of projections are permitted from walls of Type I and II buildings? Type III, IV and V buildings?
- When must combustible projections be protected? What other options are available?
- How are multiple buildings on the same lot addressed in regard to exterior wall and opening protection?
- Under which conditions must fire-resistance-rated exterior walls be rated for fire exposure from both sides?

KEY POINTS:
(Cont'd)

- Under which conditions are openings in exterior walls prohibited? What are the limitations where protected openings are provided? Unprotected openings?
- How are protected and unprotected openings regulated in the same exterior wall?
- How does the presence of a sprinkler system affect the amount of unprotected openings?
- Which special provisions apply to openings in the first story of exterior walls?
- When must exterior openings in adjacent stories be protected? Where required, what methods of protection are available?
- What is a parapet? Where are parapets required?
- What fire-resistance rating is mandated for required parapets?
- What is the minimum required height of a parapet? How does the slope of the roof affect the minimum required height?

Code Text: *The provisions of Chapter 7 shall govern the materials and assemblies used for structural fire resistance and fire-resistance-rated construction separation of adjacent spaces to safeguard against the spread of fire and smoke within a building and the spread of fire to or from buildings.*

Discussion and Commentary: There are basically two reasons for the protection of various building elements with construction resistant to fire. One, structural elements such as columns, girders, bearing walls and other loadbearing members are often required by the code to maintain their structural integrity under fire conditions for a prescribed time period. Two, horizontal and vertical assemblies are used to create compartments, including control areas, or to isolate portions of the building, such as exitways, through fire-resistant construction.

FIRE RESISTANCE. That property of materials or their assemblies that prevents or retards the passage of excessive heat, hot gases or flames under conditions of use.

In addition to limiting or resisting the spread of fire and heat within a building, certain provisions are intended to provide protection for adjoining structures. The code also uses fire-resistance-rated construction to restrict the passage of smoke to specific areas.

Code Text: *Fire-resistance rating is the period of time a building element, component or assembly maintains the ability to confine a fire, continues to perform a given structural function, or both as determined by the tests, or the methods based on tests, prescribed in Section 703. The fire-resistance rating of building elements shall be determined in accordance with the test procedures set forth in ASTM E 119 or in accordance with Section 703.3 (Alternative methods for determining fire resistance).*

Discussion and Commentary: ASTM E 119 is the referenced standard, *Standard Test Methods for Fire Tests of Building Construction and Materials*. These test methods are used for the great majority of building components or assemblies that are mandated by the code to have a fire-resistance rating.

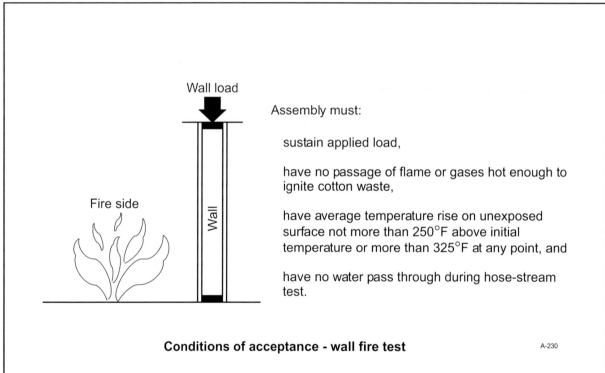

Wall load

Fire side

Wall

Assembly must:

sustain applied load,

have no passage of flame or gases hot enough to ignite cotton waste,

have average temperature rise on unexposed surface not more than 250°F above initial temperature or more than 325°F at any point, and

have no water pass through during hose-stream test.

Conditions of acceptance - wall fire test

A-230

For nonsymmetrical wall construction, where interior walls and partitions are provided with differing membranes on opposing sides, the IBC mandates that tests be performed from both sides. The side with the shortest test duration is the basis for the fire-resistance rating.

Code Text: *The application of any of the alternative methods listed in Section 703.3 shall be based on the fire exposure and acceptance criteria specified in ASTM E 119. The required fire resistance of a building element shall be permitted to be established by any of the following methods or procedures: (1) fire-resistance designs documented in approved sources, (2) prescriptive designs of fire-resistance-rated building elements as prescribed in Section 720, (3) calculations in accordance with Section 721, (4) engineering analysis based on a comparison of building element designs having fire-resistance ratings as determined by the test procedures set forth in ASTM E 119, or (5) alternative protection methods as allowed by Section 104.11.*

Discussion and Commentary: Prescriptive details of fire-resistance-rated building elements are contained in Section 720. Generic listings for structural parts, walls, partitions, floor systems and roof systems are addressed.

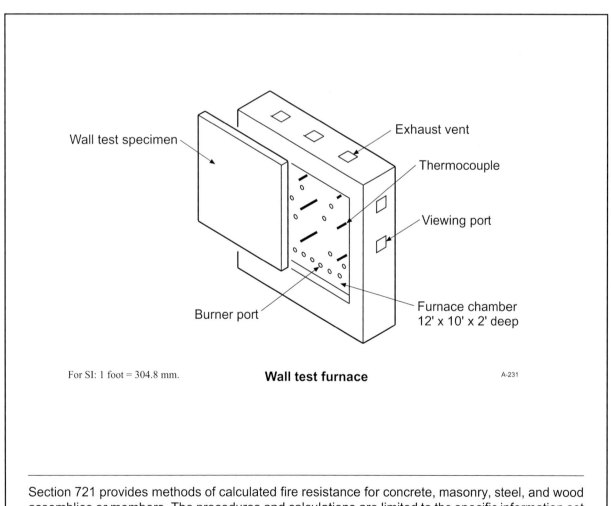

For SI: 1 foot = 304.8 mm.

Wall test furnace

A-231

Section 721 provides methods of calculated fire resistance for concrete, masonry, steel, and wood assemblies or members. The procedures and calculations are limited to the specific information set forth in this section and are not to be used in any other manner.

Code Text: *The provisions of Section 720 contain prescriptive details of fire-resistance-rated building elements. The materials of construction listed in Tables 720.1(1), 720.1(2) and 720.1(3) shall be assumed to have the fire-resistance ratings prescribed therein. Where materials that change the capacity for heat dissipation are incorporated into a fire-resistance-rated assembly, fire test results or other substantiating data shall be made available to the building official to show that the required fire-resistance rating time period is not reduced.*

Discussion and Commentary: The tables in Section 720 provide the details for obtaining desired fire-resistance ratings for structural parts, walls and partitions, and floor and roof systems. The methods and materials found in the tables are to be used in the same manner as any listed assembly.

Table 720.1(2)
Interior partition: Item 14-1.5

2 in. by 4 in. wood studs 16 in. on center with two layers $^5/_8$ in. Type X gypsum wallboard each side. Base layers applied vertically and nailed with 6d cooler or wallboard nails at 9 in. on center. Face layer applied vertically or horizontally and nailed with 8d cooler or wallboard nails at 7 in. on center. For nail-adhesive application, base layers are nailed 6 in. on center. Face layers applied with coating of approved wallboard adhesive and nailed 12 in. on center.

For SI: 1 inch = 25.4 mm.

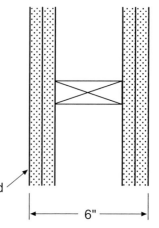

2 hour rated

6"

A-232

When insulation or a similar material is added to a fire-resistance-rated assembly, it may change the assembly's capacity to dissipate heat. Particularly in noncombustible horizontal assemblies, the fire-resistance rating may be diminished to some degree.

Topic: Scope

Category: Fire-resistance-rated Construction

Reference: IBC 721.1

Subject: Calculated Fire Resistance

Code Text: *The provisions of Section 721 contain procedures by which the fire resistance of specific materials or combinations of materials is established by calculations. The procedures apply only to the information contained in Section 721 and shall not be otherwise used. The calculated fire resistance of concrete, concrete masonry, and clay masonry assemblies shall be permitted in accordance with ACI 216.1/TMS 0216. The calculated fire resistance of steel assemblies shall be permitted in accordance with Chapter 5 of ASCE 29.*

Discussion and Commentary: Another method used to obtain the necessary fire-resistance ratings mandated by the code is calculation. The provisions for calculating fire resistance are applicable to concrete assemblies, concrete masonry, clay brick and tile masonry, steel assemblies and wood assemblies.

TABLE 721.6.2(1)
TIME ASSIGNED TO WALLBOARD MEMBRANES[a,b,c,d]

DESCRIPTION OF FINISH	TIME[e] (minutes)
$^3/_8$-inch wood structural panel bonded with exterior glue	5
$^{15}/_{32}$-inch wood structural panel bonded with exterior glue	10
$^{19}/_{32}$-inch wood structural panel bonded with exterior glue	15
$^3/_8$-inch gypsum wallboard	10
$^1/_2$-inch gypsum wallboard	15
$^5/_8$-inch gypsum wallboard	30
$^1/_2$-inch Type X gypsum wallboard	25
$^5/_8$-inch Type X gypsum wallboard	40
Double $^3/_8$-inch gypsum wallboard	25
$^1/_2$- + $^3/_8$-inch gypsum wallboard	35
Double $^1/_2$-inch gypsum wallboard	40

For SI: 1 inch = 25.4 mm.

a. These values apply only when membranes are installed on framing members which are spaced 16 inches o.c.

b. Gypsum wallboard installed over framing or furring shall be installed so that all edges are supported, except $^5/_8$-inch Type X gypsum wallboard shall be permitted to be installed horizontally with the horizontal joints staggered 24 inches each side and unsupported but finished.

c. On wood frame floor/ceiling or roof/ceiling assemblies, gypsum board shall be installed with the long dimension perpendicular to framing members and shall have all joints finished.

d. The membrane on the unexposed side shall not be included in determining the fire resistance of the assembly. When dissimilar membranes are used on a wall assembly, the calculation shall be made from the least fire-resistant (weaker) side.

e. The time assigned is not a finished rating.

TABLE 721.6.2(2)
TIME ASSIGNED FOR CONTRIBUTION OF WOOD FRAME [a,b,c]

DESCRIPTION	TIME ASSIGNED TO FRAME (minutes)
Wood studs 16 inches o.c.	20
Wood floor and roof joists 16 inches o.c.	10

For SI: 1 inch = 25.4 mm.

a. This table does not apply to studs or joists spaced more than 16 inches o.c.

b. All studs shall be nominal 2 × 4 and all joists shall have a nominal thickness of at least 2 inches.

c. Allowable spans for joists shall be determined in accordance with Sections 2308.8, 2308.10.2 and 2308.10.3.

Applicable to both loadbearing and nonloadbearing assemblies, the calculated fire resistance for wood-framed walls, floor/ceiling assemblies and roof/ceiling assemblies is limited to a 1-hour rating.

Code Text: *Materials required to be noncombustible shall be tested in accordance with ASTM E 136. The term "noncombustible" does not apply to the flame spread characteristics of interior finish or trim materials. A material shall not be classified as a noncombustible building construction material if it is subject to an increase in combustibility or flame spread beyond the limitations herein established through the effects of age, moisture, or other atmospheric conditions.*

Discussion and Commentary: In buildings of Types I, II, III and IV construction, specific elements are required to be constructed of noncombustible materials. Such materials are desirable because they do not aid combustion, nor do they add appreciable heat to an ambient fire. Under conditions of the test, a material may have a limited amount of combustible content and still qualify as noncombustible.

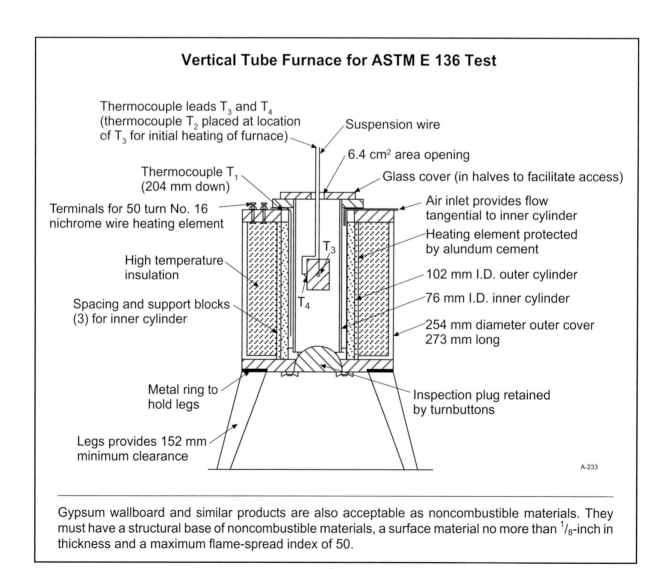

Vertical Tube Furnace for ASTM E 136 Test

Thermocouple leads T_3 and T_4 (thermocouple T_2 placed at location of T_3 for initial heating of furnace)

Suspension wire

6.4 cm^2 area opening

Thermocouple T_1 (204 mm down)

Glass cover (in halves to facilitate access)

Terminals for 50 turn No. 16 nichrome wire heating element

Air inlet provides flow tangential to inner cylinder

Heating element protected by alundum cement

High temperature insulation

T_3

102 mm I.D. outer cylinder

76 mm I.D. inner cylinder

Spacing and support blocks (3) for inner cylinder

T_4

254 mm diameter outer cover 273 mm long

Metal ring to hold legs

Inspection plug retained by turnbuttons

Legs provides 152 mm minimum clearance

A-233

Gypsum wallboard and similar products are also acceptable as noncombustible materials. They must have a structural base of noncombustible materials, a surface material no more than $1/8$-inch in thickness and a maximum flame-spread index of 50.

Topic: Extent of Projections

Reference: IBC 704.2

Category: Fire-resistance-rated Construction

Subject: Exterior Walls

Code Text: *Projections shall not extend beyond the distance determined by the following two methods, whichever results in the lesser projection: 1) a point one-third the distance to the lot line from an assumed vertical plane located where protected openings are required in accordance with Section 704.8, or 2) more than 12 inches (305 mm) into areas where openings are prohibited.*

Discussion and Commentary: Where projections extend beyond exterior walls that are in close proximity to a lot line, they create problems due to the trapping of convected heat from a fire in an adjacent building. By providing for some degree of separation between the lot line and the edge of the projection, the code allows for open space to assist in the heat's dissipation.

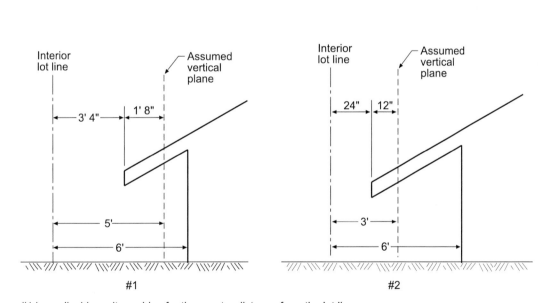

#1 is applicable as it provides for the greater distance from the lot line

Projections beyond the exterior wall

For SI: 1 inch = 25.4 mm, 1 foot = 304.8 mm.

Rather than regulate the maximum length a projection may extend beyond the building face, the projection limitation is established as the distance between the lot line and the edge of the projection. Two methods of measurement are described, with the applicable method resulting in the greater separation.

Code Text: *Projections from walls of Type I or II construction shall be of noncombustible materials or combustible materials as allowed by Sections 1406.3 and 1406.4. Projections from walls of Type III, IV or V construction shall be of any approved material. Combustible projections located where openings are not permitted or where protection of openings is required shall be of at least 1-hour fire-resistance-rated construction, Type IV construction, fire-retardant-treated wood or as required by Section 1406.3.*

Discussion and Commentary: Cornices, eave overhangs, exterior balconies and similar architectural appendages extending beyond the floor area are considered projections. Projections from noncombustible buildings are regulated to prevent a fire hazard created by inappropriate use of combustible materials at an exterior wall.

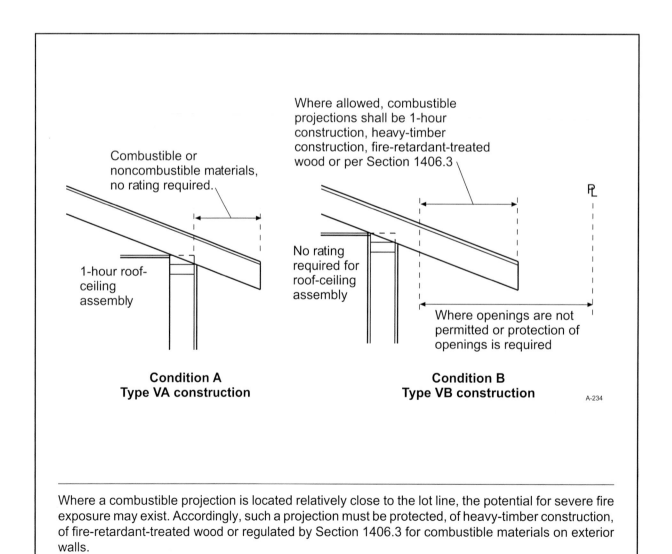

Combustible or noncombustible materials, no rating required.

1-hour roof-ceiling assembly

Condition A
Type VA construction

Where allowed, combustible projections shall be 1-hour construction, heavy-timber construction, fire-retardant-treated wood or per Section 1406.3

No rating required for roof-ceiling assembly

Where openings are not permitted or protection of openings is required

Condition B
Type VB construction

A-234

Where a combustible projection is located relatively close to the lot line, the potential for severe fire exposure may exist. Accordingly, such a projection must be protected, of heavy-timber construction, of fire-retardant-treated wood or regulated by Section 1406.3 for combustible materials on exterior walls.

Topic: Buildings on the Same Lot	**Category:** Fire-resistance-rated Construction
Reference: IBC 704.3	**Subject:** Exterior Walls

Code Text: *For the purposes of determining the required wall and opening protection and roof-covering requirements, buildings on the same lot shall be assumed to have an imaginary line between them.* See exception where aggregate area of multiple buildings is within limits specified in Chapter 5 for a single building.

Discussion and Commentary: Where two or more buildings are placed on the same piece of property, their exterior walls and openings must be regulated in the same manner as if they were on separate lots. However, if the buildings could be constructed as a single structure under one roof and meet the size requirements based on occupancy and type of construction, then an assumed imaginary line is not required.

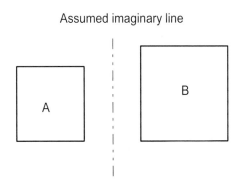

Assumed imaginary line

A   B

Case I: an assumed imaginary line between buildings

    A. Fire resistance and opening protection for walls adjacent to the lot line must comply with the code

    B. Lot line may be placed to take best advantage of wall and opening protection

Case II: as a single building

    A. Allowable area and type of construction are based on the most restrictive requirements for the occupancies housed

    B. Total floor area may not exceed that allowed for a single building

Buildings on the same property

A-235

Where a new building is to be constructed on the same lot as an existing building, the assumed imaginary line must be placed in a location where it will not cause the exterior wall and opening protection of the existing building to become noncompliant.

Code Text: *Exterior walls shall be fire-resistance rated in accordance with Tables 601 and 602. The fire-resistance rating of exterior walls with a fire separation distance of greater than 5 feet (1524 mm) shall be rated for exposure to fire from the inside. The fire-resistance rating of exterior walls with a fire separation distance of 5 feet (1524 mm) or less shall be rated for exposure to fire from both sides.*

Discussion and Commentary: Exposure of the exterior wall to an interior fire does not vary based on the distance of the wall from the lot line. However, exterior fire exposure decreases with an increase in the distance between the lot line and the exterior wall (fire separation distance). A fire separation distance of 5 feet is considered by the IBC to be a reasonable limit of flame impingement (direct exterior fire exposure) from an adjacent building.

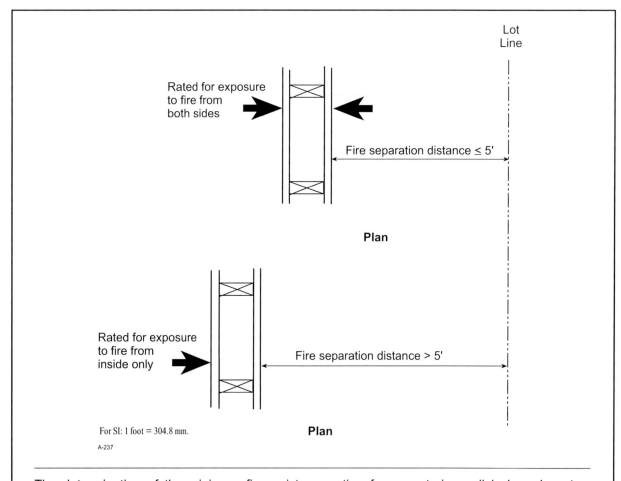

The determination of the minimum fire-resistance-rating for an exterior wall is based on two conditions: 1) the type of construction of the building, and 2) the fire separation distance. The higher of the two ratings regulates the minimum level of fire resistance.

Topic: Allowable Area of Openings	**Category:** Fire-resistance-rated Construction
Reference: IBC 704.8	**Subject:** Exterior Walls

Code Text: *The maximum area of unprotected or protected openings permitted in an exterior wall in any story shall not exceed the values set forth in Table 704.8. Where both unprotected and protected openings are located in the exterior wall in any story, the total area of the openings shall comply with the following formula:* $A/a + A_u/a_u < 1.0$.

Discussion and Commentary: Based on the fire separation distance, the amount of openings in an exterior wall is regulated on a floor-by-floor basis. Where all of the exterior openings are protected, a higher percentage of the exterior wall surface may be provided with openings, whereas a lesser amount is permitted if all openings are unprotected. The IBC also permits both protected and unprotected openings in an exterior wall, provided they comply with the unity formula. In a fully-sprinklered building, the maximum allowable area of unprotected openings is the same as that allowed for protected openings.

Given: A nonsprinklered building of Type VA construction. The 1-hour exterior wall shown is located 12 feet from an interior lot line. It has protected openings as shown.

Determine: the maximum area permitted for unprotected openings.

Solution:

$$\frac{A}{a} + \frac{A_u}{a_u} \le 1.0$$

$$\frac{83}{(45\%)(18 \times 40)} + \frac{A_u}{(15\%)(18 \times 40]} = 1.0$$

$$\frac{83}{324} + \frac{A_u}{108} = 1.0$$

$$0.25 + \frac{81}{108} = 1.0$$

81 square feet of unprotected openings are permitted

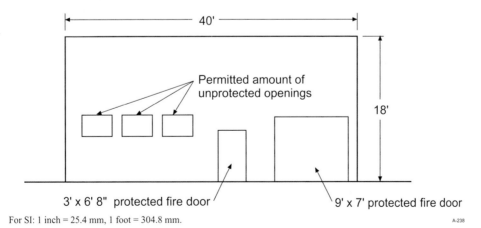

3' x 6' 8" protected fire door   9' x 7' protected fire door

For SI: 1 inch = 25.4 mm, 1 foot = 304.8 mm.   A-238

If a building's exterior bearing wall, exterior nonbearing wall or exterior structural frame is not required to be fire-resistance rated by Table 601 or 602, then an unlimited percentage of unprotected openings is permitted regardless of fire separation distance.

Code Text: *In occupancies other than Group H, unlimited unprotected openings are permitted in the first story of exterior walls of the first story above grade facing a street that have a fire separation distance of greater than 15 feet (4572 mm), or facing an unoccupied space. The unoccupied space shall be on the same lot or dedicated for public use, shall not be less than 30 feet (9144 mm) in width and shall have access from a street by a posted fire lane in accordance with the International Fire Code.*

Discussion and Commentary: Because the first story of a building is generally readily available for fire department access and manual suppression efforts, unprotected openings are not restricted, provided a moderate amount of open space is provided adjacent to the exterior wall. It is expected that the fire department can quickly mitigate the potential radiant heat exposure to surrounding buildings or structures.

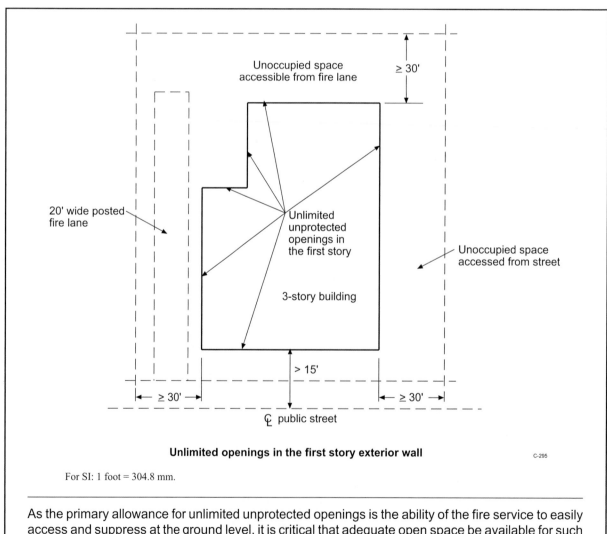

Unlimited openings in the first story exterior wall

C-295

For SI: 1 foot = 304.8 mm.

As the primary allowance for unlimited unprotected openings is the ability of the fire service to easily access and suppress at the ground level, it is critical that adequate open space be available for such purposes. Where not accessed directly from a street, a fire lane must be provided.

Topic: Vertical Separation of Openings

Category: Fire-resistance-rated Construction

Reference: IBC 704.9

Subject: Exterior Walls

Code Text: *Openings in exterior walls in adjacent stories shall be separated vertically to protect against fire spread on the exterior of the buildings where the openings are within 5 feet (1524 mm) of each other horizontally and the opening in the lower story is not a protected opening with a fire protection rating of not less than $^3/_4$ hour. Such openings shall be separated vertically at least 3 feet (914 mm) by spandrel girders, exterior walls or other similar assemblies that have a fire-resistance rating of at least 1 hour or by flame barriers than extend horizontally at least 30 inches (762 mm) beyond the exterior wall.* See exceptions for buildings no more than three stories in height, buildings that are fully sprinklered and open parking garages.

Discussion and Commentary: Where unprotected openings occur in adjacent stories, a fire that breaks out of an opening in a lower story can spread vertically to upper stories of the building.

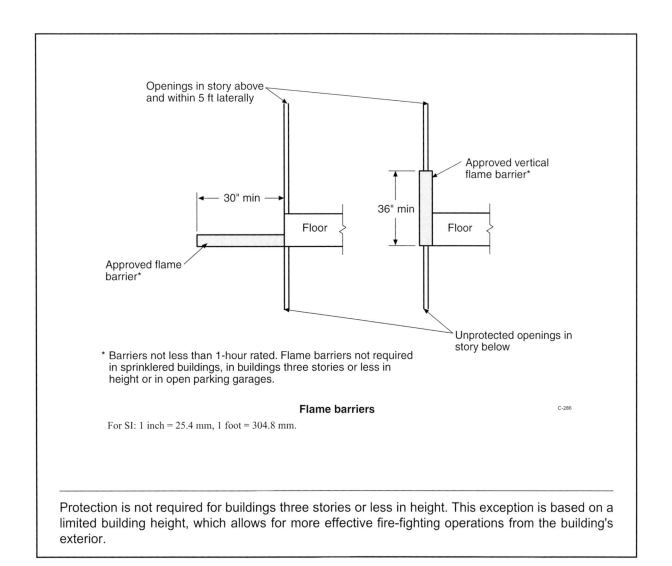

Openings in story above and within 5 ft laterally

Approved vertical flame barrier*

30" min

36" min

Floor

Floor

Approved flame barrier*

Unprotected openings in story below

* Barriers not less than 1-hour rated. Flame barriers not required in sprinklered buildings, in buildings three stories or less in height or in open parking garages.

Flame barriers

C-286

For SI: 1 inch = 25.4 mm, 1 foot = 304.8 mm.

Protection is not required for buildings three stories or less in height. This exception is based on a limited building height, which allows for more effective fire-fighting operations from the building's exterior.

Code Text: *Parapets shall be provided on exterior walls of buildings.* See six exceptions for construction methods or locations that would eliminate the requirement for parapets. *Parapets shall have the same fire-resistance rating as that required for the supporting wall, and on any side adjacent to a roof surface, shall have noncombustible faces for the uppermost 18 inches (457 mm), including counterflashing and coping materials. The height of the parapet shall not be less than 30 inches (762 mm) above the point where the roof surface and the wall intersect.*

Discussion and Commentary: A parapet wall is defined as the part of any wall entirely above the roof line. Its purpose is to prevent the spread of fire from the roof of the subject building to an adjacent building and to protect the roof of a building from exposure to a fire in an adjacent building.

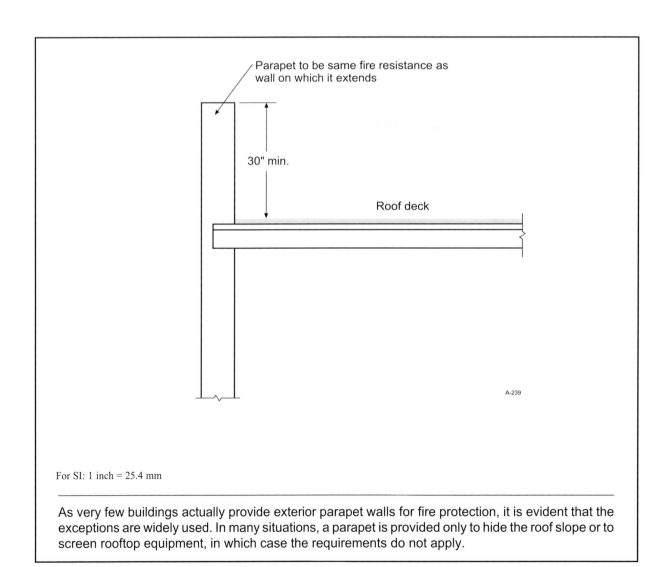

Parapet to be same fire resistance as wall on which it extends

30" min.

Roof deck

A-239

For SI: 1 inch = 25.4 mm

As very few buildings actually provide exterior parapet walls for fire protection, it is evident that the exceptions are widely used. In many situations, a parapet is provided only to hide the roof slope or to screen rooftop equipment, in which case the requirements do not apply.

Code Text: *A parapet need not be provided on an exterior wall where 1-hour fire-resistance-rated exterior walls that terminate at the underside of the roof sheathing, deck or slab, provided:* four conditions are met.

Discussion and Commentary: Limited to walls having a 1-hour fire-resistance rating, exception 4 permits exterior walls to terminate at the underside of the roof sheathing as an alternative to the use of parapets. Protection of the roof construction is provided from the interior of the building rather than at the exterior side. In addition to the restrictions on roof covering materials and openings in the roof, roof framing elements are addressed where installed parallel or perpendicular to their supporting walls.

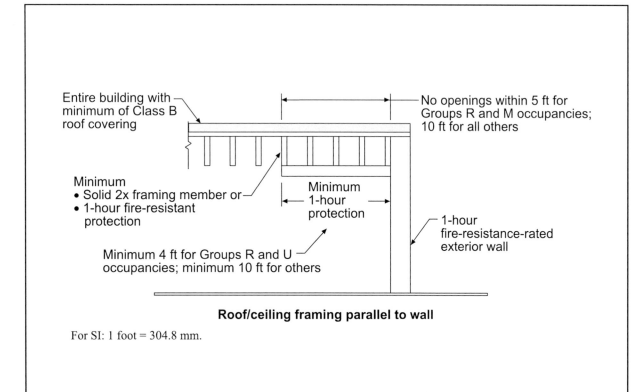

Roof/ceiling framing parallel to wall

For SI: 1 foot = 304.8 mm.

Other parapet exceptions are applicable where 1) the wall is not required to be fire-resistance-rated by Table 602; 2) no story exceeds 1,000 square feet in floor area; 3) the wall terminates at a minimum 2-hour fire-resistance-rated roof; 4) the roof is constructed of noncombustible materials, including the deck and supporting construction; 5) the wall is permitted to have a minimum of 25 percent unprotected openings per Table 708.4; or 6) in Groups R-2 and R-3, a number of special conditions are met.

Quiz

Study Session 5
IBC Sections 701 – 704

1. An opening around a penetrating item is a(n) _____.

    a. annular space   b. penetration

    c. through penetration   d. joint

    Reference_____

2. A _____ is a listed device designed to limit automatically the radiative heat transfer through an air inlet/outlet opening in the ceiling membrane of a fire-resistance-rated floor/ceiling or roof/ceiling assembly.

    a. horizontal fire damper

    b. ceiling radiation damper

    c. combination fire/smoke damper

    d. horizontal access door

    Reference_____

3. The time period that a through-penetration firestop system limits the spread of fire through a penetration is considered _____.

    a. a fire protection rating   b. an F rating

    c. a T rating   d. a fire-resistance rating

    Reference_____

4. All of the following building components may be used to create a fire area except _____.

    a. fire walls   b. fire barriers

    c. exterior walls   d. fire partitions

Reference_____

5. The measurement between the face of an exterior wall and the closest interior lot line is described as the _____.

    a. fire separation distance   b. fire exposure setback

    c. clearance to construction   d. exterior fire exposure

Reference_____

6. A _____ must have sufficient structural stability under fire conditions to allow collapse of construction on either side without collapse of the wall.

    a. fire wall   b. fire separation wall

    c. fire barrier   d. smoke barrier

Reference_____

7. A smoke compartment is created by enclosure of a space on all sides by _____.

    a. smoke partitions   b. smoke barriers

    c. shaft enclosure construction   d. fire partitions

Reference_____

8. The T rating for a penetration firestop system is based on a maximum temperature rise of _____ °F above its initial temperature through the penetration on the nonfire side.

    a. 250   b. 325

    c. 375   d. 400

Reference_____

9. Unless alternate methods are used for determining the fire-resistance ratings of building elements, the procedures set forth in _____ shall be applicable.

    a. ANSI Z 97.1   b. ASCE 5

    c. ASTM E 119   d. UL 555

Reference_____

10. Noncombustible building materials are required to some degree in all types of construction except for _____.

    a. Type II   b. Type III

    c. Type IV   d. Type V

Reference_____

11. Composite materials with a noncombustible structural base are considered noncombustible where the surface material is limited to _____ inch in thickness and has a maximum flame spread index of _____.

    a. $^1/_{28}$, 25   b. $^1/_{16}$, 50

    c. $^1/_8$, 25   d. $^1/_8$, 50

Reference_____

12. A projection is considered a building element that extends beyond the _____ of a building.

    a. exterior wall   b. fire area

    c. building area   d. floor area

Reference_____

13. Projections shall never extend more than _____ inches into areas where openings are prohibited.

    a. 0   b. 12

    c. 30   d. 36

Reference_____

14. In a building of Type III construction, projections shall be _____.

    a. of noncombustible construction

    b. of combustible construction

    c. of minimum one-hour fire-resistance-rated construction

    d. of any approved materials

Reference_____

15. For wall and opening protection and roof covering requirements, an assumed imaginary line shall be placed between _____.

    a. court walls   b. fire areas

    c. buildings on the same lot   d. different occupancies

Reference_____

16. The fire-resistance rating for an exterior wall shall be based on both interior and exterior fire exposure where the wall is located a maximum of _____ feet from the property line.

    a. 3   b. 5

    c. 10   d. 20

Reference_____

17. In a nonsprinklered Type VB building, what is the maximum area of unprotected exterior wall openings for a fire separation distance of 8 feet?

    a. 10 percent   b. 25 percent

    c. unlimited   d. unprotected openings are prohibited

Reference_____

18. In a nonsprinklered Type IIIA building, what is the maximum area of protected exterior wall openings for a fire separation distance of 20 feet?

    a. 25 percent   b. 45 percent

    c. 75 percent   d. unlimited

Reference_____

19. In a fully-sprinklered Type IIB office building, what is the maximum area of unprotected exterior wall openings for a fire separation distance of 5 feet?

    a. 10 percent   b. 15 percent

    c. 25 percent   d. unprotected openings are prohibited

Reference_____

20. In other than Group H occupancies, unlimited unprotected exterior openings are permitted in the first story above grade of exterior walls facing an unoccupied space of a least _____ feet in width.

    a. 15   b. 20

    c. 30   d. 40

Reference_____

21. Where flame barriers are required for the vertical separation of exterior openings in adjacent stories, horizontal barriers must extend at least _____ inches beyond the exterior wall.

    a. 12   b. 24

    c. 30   d. 36

Reference_____

22. Flame barriers protecting openings in exterior walls in adjacent stories shall have a minimum fire-resistance rating of _____.

    a. 20 minutes   b. 45 minutes

    c. 1 hour   d. 2 hours

Reference_____

23. A parapet is not required for those buildings having a maximum floor area of _____ square feet on any floor.

    a. 400   b. 1,000

    c. 1,500   d. 3,000

Reference_____

24. Where required, parapets must extend a minimum of _____ inches above the roof.

    a. 30   b. 32

    c. 36   d. 42

Reference_____

25. The uppermost portion of a parapet wall must have noncombustible faces for a minimum of _____ inches.

    a. 12   b. 18

    c. 24   d. 30

Reference_____

26. To obtain a fire-resistance rating, all nonsymmetrical walls except for _____ shall be tested with both faces exposed to the furnace.

    a. fire walls   b. fire barriers

    c. smoke barriers   d. exterior walls

Reference_____

27. A roof eave, where projecting into an area where protection of openings is required, may be constructed of all of the following materials, except _____.

    a. heavy-timber construction   b. noncombustible construction

    c. fire-retardant-treated wood   d. 1-half-hour fire-resistance-rated construction

Reference_____

28. In a fully sprinklered Type IIA building, the maximum allowable area of protected exterior wall openings is _____ where the fire separation distance is 12 feet.

    a. 15 percent   b. 25 percent

    c. 45 percent   d. 60 percent

Reference_____

29. A parapet is required on the exterior wall of a nonsprinklered building located a maximum of _____ feet from an interior lot line.

    a. 5   b. 10

    c. 15   d. 20

Reference_____

30. Unprotected openings are not permitted in the exterior wall of a Group H-3 warehouse where the maximum fire separation distance is _____ feet.

    a. 3   b. 5

    c. 10   d. 15

Reference_____

31. Materials required to be noncombustible shall be tested in accordance with _____.

    a. ASTM E 119   b. ASTM E 136

    c. ASTM E 84   d. ASTM C 636

Reference _____

32. The area of openings in an exterior wall of an open parking garage is permitted to be unlimited where there is a minimum fire separation distance of greater than _____ feet.

    a. 5   b. 10

    c. 20   d. 30

Reference _____

33. A parapet is not required on an exterior wall where the wall terminates at a roof with a minimum fire-resistance rating of _____ .

    a. 45 minutes   b. 1 hour

    c. $1^1/_2$ hour   d. 2 hours

Reference _____

34. A balcony projecting beyond the exterior wall on a Type IIA building is permitted to be constructed of fire-retardant-treated wood where the building is a maximum of _____ stories in height.

    a. 2   b. 3

    c. 4   d. 5

    Reference _____

35. In other than Group H occupancies, unlimited unprotected openings are permitted in a first-story exterior wall that faces a street, provided the minimum fire separation distance is greater than _____ feet.

    a. 5   b. 10

    c. 15   d. 20

    Reference _____

6

2006 IBC Sections 705 – 711
Fire-resistance-rated Construction II

OBJECTIVE: To gain an understanding of the fire-resistance-rated building components such as fire walls, fire barriers, shaft enclosures, fire partitions, smoke barriers, smoke partitions and horizontal assemblies.

REFERENCE: Sections 705 through 711, 2006 *International Building Code*

KEY POINTS:
- What is the purpose of a fire wall?
- Where is a party wall located? How is it to be constructed?
- How must a fire wall perform structurally?
- Which types of materials are permitted in a fire wall? How is the minimum fire-resistance rating of a fire wall determined?
- How must the horizontal continuity of a fire wall be accomplished? Vertical continuity?
- Which options are possible where a fire wall serves a stepped building?
- How shall the penetration of combustible framing members entering into a masonry or concrete fire wall be addressed?
- What are the limitations on openings in a fire wall?
- Where are fire barriers utilized?
- How must a fire barrier be constructed? What restrictions are placed on openings?
- What is a fire area? What is its purpose? How is the minimum fire-resistance rating for fire barrier assemblies separating fire areas determined?
- What is the purpose of a shaft enclosure? Where are such enclosures required?
- Which types of conditions permit vertical openings without enclosure?
- How does the height of a building affect the fire-resistance ratings of shaft enclosures?
- Which methods are mandated for construction of a shaft enclosure?
- Which types of openings are permitted to penetrate a shaft enclosure? How must they be protected?
- How must shafts be enclosed at the top? At the bottom?
- Which special provisions govern refuse and laundry chutes?
- Under which conditions must elevator lobbies be provided?
- Where are fire partitions required to be installed?

KEY POINTS:
(Cont'd)

- What minimum fire-resistance rating is required for fire partitions? What rating is required in a sprinklered hotel or apartment building?
- To what extent must a fire partition extend above a ceiling?
- What is the function of a smoke barrier?
- What fire-resistance rating is mandated for smoke barriers?
- How must openings in a smoke barrier be protected?
- Where are smoke barriers required?
- How is a smoke partition to be constructed? How are door openings, penetrations, ducts and air transfer openings regulated?
- How is the required rating of a horizontal assembly determined?
- What is the minimum fire-resistance rating for floor assemblies separating dwelling units in apartment buildings? Sleeping units in Group R-1 occupancies?
- What weight of ceiling panel requires no additional devices for the prevention of lateral displacement?
- When may the ceiling membrane in a fire-resistance-rated horizontal assembly be omitted? Floor membrane?
- How are skylights regulated in fire-resistance-rated roof construction?

| **Topic:** Definition and Scope | **Category:** Fire-resistance-rated Construction |
| **Reference:** IBC 702, 705.1 | **Subject:** Fire Walls |

Code Text: *A fire wall is a fire-resistance-rated wall having protected openings, which restricts the spread of fire and extends continuously from the foundation to or through the roof, with sufficient structural stability under fire conditions to allow collapse of construction on either side without collapse of the wall. Each portion of a building separated by one or more fire walls that comply with the provisions of Section 705 shall be considered a separate building.*

Discussion and Commentary: By placing one or more fire walls in a large-area building, multiple smaller-area buildings are created. Each of these smaller spaces can then be considered a unique building for code purposes. Under various conditions, fire walls can serve to reduce a structure's type of construction, eliminate a required sprinkler system or increase the permitted size of a facility under one roof.

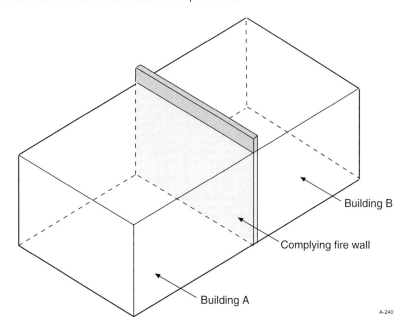

Fire wall to have sufficient structural stability under fire conditions to allow collapse of construction on either side without collapse of wall

Building B

Complying fire wall

Building A

A-240

In a situation where a fire wall separates distinct occupancy groups that are required to be separated by a fire barrier wall, the most restrictive requirements of each separation apply. This includes both the wall's continuity and the required fire-resistance rating.

Code Text: *Fire walls shall be of any approved noncombustible materials.* See exception for Type V construction. *Fire walls shall have a fire-resistance rating of not less than that required by Table 705.4.*

Discussion and Commentary: A fire wall is designed to act in a manner similar to an exterior wall, as a barrier to prevent a fire in one building from spreading to the other building. Accordingly, construction of the fire wall must be commensurate with the exterior wall requirements for the construction type. In addition, the fire-resistance rating of the wall must be considerable in order to provide the necessary level of protection based on the anticipated fire loading that is due to the uses of the separate buildings. The required ratings vary based on occupancy and, to some degree, type of construction.

TABLE 705.4
FIRE WALL FIRE-RESISTANCE RATINGS

GROUP	FIRE-RESISTANCE RATING (hours)
A, B, E, H-4, I, R-1, R-2, U	3[a]
F-1, H-3[b], H-5, M, S-1	3
H-1, H-2	4[b]
F-2, S-2, R-3, R-4	2

a. Walls shall be not less than 2-hour fire-resistance rated where separating buildings of Type II or V construction.
b. For Group H-1, H-2 or H-3 buildings, also see Sections 415.4 and 415.5.

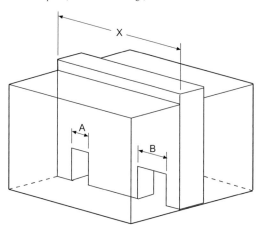

A + B ≤ 25% of X

Each opening limited to 120 square feet unless both buildings are sprinklered

Fire-protection rating based on Tables 705.4 and 715.4

For SI: 1 square foot = $0.093m^2$

A-241

Per Section 705.8, the total width of all openings in a fire wall is limited to 25 percent of the length of the wall in each story. There is no limit on the amount of total wall area containing openings; however, each opening is limited to 120 square feet in nonsprinklered buildings.

Code Text: *Fire walls shall be continuous from exterior wall to exterior wall and shall extend at least 18 inches (457 mm) beyond the exterior surface of exterior walls.* See exceptions for various methods of terminating the fire wall at the interior surface of the exterior sheathing or finish materials.

Discussion and Commentary: Historically, the code has addressed the hazards of fire exposure at the fire wall from a vertical perspective, at the roof. There is also concern of a similar hazard from the horizontal perspective, at the intersection of the fire wall and the exterior wall. The 18-inch extension is intended to abate the potential for fire to travel from one building to the other around the fire wall. The 18-inch extension must extend the full height of the fire wall.

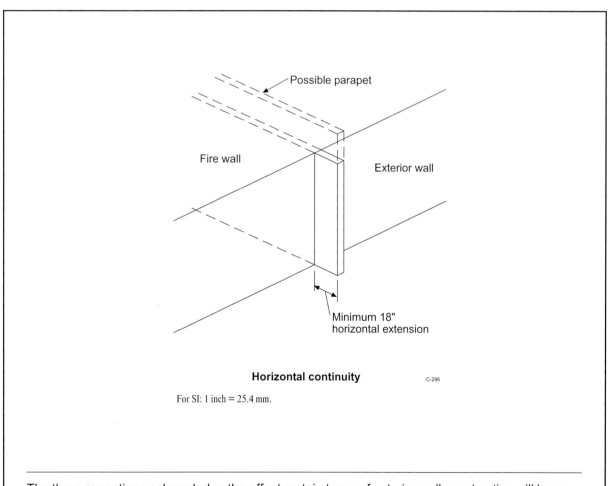

Horizontal continuity   C-296

For SI: 1 inch = 25.4 mm.

The three exceptions acknowledge the effect certain types of exterior wall construction will have on fire breaching the exterior of the building and exposing the adjacent building. These methods of protection are similar to those used at the roof construction where a parapet is not provided.

Code Text: *Fire walls shall extend from the foundation to a termination point at least 30 inches (762 mm) above both adjacent roofs.* See exceptions for buildings with different roof levels, those with noncombustible roof construction, and those constructed under special provisions.

Discussion and Commentary: To ensure the *separate building* concept, a fire wall must be continuous vertically with no horizontal offsets from the foundation, through the roof to a point at least 30 inches above. Various exceptions to the parapet requirement allow the fire wall to terminate at the bottom of the roof deck or sheathing. According to many of the exceptions, the roof covering must be minimum Class B, and no openings in the roof are permitted within 4 feet of the fire wall.

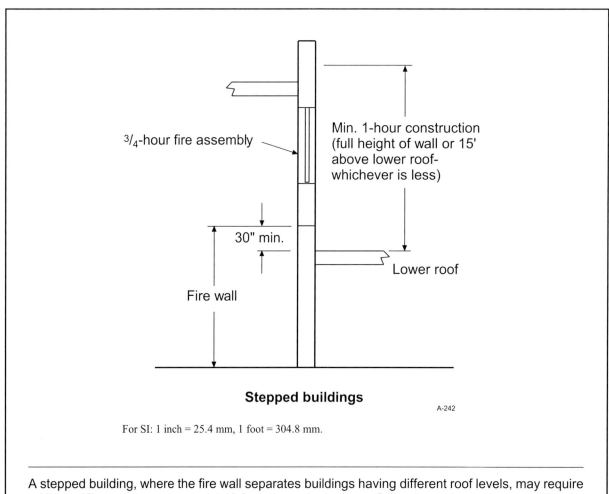

Stepped buildings

A-242

For SI: 1 inch = 25.4 mm, 1 foot = 304.8 mm.

A stepped building, where the fire wall separates buildings having different roof levels, may require additional fire resistance to a point 15 feet above the lower roof. An alternative method provides for minimum 1-hour horizontal protection of the lower roof assembly.

Code Text: *A fire barrier is a fire-resistance-rated wall assembly of materials designed to restrict the spread of fire in which continuity is maintained. Fire barriers installed as required elsewhere in* the International Building Code *or the* International Fire Code *shall comply with* Section 706.

Discussion and Commentary: The term *fire barrier* is specific in the IBC and is used to describe a unique type of vertical fire separation element. Many of the building elements required to be constructed as fire barriers are listed in Section 706.3, including shaft enclosures, exit enclosures, exit passageways, horizontal exits, atriums, incidental use areas, control areas, separation of mixed occupancies and seperation of fire areas.

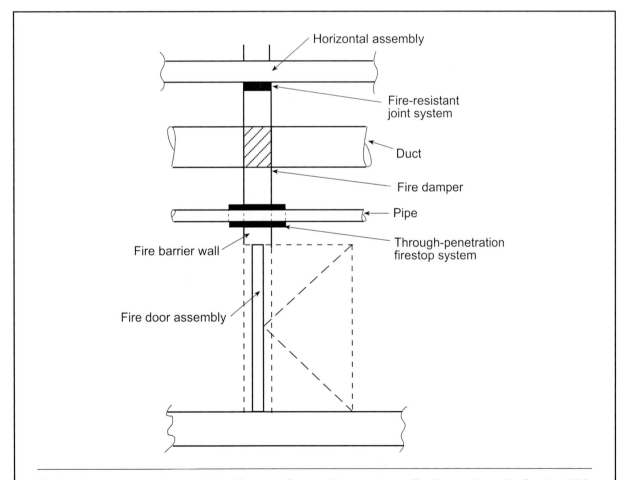

Fire barriers may also be mandated for specific conditions not specifically mentioned in Section 706. Throughout the IBC, as well as the other *International Codes*, fire barriers are identified as the element used to provide the necessary fire separation for compartmentation of building spaces.

Topic: Fire-resistance-rated Glazing
Reference: IBC 706.2.1

Category: Fire-resistance-rated Construction
Subject: Fire Barriers

Code Text: *Fire-resistance-rated glazing, when tested in accordance with ASTM E 119 and complying with the requirements of Section 706, shall be permitted. Fire-resistance-rated glazing shall bear a label or other identification showing the name of the manufacturer, the test standard and the identifier "W-XXX," where the "XXX" is the fire-resistance rating in minutes. Such label or identification shall be issued by an approved agency and shall be permanently affixed to the glazing.*

Discussion and Commentary: Under the provisions of Table 715.5, fire windows are not permitted in fire barriers having a required fire-resistance rating of more than 1 hour. Therefore, glazing in such fire barriers must either be protected by complying fire shutters or be in compliance with the requirements of Section 706.2.1. The fire-resistance-rated glazing permitted by this section is acceptable because it is tested to the same criteria as any fire barrier wall.

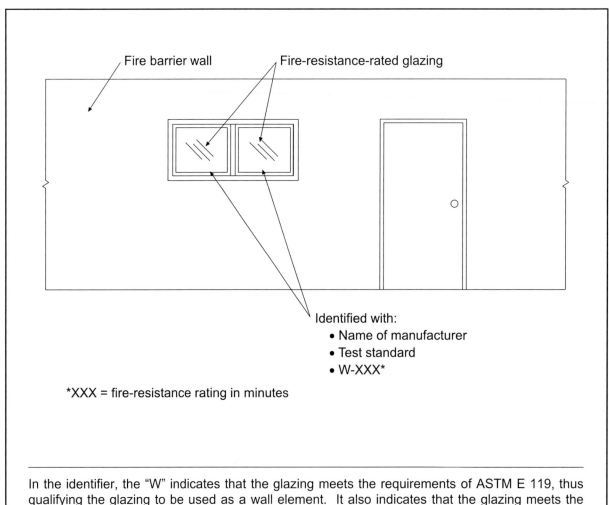

Fire barrier wall   Fire-resistance-rated glazing

Identified with:
- Name of manufacturer
- Test standard
- W-XXX*

*XXX = fire-resistance rating in minutes

In the identifier, the "W" indicates that the glazing meets the requirements of ASTM E 119, thus qualifying the glazing to be used as a wall element. It also indicates that the glazing meets the fire-resistance, hose-stream and temperature-rise requirements of the test standard.

Topic: Single-occupancy Fire Areas **Category:** Fire-resistance-rated Construction
Reference: IBC 706.3.9, Table 706.3.9 **Subject:** Fire Barriers

Code Text: *The fire barrier or horizontal assembly, or both, separating a single occupancy into different fire areas shall have a fire-resistance rating of not less than that indicated in Table 706.3.9.*

Discussion and Commentary: The code recognizes that in many buildings there are two methods to limit the spread of fire, either 1) the use of an automatic sprinkler system or 2) the creation of fire-resistive compartments that contain a fire's movement (fire areas). Section 903.2 identifies those occupancies where compartmentation is an acceptable alternative to a sprinkler system. Table 706.3.9 then mandates the minimum level of fire resistance of the fire barriers utilized to separate the building into two or more compartments (fire areas). As a result, the use of Table 706.3.9 is only applicable in buildings not protected by an automatic sprinkler system.

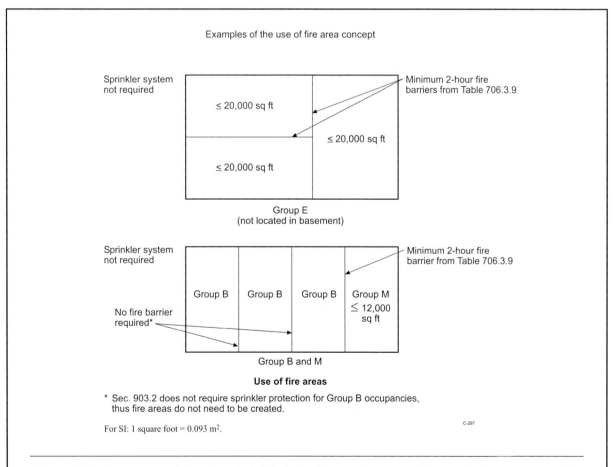

Examples of the use of fire area concept

Use of fire areas

* Sec. 903.2 does not require sprinkler protection for Group B occupancies, thus fire areas do not need to be created.

For SI: 1 square foot = 0.093 m².

C-297

Although the language of Section 706.3.9 limits its use to separating a single occupancy into different fire areas, it must also be used in mixed-occupancy conditions. The controlling fire-resistance rating of the fire barrier or horizontal assembly separating the occupancies is based on the higher of the ratings as established by Table 706.3.9 for the occupancies involved.

Code Text: *Fire barrier walls shall extend from the top of the floor/ceiling assembly below to the underside of the floor or roof slab or deck above and shall be securely attached thereto. Such fire barriers shall be continuous through concealed spaces, such as the space above a suspended ceiling. The supporting construction for fire barrier walls shall be protected to afford the required fire-resistance rating of the fire barrier supported, except for 1-hour fire-resistance-rated incidental use area separations as required by Table 508.2 in buildings of Types IIB, IIIB and VB construction.*

Discussion and Commentary: Where a wall is required to serve as a fire barrier, it must be tight from deck to deck in order to provide a full separation. A fire barrier is often used with a horizontal assembly in a multistory building to provide a complete separation.

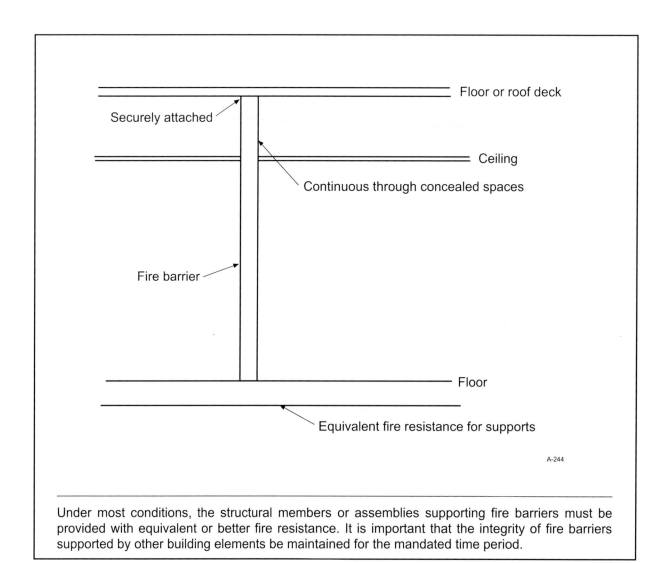

Under most conditions, the structural members or assemblies supporting fire barriers must be provided with equivalent or better fire resistance. It is important that the integrity of fire barriers supported by other building elements be maintained for the mandated time period.

Code Text: *Openings in a fire barrier wall shall be protected in accordance with Section 715. Openings shall be limited to a maximum aggregate width of 25 percent of the length of the wall, and the maximum area of any single opening shall not exceed 156 square feet (15 m^2). Openings in exit enclosures and exit passageways shall also comply with Sections 1020.1.1 and 1021.4, respectively.* See exceptions for 1) sprinklered adjoining fire areas, 2) fire doors serving an exit enclosure, 3) openings tested per ASTM E 119, and 4) fire windows in atrium separation walls.

Discussion and Commentary: As openings in a fire barrier create a potential breach in the integrity of the fire-resistive separation, a limit is placed on the amount of permitted openings. The limitation allows for design flexibility without compromising the necessary level of fire separation. The aggregate area of such openings is not limited; however, each opening is limited to 156 square feet.

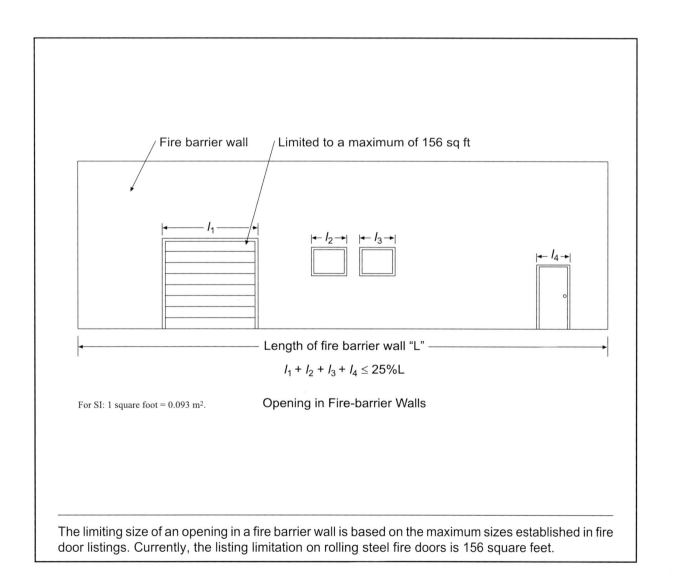

Fire barrier wall   Limited to a maximum of 156 sq ft

l_1 l_2 l_3 l_4

Length of fire barrier wall "L"

$$l_1 + l_2 + l_3 + l_4 \leq 25\%L$$

For SI: 1 square foot = 0.093 m².   Opening in Fire-barrier Walls

The limiting size of an opening in a fire barrier wall is based on the maximum sizes established in fire door listings. Currently, the listing limitation on rolling steel fire doors is 156 square feet.

Code Text: *A shaft is an enclosed space extending through one or more stories of a building, connecting vertical openings in successive floors, or floors and roof. A shaft enclosure is the walls or construction forming the boundaries of a shaft. Openings through a floor/ceiling assembly shall be protected by a shaft enclosure complying with Section 707. See 13 exceptions identifying where a shaft enclosure is not required.*

Discussion and Commentary: It is not uncommon in multistory buildings to have openings that are provided to accommodate elevators, mechanical equipment or similar devices, and to transmit light or ventilation air. Because of the potential for the rapid spread of fire, smoke and gases vertically through buildings, such openings must be protected with fire-resistance-rated shaft enclosures. There are many special and specific provisions that modify the general requirements.

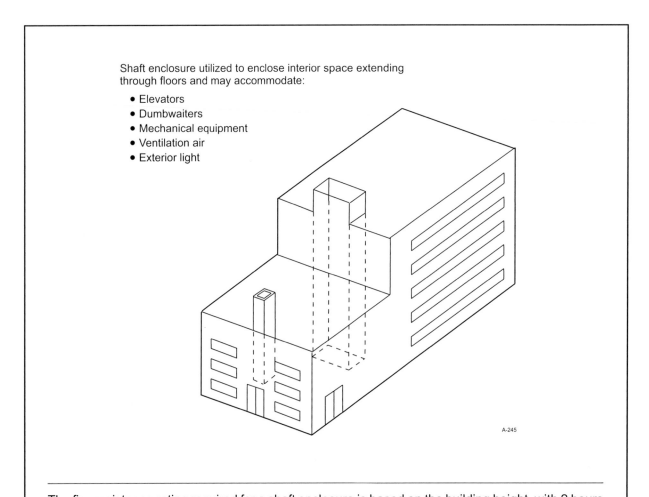

Shaft enclosure utilized to enclose interior space extending through floors and may accommodate:

- Elevators
- Dumbwaiters
- Mechanical equipment
- Ventilation air
- Exterior light

A-245

The fire-resistance rating required for a shaft enclosure is based on the building height, with 2 hours being required where four stories or more are connected. Where less than four stories are connected, 1 hour is required. The enclosure rating cannot be less than that of any floor penetrated.

Code Text: *Shaft enclosures shall be constructed as fire barriers in accordance with Section 706 or horizontal assemblies constructed in accordance with Section 711, or both, and shall have continuity in accordance with Section 706.5 for fire barriers or Section 711.4 for horizontal assemblies as applicable.*

Discussion and Commentary: The general provisions dictate that a shaft be completely enclosed with fire-resistance-rated construction. However, there are conditions that modify this rule. The protection of exterior shaft walls is often unnecessary. Additionally, for those shafts that do not extend to the bottom of the building, the code provides three methods of maintaining the integrity of the shaft enclosure.

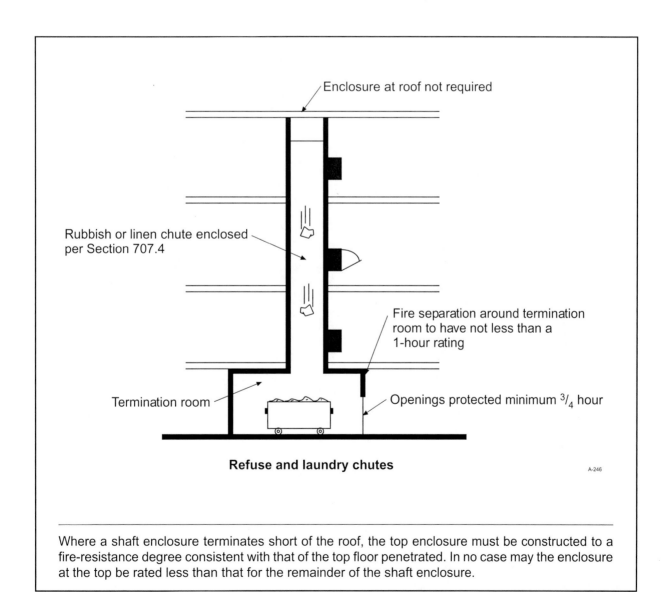

Enclosure at roof not required

Rubbish or linen chute enclosed per Section 707.4

Fire separation around termination room to have not less than a 1-hour rating

Termination room

Openings protected minimum $^3/_4$ hour

Refuse and laundry chutes

A-246

Where a shaft enclosure terminates short of the roof, the top enclosure must be constructed to a fire-resistance degree consistent with that of the top floor penetrated. In no case may the enclosure at the top be rated less than that for the remainder of the shaft enclosure.

Code Text: *An enclosed elevator lobby shall be provided at each floor where an elevator shaft enclosure connects more than three stories. The lobby shall separate the elevator shaft enclosure doors from each floor by fire partitions equal to the fire-resistance rating of the corridor and the required opening protection.* See exceptions for 1) enclosed elevator lobbies, 2) elevators not required to be located in a shaft, 3) additional doors provided at the hoistway opening, 4) sprinklered buildings with no occupied floor more than 75 feet above the lowest level of fire department vehicle access, 5) the use of smoke partitions in fully sprinklered buildings, and 6) elevator hoistways that are pressurized.

Discussion and Commentary: The purpose of an elevator lobby under this provision is to reduce the potential for smoke to travel from the floor of fire origin to any other floor of the building by way of an elevator shaft enclosure. An allowance is provided for those low-rise buildings where the elevator shaft connects only two or three stories.

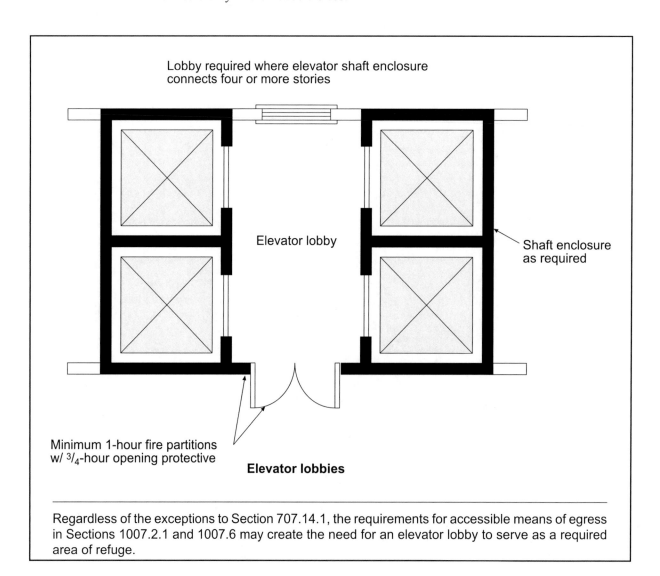

Lobby required where elevator shaft enclosure connects four or more stories

Elevator lobby

Shaft enclosure as required

Minimum 1-hour fire partitions w/ $^3/_4$-hour opening protective

Elevator lobbies

Regardless of the exceptions to Section 707.14.1, the requirements for accessible means of egress in Sections 1007.2.1 and 1007.6 may create the need for an elevator lobby to serve as a required area of refuge.

Code Text: *A fire partition is a vertical assembly of materials designed to restrict the spread of fire in which openings are protected. The following wall assemblies shall comply with Section 708: 1) walls separating dwelling units in the same building, 2) walls separating sleeping units in occupancies in Group R-1 hotel, R-2 and I-1 occupancies, 3) walls separating tenant spaces in covered mall buildings as required by Section 402.7.2, 4) corridor walls as required by Section 1017.1, 5) elevator lobby separation as required by Section 707.14.1, and 6) residentail aircraft hangars. Fire partitions shall have a fire-resistance rating of not less than 1 hour. See exceptions for corridor walls and sprinklered buildings.*

Discussion and Commentary: Required to have a fire-resistance-rating of one-hour, fire partitions provide a moderate level of separation that is necessary under certain conditions. Although fire partitions have limited applications, they are important elements in the specific uses and areas in which they are mandated.

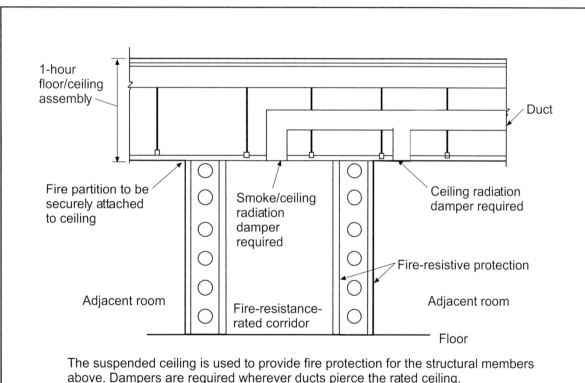

The suspended ceiling is used to provide fire protection for the structural members above. Dampers are required wherever ducts pierce the rated ceiling.

Corridor fire partitions

A-248

In sprinklered buildings of Types IIB, IIIB and VB construction, the 1-hour fire-resistance rating for dwelling unit and guestroom separations may be reduced to $^1/_2$ hour. For a typical wood-stud wall system, this separation could be satisfied with $^1/_2$-inch gypsum board on each side.

Code Text: *Fire partitions shall extend from the top of the foundation or floor/ceiling assembly below to the underside of the floor or roof sheathing, slab or deck above or to the fire-resistance rated floor/ceiling or roof/ceiling assembly above, and shall be securely attached thereto. If the partitions are not continuous to the sheathing, deck or slab, and where constructed of combustible construction, the space between the ceiling and the sheathing, deck or slab above shall be fireblocked or draftstopped in accordance with Sections 717.2 and 717.3 at the partition line.* See multiple exceptions.

Discussion and Commentary: The method of continuity is a primary difference between fire barriers and fire partitions. Fire partitions need not extend through a concealed space, such as the one above a suspended ceiling, provided that the ceiling is a portion of a fire-resistance-rated floor/ceiling or roof/ceiling assembly.

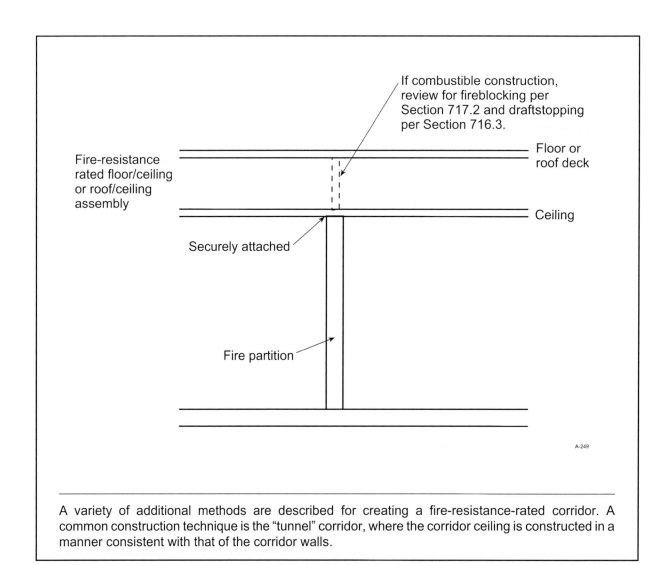

If combustible construction, review for fireblocking per Section 717.2 and draftstopping per Section 716.3.

Floor or roof deck

Fire-resistance rated floor/ceiling or roof/ceiling assembly

Ceiling

Securely attached

Fire partition

A-249

A variety of additional methods are described for creating a fire-resistance-rated corridor. A common construction technique is the "tunnel" corridor, where the corridor ceiling is constructed in a manner consistent with that of the corridor walls.

Topic: Definition and Continuity	Category: Fire-resistance-rated Construction
Reference: IBC 702, 709.4	Subject: Smoke Barriers

Code Text: *A smoke barrier is a continuous membrane, either vertical or horizontal, such as a wall, floor, or ceiling assembly, that is designed and constructed to restrict the movement of smoke. Smoke barriers shall form an effective membrane continuous from outside wall to outside wall and from the top of the foundation or floor/ceiling assembly below to the underside of the floor or roof sheathing, deck or slab above, including continuity through concealed spaces, such as those found above suspended ceilings, and including interstitial structural and mechanical spaces.* See exception for fire/smoke-resistant ceilings.

Discussion and Commentary: Where the primary concern of the code is the containment of smoke, the use of a smoke barrier is mandated. The locations for smoke barriers are found in various provisions in the IBC, including Section 407.4 for Group I-2 occupancies and Section 1007.6.2 for areas of refuge.

Required Use of Smoke Barriers

- Compartmentation of underground buildings (Sec. 405.4.2)
- Compartmentation of Group I-2 (Sec. 407.4)
- Compartmentation of Group I-3 (Sec. 408.6)
- Smoke control systems (Sec. 402.9, 404.4 and 909.5)
- Areas of refuge (Sec. 1007.6.2)

A smoke barrier must have a minimum 1-hour fire-resistance rating. Openings, penetrations, joints, ducts and transfer openings must also be protected to minimize the passage of smoke through the barrier. Opening protectives must have a minimum 20-minute fire-protection rating.

Code Text: *Smoke partitions installed as required elsewhere in the IBC shall comply with Section 710. The walls shall be of materials permitted by the building type of construction. Unless required elsewhere in the IBC, smoke partitions are not required to have a fire-resistance rating. Smoke partitions shall extend from the top of the foundation or floor below to the underside of the floor or roof sheathing, deck or slab above or to the underside of the ceiling above where the ceiling membrane is constructed to limit the transfer of smoke.*

Discussion and Commentary: A smoke partition is designed for a singular purpose, to limit the movement of smoke from one area to another. Therefore, windows in smoke partitions must be sealed, penetrations and joints must be adequately filled, and smoke dampers used to protect air transfer openings.

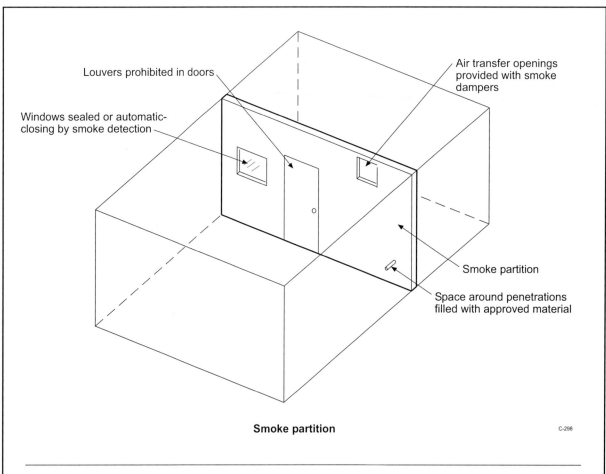

Louvers prohibited in doors

Air transfer openings provided with smoke dampers

Windows sealed or automatic-closing by smoke detection

Smoke partition

Space around penetrations filled with approved material

Smoke partition

C-298

Smoke partition is a specific element with specific requirements, much like smoke barriers, fire barriers, fire partitions and fire walls. Only where the code specifically mandates smoke partitions are the requirements of Section 710 applicable.

Code Text: *The fire-resistance rating of floor and roof assemblies shall not be less than that required by the building type of construction. Where the floor assembly separates mixed occupancies, the assembly shall have a fire-resistance rating of not less than that required by Section 508.3.3 based on the occupancies being separated. Where the floor assembly separates a single occupancy into different fire areas, the assembly shall have a fire-resistance rating of not less than that required by Section 706.3.9.*

Discussion and Commentary: Table 601 regulates the minimum fire-resistance ratings for floor construction based on the building's type of construction. This minimum level of fire-resistance must always be maintained. Other provisions of the code must also be considered where a horizontal separation is needed within a multistory building, such as for control areas or occupancy separations.

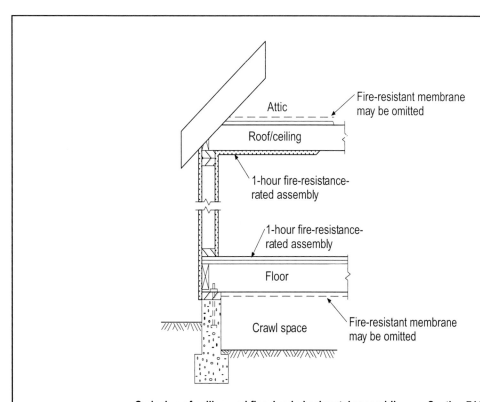

Omission of ceiling and flooring in horizontal assemblies per Section 711.3.3

A-250

Other than permitted openings, penetrations or joints, horizontal assemblies must be continuous in order to isolate totally one floor from another. An allowance is permitted for fire-resistance-rated roof construction, where skylights and other penetrations may be unprotected.

Quiz

Study Session 6

IBC Sections 705 – 711

1. Fire walls must be constructed of noncombustible materials unless separating buildings of Type _____ construction.

   a. I   b. III

   c. IV   d. V

   Reference_____

2. A fire wall in a Type VA building housing a Group M occupancy must have a minimum fire-resistance rating of _____ hour(s).

   a. 1   b. 2

   c. 3   d. 4

   Reference_____

3. In a Type IIB building containing a Group B occupancy, what is the minimum required fire- resistance rating for a fire wall?

   a. 1 hour   b. 2 hour

   c. 3 hour   d. 4 hour

   Reference_____

4.	As a general provision, what minimum distance must a fire wall extend horizontally beyond the exterior surface of exterior walls?

    a.	18 inches

    b.	20 inches

    c.	30 inches

    d.	4 feet

Reference_____

5.	In general, what minimum distance above the roof must a fire wall extend?

    a.	18 inches

    b.	30 inches

    c.	3 feet

    d.	4 feet

Reference_____

6.	Where a fire wall occurs at a location where the roof heights differ, the fire wall may terminate at the underside of the lower roof slab, provided _____.

    a.	the lower roof is of noncombustible construction

    b.	the first 10 feet of the lower roof assembly has a minimum 1-hour rating

    c.	the exterior wall of the higher portion is 1-hour to a height of 10 feet

    d.	openings in the lower roof have a minimum 45-minute fire-protective rating

Reference_____

7.	Embedded ends of combustible members entering masonry or concrete fire walls shall be separated a minimum distance of _____ inch(es).

    a.	1

    b.	2

    c.	4

    d.	6

Reference_____

8.	The aggregate width of openings in a fire wall is limited to a maximum of _____ of the length of the wall.

    a.	10 percent

    b.	25 percent

    c.	$33^{1}/_{3}$ percent

    d.	50 percent

Reference_____

9. Fire barriers are required for the separation of all of the following building elements, except _____.

    a. shaft enclosures   b. exit passageways

    c. incidental use areas   d. fire-resistance-rated corridors

Reference_____

10. In nonrated buildings, construction supporting fire barriers need not be protected by equivalent fire resistance where the fire barriers are used for _____.

    a. 1-hour occupancy separations

    b. 1-hour vertical exit enclosures

    c. 1-hour incidental use separations

    d. 2-hour horizontal exits

Reference_____

11. In a nonsprinklered building, any single opening in a fire barrier is limited to a maximum area of _____ square feet.

    a. 20   b. 100

    c. 120   d. 156

Reference_____

12. Fire doors in a fire barrier are permitted to be unlimited in area and aggregate width where the fire barrier is utilized as a(n) _____ separation.

    a. exit enclosure   b. horizontal exit

    c. incidental use area   d. exit access corridor

Reference_____

13. In a fully sprinklered hotel, under specific conditions, a stairway not considered a means of egress need not be protected by a shaft enclosure where connecting a maximum of _____ stories.

    a. two   b. three

    c. four   d. six

Reference_____

14. A floor opening is permitted between a maximum of two stories without a shaft enclosure, under limited conditions, in all but which one of the following occupancies?

    a. Group A-2   b. Group H-2

    c. Group I-2   d. Group S-2

    Reference_____

15. A shaft enclosure shall have a minimum 2-hour fire-resistance rating where connecting a minimum of _____ stories.

    a. two   b. three
    c. four   d. six

    Reference_____

16. Where an elevator lobby is required in a nonsprinklered building, it shall be separated from the floor level through the use of _____.

    a. smoke barriers   b. fire barriers

    c. fire partitions   d. smoke partitions

    Reference_____

17. A fire partition is not the appropriate wall assembly in which of the following locations?

    a. walls separating sleeping units in a Group R-1 hotel

    b. walls separating tenant spaces in a covered mall building

    c. walls separating control areas in a manufacturing occupancy

    d. walls separating dwelling units in an apartment building

    Reference_____

18. What is the minimum required fire-resistance rating for a fire partition separating sleeping units in a Group I-1 occupancy housed in a Type IIA building?

    a. 20 minutes   b. 30 minutes

    c. 1 hour   d. no rating is required

    Reference_____

19. Where a fire-resistance-rated corridor ceiling is constructed as required for the corridor walls, the walls shall not terminate before reaching _____.

    a. the lower membrane of the ceiling assembly

    b. the upper membrane of the ceiling assembly

    c. the underside of the floor or roof deck above

    d. an approved fire-resistant joint system

Reference_____

20. In other than a Group I-3 occupancy, what is the minimum required fire-resistance rating for a smoke barrier?

    a. no rating is required   b. 30 minutes

    c. 45 minutes   d. 1 hour

Reference_____

21. In which one of the following types of construction must a smoke barrier be supported by fire-resistance-rated construction having at least an equivalent rating to the smoke barrier supported?

    a. Type IIB   b. Type IIIA

    c. Type IIIB   d. Type VB

Reference_____

22. What is the minimum required fire-protection rating for doors installed in a smoke barrier?

    a. No rating is required   b. 20 minutes

    c. 45 minutes   d. 1 hour

Reference_____

23. Wire or other approved devices shall be installed above lay-in ceiling panels in a fire-resistance-rated floor/ceiling assembly to prevent vertical displacement where the weight of the panels is not adequate to resist a minimum upward force of _____.

    a. 1 psf   b. 2 psf

    c. 5 psf   d. 15 psf

Reference_____

24. Under which of the following conditions may the ceiling membrane of a fire-resistance-rated horizontal assembly be omitted?

    a.   where usable attic space occurs above

    b.   where the assembly has a minimum fire-resistance rating of 2 hours

    c.   where the floor construction is limited to combustible construction

    d.   where an unusable crawl space occurs below

Reference_____

25. Unprotected skylights are permitted through a fire-resistance-rated roof deck, provided _____.

    a.   the skylights were tested as a part of the roof assembly

    b.   the roof construction has not less than a 1-hour fire-resistance rating

    c.   the structural integrity of the roof construction is maintained

    d.   they are limited to 10 percent of the total roof area

Reference_____

26. Where a fire barrier is utilized to separate a Group M occupancy into multiple fire areas, the minimum fire-resistance rating of the fire barrier shall be _____ hour(s).

    a. 1   b. 2

    c. 3   d. 4

Reference_____

27. The termination room for a laundry chute shall be separated from the remainder of the building by a fire barrier having a minimum fire-resistance rating of _____.

    a. 0, no rating is required   b. $^1/_2$ hour

    c. 1 hour   d. 2 hours

Reference_____

28. What hourly rating is mandated for the construction of smoke partitions?

    a. 0, no rating is required   b. $^1/_2$ hour

    c. $^3/_4$ hour   d. 1 hour

Reference_____

29. Openings in smoke partitions shall have a minimum fire-protection rating of _____.

    a. 0, no rating is required   b. 20 minutes

    c. 45 minutes   d. 1 hour

    Reference_____

30. In a Type IIB hotel that is sprinklered in accordance with NFPA 13, what is the minimum fire-resistance rating for the fire partitions and horizontal assemblies that separate sleeping units?

    a. 0, no rating is required   b. $^1/_2$ hour

    c. 1 hour   d. 2 hours

    Reference_____

31. In general, fire walls shall extend to the outer edge of horizontal projecting elements that are located within _____ feet of the fire wall.

    a. 4   b. 5

    c. 8   d. 10

    Reference_____

32. Unless exempted, an enclosed elevator lobby is required at each floor level where an elevator shaft enclosure connects a minimum of _____ stories.

    a. 3   b. 4

    c. 5   d. 6

    Reference_____

33. Where elevator hoistway pressurization is provided in lieu of required enclosed elevator lobbies, the hoistways shall be pressurized to maintain a minimum positive pressure of _____ inches of water column and a maximum positive pressure of _____ inches of water column with respect to adjacent occupied space on all floors.

    a. 0.03, 0.05   b. 0.03, 0.06

    c. 0.04, 0.06   d. 0.04, 0.08

    Reference_____

34. An opening into a laundry chute access room shall be protected by opening protectives having a minimum fire protection rating of _____.

    a. 20 minutes   b. 45 minutes

    c. 1 hour   d. $1\frac{1}{2}$ hours

Reference _____

35. Where a membrane penetration of a fire-resistance-rated wall is made by a listed electrical box, the annular space between the wall membrane and the box shall be a maximum of _____ inch unless otherwise listed.

    a. $\frac{1}{16}$ b. $\frac{1}{8}$

    c. $\frac{3}{16}$ d. $\frac{1}{4}$

Reference _____

Study Session

7

2006 IBC Sections 712 – 719
Fire-resistance-rated Construction III

OBJECTIVE: To gain an understanding of the installation of fireblocking and draftstopping in combustible construction as well as the methods of protecting fire-resistance-rated building components where they contain doors, windows, ducts, air transfer openings and penetrations.

REFERENCE: Sections 712 through 719, 2006 *International Building Code*

KEY POINTS:
- Which two types of penetrations are regulated by the code?
- What are the appropriate installation details where sleeves are used in penetrating a fire-resistance-rated assembly?
- What is a through penetration? Membrane penetration?
- Penetrating items of steel, ferrous or copper may be protected in what manner?
- What is an F rating and a T rating? What minimum fire-resistance rating is required of a through-penetration firestop system?
- How can the membrane penetration by a steel outlet box be addressed? A listed electrical box? A fire sprinkler?
- Which limitations are placed on noncombustible penetrating items connecting to combustible items?
- How are fire-resistance-rated horizontal assemblies penetrated? Nonfire-resistance-rated horizontal assemblies?
- What is the purpose of a fire-resistance-rated joint system? When is such a system required?
- How should the intersection of an exterior curtain wall and a fire-resistance-rated floor assembly be accomplished?
- Where must structural frame members be individually protected? How are columns to be protected above a ceiling?
- How is the required fire-protection rating of an opening protective determined?
- What are the characteristics of a fire-resistance-rated corridor door or smoke barrier door?
- In what locations must a fire door be rated to a maximum transmission temperature end point?

KEY POINTS:
(Cont'd)

- What are the labeling requirements for fire doors? Smoke and draft control doors? Fire-protection-rated glazing?
- Where are self-closing door assemblies mandated? Automatic-closing assemblies?
- What is the maximum fire-protection rating assigned to fire-protection-rated glazing?
- Under which conditions is wired glass considered equivalent to $^3/_4$-hour fire-protection-rated glazing?
- What is the maximum size of a fire window assembly containing a wired glass panel?
- What are the different types of dampers? How should they be actuated?
- How must fire and smoke dampers be identified and accessed?
- Where are fire dampers required? Smoke dampers? Ceiling radiation dampers?
- In what specific locations are fire dampers not required for penetrations of shaft enclosures? Where smoke dampers are not required?
- Why is draftstopping and fireblocking unnecessary in noncombustible construction?
- Where is fireblocking required to be installed? Draftstopping?
- Which materials are acceptable as fireblocks? Draftstops?

Code Text: *A through penetration is an opening that passes through an entire assembly. A membrane penetration is an opening made through one side (wall, floor or ceiling membrane) of an assembly. The provisions of Section 712 shall govern the materials and methods of construction used to protect through penetrations and membrane penetrations.*

Discussion and Commentary: Fire-resistance-rated walls and horizontal assemblies are usually penetrated, both fully and partially, with piping, conduit, outlet boxes, cable, vents and similar penetrating items. The IBC regulates both the materials and the methods of penetration based on the specific conditions that exist. Where sleeves are used, they must be fastened securely in place, and all open space within and around the sleeve must be appropriately protected.

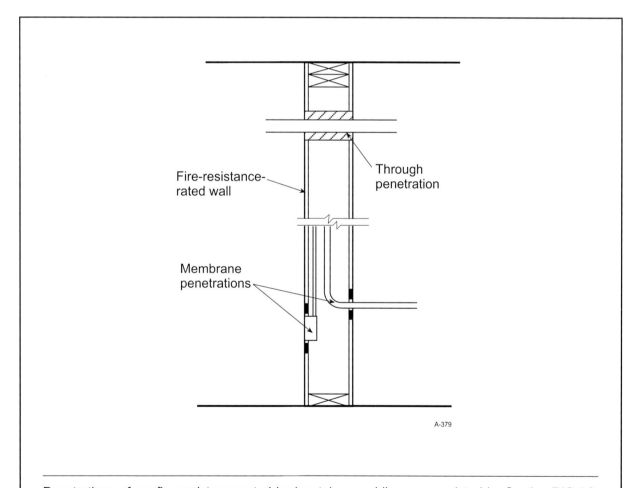

A-379

Penetrations of nonfire-resistance-rated horizontal assemblies are regulated by Section 712.4.2. Although some horizontal assemblies may not require a fire-resistance rating, the code intends that some degree of separation (compartmentalization) be provided from one story to another.

Code Text: *Penetrations into or through fire walls, fire barrier walls, smoke barrier walls, and fire partitions shall comply with Sections 712.3.1 through 712.3.4.*

Discussion and Commentary: In general, penetrations into or through fire-resistance-rated walls must be either protected with an approved through-penetration firestop system or installed as a tested component of an approved fire-resistance-rated assembly. These methods are considered proprietary, with each penetration being regulated by the specifics of the installation. Two generic methods are identified as exceptions to the general requirements; however, both methods are based on the penetration only of steel, ferrous or copper pipes or steel conduits. Under such conditions, the annular space around the penetrating items shall be filled with an appropriate material.

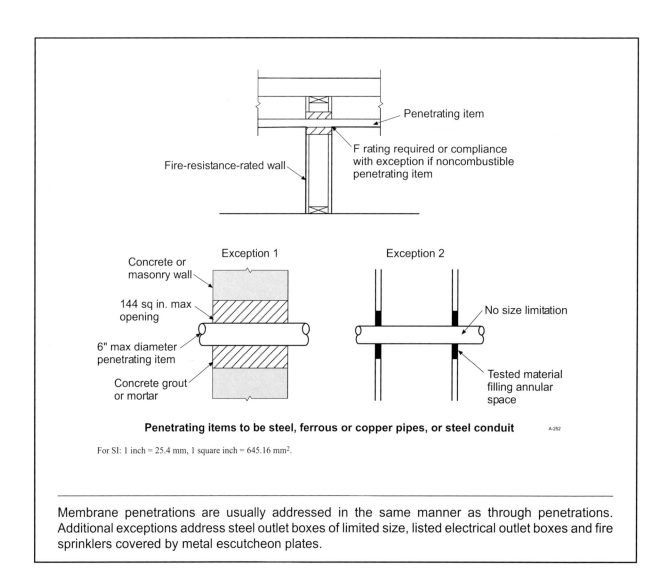

Penetrating items to be steel, ferrous or copper pipes, or steel conduit   A-252

For SI: 1 inch = 25.4 mm, 1 square inch = 645.16 mm².

Membrane penetrations are usually addressed in the same manner as through penetrations. Additional exceptions address steel outlet boxes of limited size, listed electrical outlet boxes and fire sprinklers covered by metal escutcheon plates.

Code Text: *Penetrations of fire-resistance-rated walls by ducts that are not protected with fire dampers shall comply with the provisions for penetrations. Noncombustible penetrating items shall not connect to combustible items beyond the point of firestopping unless it can be demonstrated that the fire-resistance integrity of the wall is maintained.*

Discussion and Commentary: Duct penetrations of fire-resistance-rated wall assemblies are typically protected with fire dampers in accordance with Section 716.5. However, in those locations where dampers are not required, it is still necessary to address the structural integrity of the fire-resistive-rated wall where it is penetrated. Thus, the space between the duct and the wall must be protected in a manner consistent with that used for pipes, conduits and similar items.

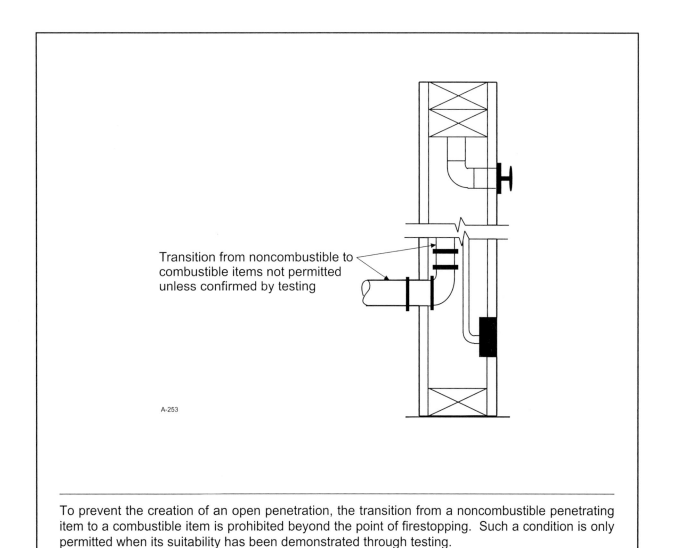

Transition from noncombustible to combustible items not permitted unless confirmed by testing

A-253

To prevent the creation of an open penetration, the transition from a noncombustible penetrating item to a combustible item is prohibited beyond the point of firestopping. Such a condition is only permitted when its suitability has been demonstrated through testing.

Code Text: *Penetrations of a floor, floor/ceiling assembly or the ceiling membrane of a roof/ceiling assembly shall be protected in accordance with Section 707. Penetrations of the fire-resistance rated floor, floor/ceiling assembly or the ceiling membrane of a roof/ceiling assembly shall comply with Sections 712.4.1.1.1 or 712.4.1.1.2.*

Discussion and Commentary: Where horizontal construction is penetrated by a duct, pipe, tube, wire, conduit, cable, vent or similar item, the primary requirements are based on Section 707 for shaft enclosures. However, Exceptions 3 and 4 permit the use of Section 712.4 for both through penetrations and membrane penetrations. The provisions for horizontal assemblies are very similar to those for walls, with special allowances for steel, copper or ferrous penetrating items. In addition to the typical penetrations, the code requires that recessed fixtures in fire-resistance-rated floor/ceiling assemblies be installed so that the required fire resistance is not reduced.

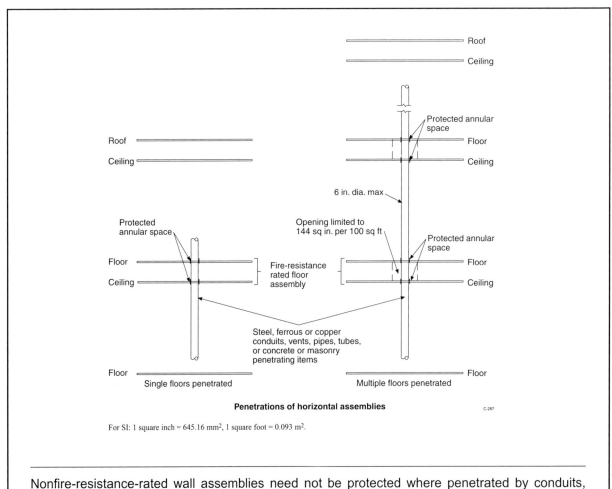

Penetrations of horizontal assemblies

C-287

For SI: 1 square inch = 645.16 mm², 1 square foot = 0.093 m².

Nonfire-resistance-rated wall assemblies need not be protected where penetrated by conduits, piping and similar penetrating items. However, such protection is mandated for the penetration of nonfire-resistance-rated horizontal assemblies to limit vertical fire spread.

Topic: General Provisions	**Category:** Fire-resistance-rated Construction
Reference: IBC 713.1, 713.2	**Subject:** Fire-resistant Joint Systems

Code Text: *Joints installed in or between fire-resistance-rated walls, floor or floor/ceiling assemblies and roofs or roof/ceiling assemblies shall be protected by an approved fire-resistant joint system designed to resist the passage of fire for a time period not less than the required fire-resistance rating of the wall, floor or roof in or between which it is installed. Fire-resistant joint systems shall be securely installed in or on the joint for its entire length so as not to dislodge, loosen or otherwise impair its ability to accommodate expected building movements and to resist the passage of fire and hot gases.*

Discussion and Commentary: Joints are created where the structural design of a building necessitates a separation between building components in order to accommodate anticipated structural displacements caused by thermal expansion and contraction, seismic activity, wind or other loads. The integrity of the fire-resistant separation must be maintained where such joints occur.

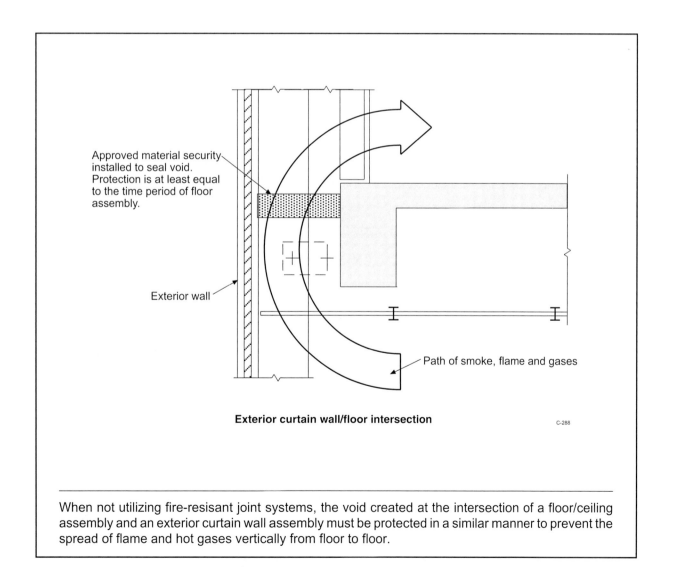

Approved material security installed to seal void. Protection is at least equal to the time period of floor assembly.

Exterior wall

Path of smoke, flame and gases

Exterior curtain wall/floor intersection

C-288

When not utilizing fire-resisant joint systems, the void created at the intersection of a floor/ceiling assembly and an exterior curtain wall assembly must be protected in a similar manner to prevent the spread of flame and hot gases vertically from floor to floor.

Code Text: *Columns, girders, trusses, beams, lintels or other structural members that are required to have a fire-resistance rating and that support more than two floors or one floor and roof, or support a load-bearing wall or a nonload-bearing wall more than two stories high, shall be individually protected on all sides for the full length with materials having the required fire-resistance rating. Other structural members required to have a fire-resistance rating shall be protected by individual encasement, by a membrane or ceiling protection as specified in Section 711, or by a combination of both.*

Discussion and Commentary: Because of the differences in both the testing procedure and the conditions of acceptance, structural frame members carrying significant portions of the structure cannot simply be protected by enclosure within a fire-resistance-rated wall, floor/ceiling or roof/ceiling assembly. Therefore, under specific conditions, individual encasement is mandated.

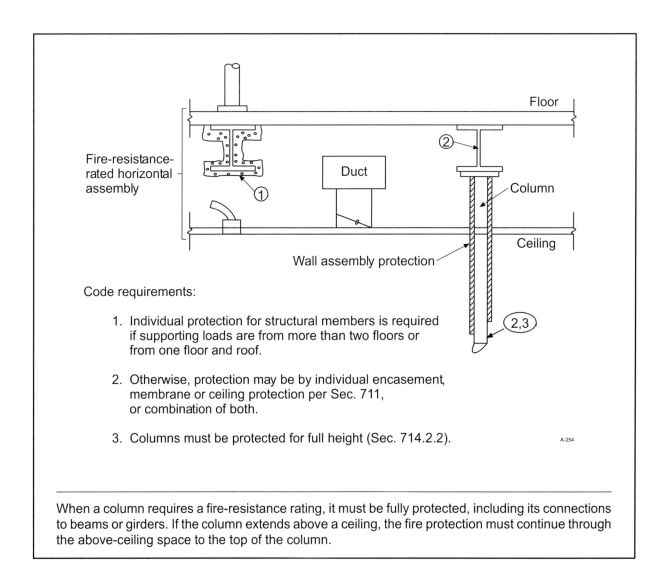

Fire-resistance-rated horizontal assembly

Floor

Duct

Column

Ceiling

Wall assembly protection

Code requirements:

1. Individual protection for structural members is required if supporting loads are from more than two floors or from one floor and roof.

2. Otherwise, protection may be by individual encasement, membrane or ceiling protection per Sec. 711, or combination of both.

3. Columns must be protected for full height (Sec. 714.2.2).

A-254

When a column requires a fire-resistance rating, it must be fully protected, including its connections to beams or girders. If the column extends above a ceiling, the fire protection must continue through the above-ceiling space to the top of the column.

Code Text: *Approved fire door and fire shutter assemblies shall be constructed of any material or assembly of component materials that conforms to the test requirements of Section 715.4.1 (side-hinged or pivoted swinging doors), 715.4.2 (other types of doors) or 715.4.3 (doors in corridors and smoke barriers) and the fire-protection rating indicated in Table 715.4.* See exceptions for tin-clad fire doors and floor fire doors.

Discussion and Commentary: The level of protection required for a fire door is commensurate with that required for the wall or partition in which it is installed. The minimum fire protection rating varies based on the wall's required rating as well as the type and use of the wall assembly under consideration.

TABLE 715.4
FIRE DOOR AND FIRE SHUTTER FIRE PROTECTION RATINGS

TYPE OF ASSEMBLY	REQUIRED ASSEMBLY RATING (hours)	MINIMUM FIRE DOOR AND FIRE SHUTTER ASSEMBLY RATING (hours)
Fire walls and fire barriers having a required fire-resistance rating greater than 1 hour	4	3
	3	3[a]
	2	$1^{1}/_{2}$
	$1^{1}/_{2}$	$1^{1}/_{2}$
Fire barriers having a required fire-resistance rating of 1 hour:		
Shaft, exit enclosure and exit passageway walls	1	1
Other fire barriers	1	$^{3}/_{4}$
Fire partitions:		
Corridor walls	1	$^{1}/_{3}$ [b]
	0.5	$^{1}/_{3}$ [b]
Other fire partitions	1	$^{3}/_{4}$
	0.5	$^{1}/_{3}$
Exterior walls	3	$1^{1}/_{2}$
	2	$1^{1}/_{2}$
	1	$^{3}/_{4}$
Smoke barriers	1	$^{1}/_{3}$ [b]

a. Two doors, each with a fire protection rating of $1^{1}/_{2}$ hours, installed on opposite sides of the same opening in a fire wall, shall be deemed equivalent in fire protection rating to one 3-hour fire door.
b. For testing requirements, see Section 715.4.3.

A fire door assembly installed in a fire-rated corridor wall or smoke barrier must have a minimum fire protection rating of 20 minutes. In addition, the door must pass an air leakage test to verify that it provides the necessary level of smoke protection.

Code Text: *Fire doors shall be labeled showing the name of the manufacturer, the name of the third-party inspection agency, the fire-protection rating, and where required for fire doors in exit enclosures and exit passageways by Section 715.4.4, the maximum transmitted temperature end point. Smoke and draft control doors complying with UL 1784 shall be labeled as such. Labels shall be approved and permanently affixed. The label shall be applied at the factory or location where fabrication and assembly are performed.*

Discussion and Commentary: To be certain that the proper protective assembly is installed in the proper location, it is critical that the assembly be listed and labeled. Field alteration of a fire door assembly is not permitted, because the assembly is usually only listed for use in the condition it was in when it left the factory.

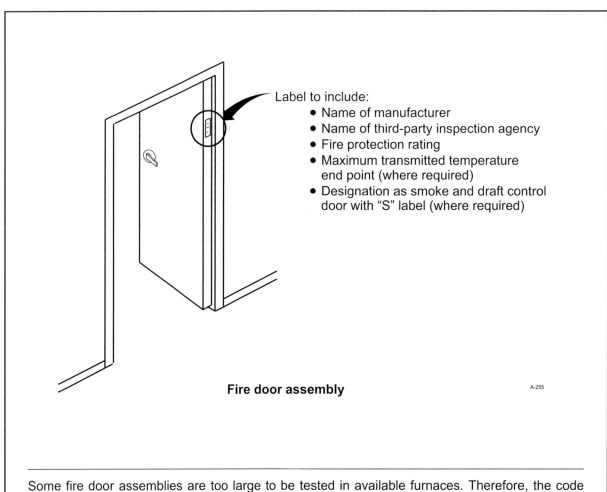

Label to include:
- Name of manufacturer
- Name of third-party inspection agency
- Fire protection rating
- Maximum transmitted temperature end point (where required)
- Designation as smoke and draft control door with "S" label (where required)

Fire door assembly

A-255

Some fire door assemblies are too large to be tested in available furnaces. Therefore, the code recognizes a certificate of inspection as proof that the oversized doors comply with the requirements for materials, design and construction for a comparable fire door.

Topic: Fire Door Glazing Identification   **Category:** Fire-resistance-rated Construction
Reference: IBC 715.4.6.3.1   **Subject:** Opening Protectives

Code Text: *For fire-protection-rated glazing in fire door assemblies, the label shall bear the following four-part identification: "D – H or NH – T or NT – XXX." "D" indicates that the glazing shall be used in fire door assemblies and that the glazing meets the fire resistance requirements of the test standard. "H" shall indicate that the glazing meets the hose stream requirements of the test standard. "NH" shall indicate that the glazing does not meet the hose stream requirements of the test. "T" shall indicate that the glazing meets the temperature requirements of Section 715.4.4.1. "NT" shall indicate that the glazing does not meet the temperature requirements of Section 715.4.4.1. The placeholder "XXX" shall specify the fire-protection rating period, in minutes.*

Discussion and Commentary: Glazing utilized in fire door assemblies can be easily identified for verification of its appropriate application. Such glazing must also be provided with the proper identification indicating its compliance as safety glazing in conformance with Section 2406.2.

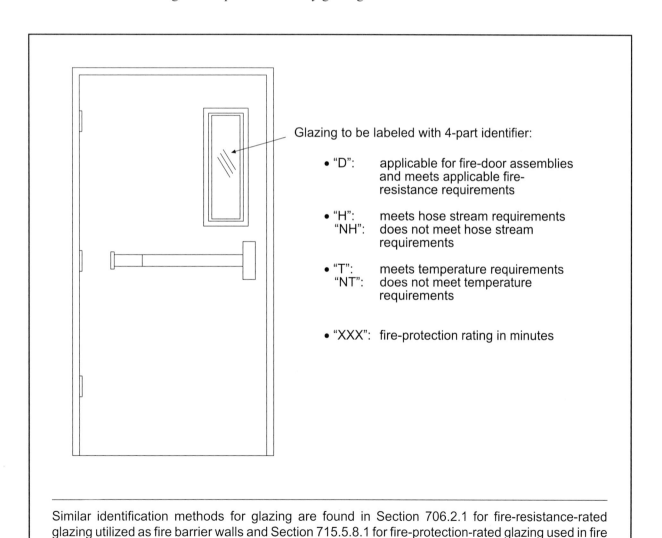

Glazing to be labeled with 4-part identifier:

- "D": applicable for fire-door assemblies and meets applicable fire-resistance requirements

- "H": meets hose stream requirements
  "NH": does not meet hose stream requirements

- "T": meets temperature requirements
  "NT": does not meet temperature requirements

- "XXX": fire-protection rating in minutes

Similar identification methods for glazing are found in Section 706.2.1 for fire-resistance-rated glazing utilized as fire barrier walls and Section 715.5.8.1 for fire-protection-rated glazing used in fire window assemblies.

Code Text: *Fire doors shall be self- or automatic-closing in accordance with Section 715.4.7.* See exceptions for fire doors in common walls between Group R-1 guestrooms and elevator car and associated hoistway doors. *Unless otherwise specifically permitted, single fire doors and both leaves of pairs of side-hinged swinging fire doors shall be provided with an active latch bolt that will secure the door when it is closed.*

Discussion and Commentary: Fire doors must close and latch to be effective during a fire. The expectation is that the doors will normally be in a closed position and that the self-closing device will cause the door to close after use. Where specifically mandated by the code, automatic-closing devices must be installed. Such devices are intended for doors normally held in an open position.

Automatic-closing doors shall be actuated by smoke detection at: (Sec. 715.4.7.3)

- Doors installed across a corridor;
- Doors protecting openings in exits or corridors required to be fire-resistance-rated;
- Doors in walls of incidental use areas required to resist the passage of smoke (Sec. 508.2.2.1);
- Doors installed in smoke barriers (Sec. 709.5);
- Doors installed in fire partitions (Sec. 708.6);
- Doors installed in fire walls (Sec. 705.8);
- Doors installed in shaft enclosures (Sec. 707.7);
- Doors installed in refuse and laundry access chutes and termination rooms (Sec.707.13.3);
- Doors installed in compartmentation walls of underground buildings (Sec. 405.4.2);
- Doors installed in elevator lobby walls of underground buildings (Sec. 405.4.3); and
- Doors installed in smoke partitions (Sec. 710.5.3).

Automatic-closing fire door assemblies are required only where specifically addressed, such as in Section 709.5 for cross-corridor doors in Group I-2 occupancies. Where automatic-closing fire doors are provided, including nonrequired locations, they must typically be smoke activated.

Code Text: *Steel window frame assemblies of 0.125-inch (3.2 mm) minimum solid section or of not less than nominal 0.048-inch-thick (1.2 mm) formed sheet steel members fabricated by pressing, mitering, riveting, interlocking or welding and having provision for glazing with $^1/_4$ inch (6.4 mm) wired glass where securely installed in the building construction and glazed with $^1/_4$ inch (6.4 mm) labeled wired glass shall be deemed to meet the requirements for a $^3/_4$-hour fire window assembly. Wired glass panels shall conform to the size limitations set forth in Table 715.5.3.*

Discussion and Commentary: Fire-protection-rated glazing (fire windows) shall typically be rated for $^3/_4$ hour. This rating is based on testing in accordance with NFPA 257. The IBC permits the use of complying $^1/_4$-inch wired glass in a steel frame in lieu of a tested assembly where a $^3/_4$-hour fire protection rating is required.

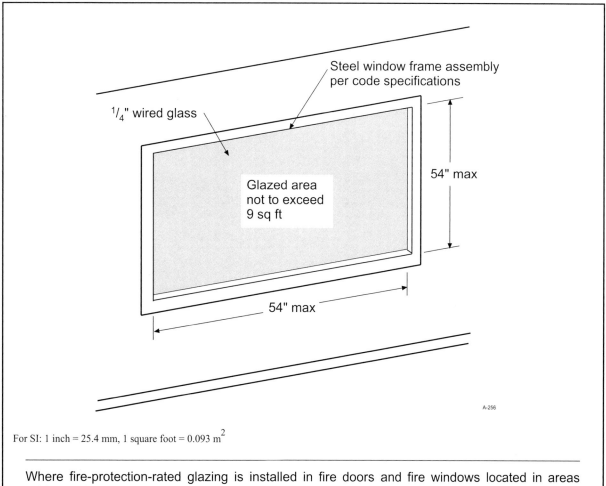

Steel window frame assembly per code specifications

$^1/_4$" wired glass

Glazed area not to exceed 9 sq ft

54" max

54" max

A-256

For SI: 1 inch = 25.4 mm, 1 square foot = 0.093 m^2

Where fire-protection-rated glazing is installed in fire doors and fire windows located in areas subject to human impact, it also shall comply with the provisions for safety glazing as identified in Section 2406.3 (Hazardous Locations).

Code Text: *A fire damper is a listed device, installed in ducts and air transfer openings designed to close automatically upon detection of heat and resist the passage of flame. A smoke damper is a listed device installed in ducts and air transfer openings that is designed to resist the passage of smoke. A ceiling radiation damper is a listed device installed in a ceiling membrane of a fire-resistance-rated floor/ceiling or roof/ceiling assembly to limit automatically the radiative heat transfer through an air inlet/outlet opening.*

Discussion and Commentary: The IBC identifies several types of dampers, each of which performs a specific function. The required type of damper is based on the function of the building element that is being penetrated by the duct or air transfer opening.

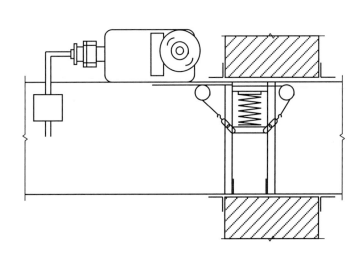

Figure courtesy
Sheet Metal and Air Conditioning Contractors National Association

Note: This illustration is not intended to exclusively endorse or
indicate preference for a combination fire and smoke damper.
Two separate dampers that satisfy the requirements for the
respective functions may also be used for fire and smoke control.

Combination fire and smoke dampers

A-258

Where both a fire and a smoke damper are mandated, the use of a combination damper is permitted. This type of listed device is designed to close automatically upon detecting heat and to resist the passage of air and smoke.

Topic: Damper Testing and Ratings

Category: Fire-resistance-rated Construction

Reference: IBC 716.3, 716.3.1

Subject: Ducts and Air Transfer Openings

Code Text: *Dampers shall be listed and bear the label of an approved testing agency indicating compliance with the standards in Section 716.3. Fire dampers shall comply with the requirements of UL 555. Smoke dampers shall comply with the requirements of UL 555S. Combination fire/smoke dampers shall comply with the requirements of both UL 555 and UL 555S. Ceiling radiation dampers shall comply with the requirements of UL 555C. Fire dampers shall have the minimum fire-protection rating specified in Table 716.3.1 for the type of penetration.*

Discussion and Commentary: Consistent with other openings that penetrate a fire-resistance-rated assembly, fire and smoke dampers must be provided where it is necessary to maintain the integrity of the assembly. The minimum damper rating is based on the rating of the assembly penetrated.

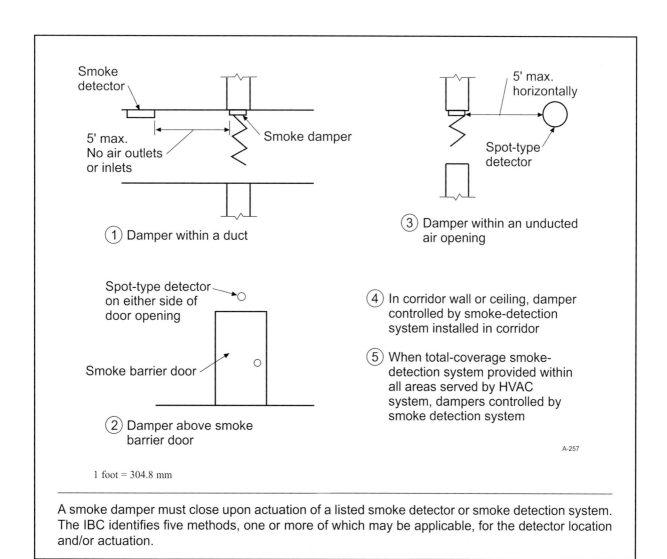

① Damper within a duct

② Damper above smoke barrier door

③ Damper within an unducted air opening

④ In corridor wall or ceiling, damper controlled by smoke-detection system installed in corridor

⑤ When total-coverage smoke-detection system provided within all areas served by HVAC system, dampers controlled by smoke detection system

A-257

1 foot = 304.8 mm

A smoke damper must close upon actuation of a listed smoke detector or smoke detection system. The IBC identifies five methods, one or more of which may be applicable, for the detector location and/or actuation.

Code Text: *Fire dampers, smoke dampers, combination fire/smoke dampers and ceiling radiation dampers shall be provided at the locations prescribed in Sections 716.5.1 through 716.5.5. Where an assembly is required to have both fire dampers and smoke dampers, combination fire/smoke dampers or a fire damper and a smoke damper shall be required.*

Discussion and Commentary: Only those specific building elements identified in the code need to be protected by fire and/or smoke dampers where penetrated by ducts or air transfer openings. As a general rule, fire dampers protect such openings in fire walls, fire barriers, shaft enclosures and fire partitions. Fire dampers may also be installed in some of those locations where a shaft enclosure is otherwise required. Smoke dampers are generally required for openings in shaft enclosures, smoke- and draft-control corridor enclosures, and smoke barriers.

DAMPER REQUIREMENTS

Location		Fire Dampers	Smoke Dampers
Fire walls		Required	
Fire barriers		Required [1, 2, 3]	
Shaft enclosures [6]		Required [1, 2, 4, 5]	Required [2, 5, 19]
Fire partitions		Required [7, 8, 20]	
Corridor enclosure		Required [7, 8]	Required [9, 17]
Smoke barriers			Required [10]
Horizontal assemblies [11]	Through penetrations	Required [12, 18]	
	Membrane penetrations	Required [13, 21]	
	Nonfire-resistance-rated assemblies	Required [14, 15, 16, 21]	

[1] Not required for penetrations tested in accordance with ASTM E 119 as part of the rated assembly.

[2] Not required for ducts used as a part of an approved smoke control system in accordance with Section 909.

[3] Not required in sprinklered building of other than Group H for 1-hour walls penetrated by ducted HVAC systems.

[4] Not required for steel exhaust subducts extending at least 22 inches vertically in exhaust shafts having continuous airflow upward to the outside.

[5] Not required in parking garage supply or exhaust shafts that are separated from other building shafts by a minimum of 2-hour fire-resistance-rated construction.

[6] See Section 1020.1.2 for permitted penetrations of exit enclosures.

[7] Not required in sprinklered buildings of other than Group H for tenant separation and corridor walls.

[8] Not required in buildings of other than Group H where duct penetration is limited to 100 square inches; is of minimum 0.0217-inch steel; does not have communicating openings between a corridor and adjacent spaces; is installed above a ceiling; does not terminate at a wall register of the fire-resistance-rated wall; and a minimum 12-inch-long steel sleeve is centered in each duct opening, approximately secured, and the annular space between the sleeve and the wall opening is filled with mineral wool batting.

[9] Not required for corridor penetrations of minimum 0.019-inch steel ducts with no openings into corridor.

[10] Not required where openings in steel ducts are limited to a single smoke compartment.

[11] General requirement mandates shaft enclosures for openings in floor and roof systems.

[12] In other than Group I-2 and I-3, fire dampers are permitted in lieu of shaft enclosures for penetration of fire-resistance-rated horizontal assembly that connects two floors.

[13] Where shaft enclosure is not provided, an approved ceiling radiation damper is required at the ceiling line of a fire-resistance-rated floor/ceiling or roof/ceiling assembly.

[14] Not required, provided the shaft enclosure does not connect more than two stories, and the annular space around the duct is filled with noncombustible material.

[15] Limited to three connected stories without shaft enclosure, provided fire dampers are installed at each floor line and annular space is filled.

[16] Not required in ducts within individual dwelling units.

[17] Not required in building with a smoke control system if not necessary for operation and control of system.

[18] Not required where penetrating up to three floors, provided the steel duct is located within a wall cavity, serves only one dwelling unit or sleeping unit, and is a maximum of 4 inches in diameter and limited to 100 square inches in 100 square feet; and provided the annular space around the duct is protected, and any grille openings in a fire-rated ceiling are protected with ceiling radiationt dampers.

[19] Not required in fully sprinklered Group B and R occupancies for kitchen, clothes dryer, bathroom and toilet room exhaust openings installed with steel subducts extending minimum of 22 inches vertically in exhaust shafts having continuous airflow upward to the outside.

[20] Not required for tenant partitions in covered mall buildings where walls not required to extend to underside of floor or roof deck above.

[21] Not required where protected by a shaft enclosure.

There will be times when a fire-resistance-rated assembly is penetrated by a duct or transfer opening that is not required to be protected by a fire or smoke damper. In such situations, the condition will be regulated and protected as a penetration, in accordance with Section 712.

Code Text: *Duct systems constructed of approved materials in accordance with the International Mechanical Code that penetrate nonfire-resistance-rated floor assemblies shall be protected by any of the following methods.* See three methods addressing 1) shaft enclosures, 2) ducts connecting only two stories, and 3) ducts connecting a maximum of three stories.

Discussion and Commentary: It is important when addressing the vertical spread of fire, smoke and gases—even in buildings where the floor assemblies are not required to have a fire-resistance rating—that some level of compartmentation is provided between stories.

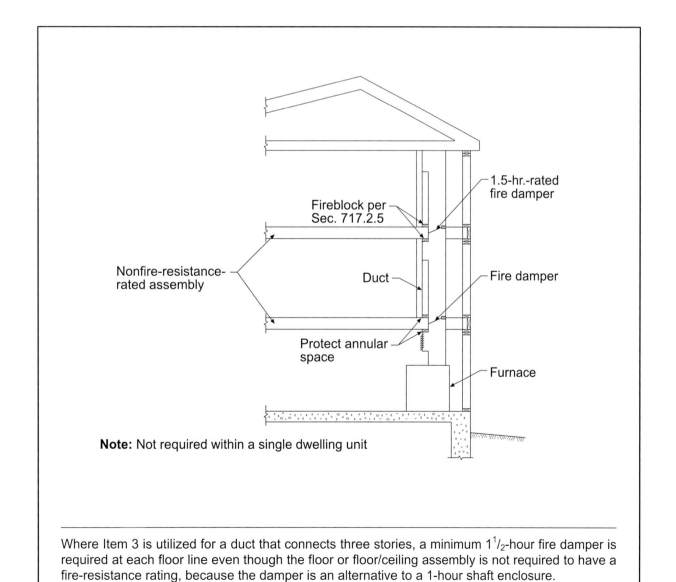

Where Item 3 is utilized for a duct that connects three stories, a minimum $1^1/_2$-hour fire damper is required at each floor line even though the floor or floor/ceiling assembly is not required to have a fire-resistance rating, because the damper is an alternative to a 1-hour shaft enclosure.

Code Text: *In combustible construction, fireblocking shall be installed to cut off concealed draft openings (both vertical and horizontal) and shall form an effective barrier between floors, between a top story and a roof or attic space. Fireblocking shall be installed in the locations specified in Sections 717.2.2 through 717.2.7.*

Discussion and Commentary: Experience has shown that the greatest fire damage to conventional light-framed wood buildings occurs when the fire travels unimpeded through concealed draft openings. Virtually any concealed air space within a building will provide an open channel through which high-temperature air and gases can spread. Fireblocking is invaluable to the control of fire prior to active fire suppression activities.

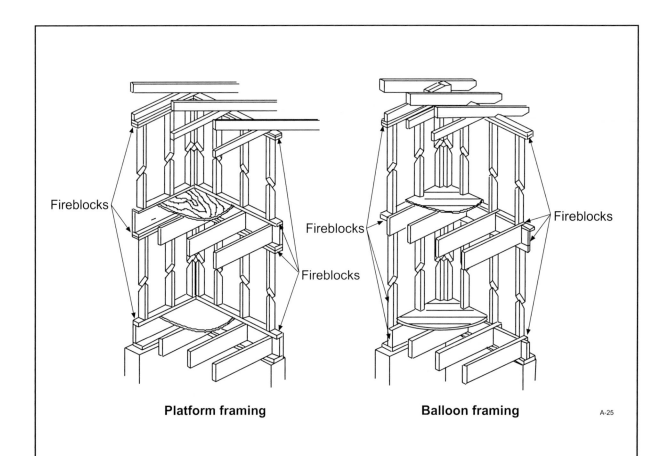

Platform framing   **Balloon framing**   A-25

In noncombustible construction, building materials located in concealed areas of the building construction do not contribute to the spread of fire. Therefore, fireblocking and draftstopping are required only in buildings of combustible construction.

Code Text: *Fireblocking shall be provided in concealed spaces of stud walls and partitions, including furred spaces, and parallel rows of studs or staggered studs, as follows: 1) vertically at the ceiling and floor levels, and 2) horizontally at intervals not exceeding 10 feet (3048 mm). Fireblocking shall be provided at interconnections between concealed vertical . . . and horizontal spaces . . . such as occur at soffits, drop ceilings, cove ceilings and similar locations.* See additional provisions for fireblocking at stairways; openings around vents, ducts and chimneys; concealed spaces of exterior architectural trim; and concealed sleeper spaces in floors.

Discussion and Commentary: The platform framing techniques that are typically used in light-frame wood construction provide adequate fireblocking between stories in the stud walls. However, furred spaces and openings for penetrating elements such as vents should be addressed carefully as avenues for fire transmission between stories or along a wall.

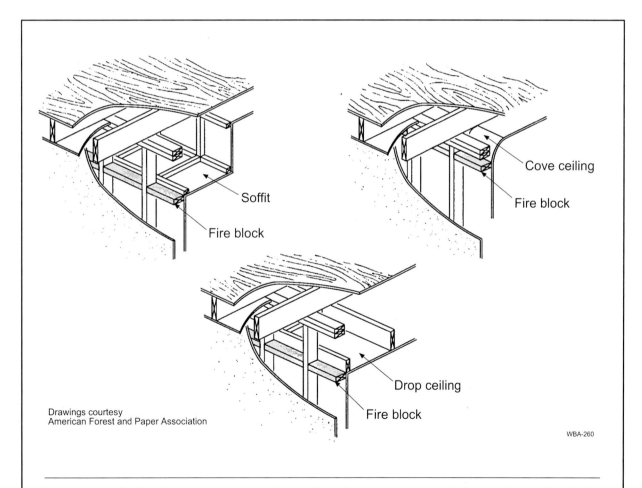

Soffit

Fire block

Cove ceiling

Fire block

Drop ceiling

Fire block

Drawings courtesy
American Forest and Paper Association

WBA-260

In general, fireblocking materials must consist of lumber or wood structural panels of the thicknesses specified, gypsum board, cement fiber board, batts or blankets of mineral wool or glass fiber, or any other approved materials securely fastened in place.

Code Text: *A draftstop is a material, device or construction installed to restrict the movement of air within open spaces of concealed areas of building components such as crawl spaces, floor/ceiling assemblies, roof/ceiling assemblies and attics. Draftstopping materials shall not be less than 0.5-inch (12.7 mm) gypsum board, 0.375-inch (9.5 mm) wood structural panel, 0.375-inch (9.5 mm) particleboard, 1-inch (25 mm) nominal lumber, cement fiberboard, batts or blankets of mineral wool or glass fiber, or other approved materials adequately supported. The integrity of draftstops shall be maintained.*

Discussion and Commentary: Draftstopping, like fireblocking, is required only in combustible construction. Although the role of draftstopping is important, it is less critical than that of fireblocking. Therefore, the protective materials used in draftstopping construction are permitted to be less substantial.

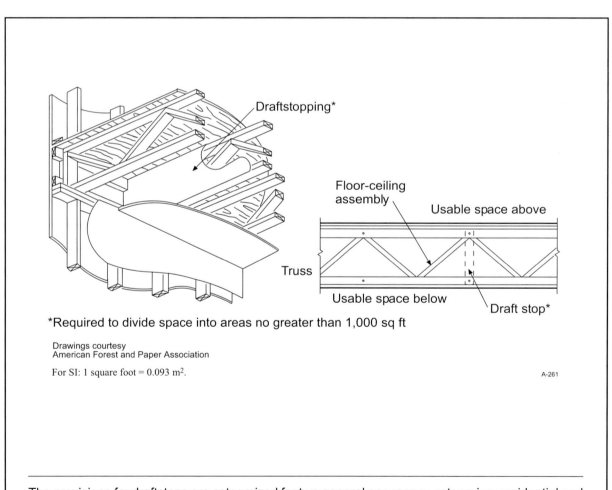

Draftstopping*

Floor-ceiling assembly

Usable space above

Truss

Usable space below

Draft stop*

*Required to divide space into areas no greater than 1,000 sq ft

Drawings courtesy
American Forest and Paper Association

For SI: 1 square foot = 0.093 m².

A-261

The provisions for draftstops are categorized for two general occupancy categories: residential and all uses other than residential. Both floor/ceiling assemblies and attics are addressed for each category. Many of the requirements are eliminated in fully sprinklered buildings.

Quiz

Study Session 7
IBC Sections 712 – 719

1. An approved through-penetration firestop system requires a T rating when which of the following conditions occurs?

    a. a fire wall is penetrated by a combustible penetrating item

    b. a fire barrier wall is penetrated above a nonrated ceiling membrane

    c. a fire partition is penetrated by a steel conduit

    d. a floor penetration of a fire-rated assembly is not contained within a wall cavity

Reference_____

2. Where the annular space at the penetration of horizontal assemblies is filled with an approved material, noncombustible piping may connect a maximum of _____ stories, provided the horizontal assemblies require no fire-resistance rating.

    a. 2   b. 3

    c. 4   d. 6

Reference_____

3. In which of the following locations is a fire-resistant joint system required to protect all joints?

    a. floors within malls   b. horizontal exit walls

    c. mezzanine floors   d. roofs where openings are permitted

Reference_____

4. Girders and beams required to have a fire-resistance rating must be individually protected on all sides where supporting a minimum of _____.

    a. a roof only   b. a floor and a roof

    c. two floors   d. two floors and a roof

Reference_____

5. The edges of lugs, brackets, rivets and bolt heads attached to fire-resistance-rated structural members shall maintain a minimum of _____ inch(es) clearance to the surface of the fire protection.

    a. $^1/_2$ b. 1

    c. $1^1/_2$ d. 2

Reference_____

6. Stirrups used for concrete reinforcement may extend a maximum of _____ inch into the required thickness of fire protection.

    a. $^1/_4$ b. $^3/_8$

    c. $^1/_2$ d. 1

Reference_____

7. A fire door assembly in a 1-hour fire barrier used in a vertical exit enclosure shall have a minimum fire protection rating of _____ hour.

    a. $^1/_3$ b. $^1/_2$

    c. $^3/_4$ d. 1

Reference_____

8. Fire door assemblies in a 2-hour exterior wall shall have a minimum fire protection rating of _____.

    a. 20 minutes   b. 45 minutes

    c. 1 hour   d. 90 minutes

Reference_____

9. Penetrations in _____ shall be tested for air leakage in accordance with the requirements of UL 1479.

    a. smoke partitions   b. smoke barriers

    c. fire barriers   d. fire walls

Reference _____

10. Where not used as a horizontal exit, what is the maximum amount of glazing permitted in a $1^1/_2$-hour swinging fire door located in a 2-hour fire-resistance-rated fire wall?

    a. 0, no glazing is permitted   b. 100 square inches

    c. 144 square inches   d. 9 square feet

Reference_____

11. In which one of the following locations may self-closing and automatic-closing devices be omitted from required fire doors?

    a. Group E corridor doors

    b. doors in common walls separating Group R-1 sleeping units

    c. doors between vertical exit enclosures and exit passageways

    d. Group I-2 smoke barrier doors

Reference_____

12. For a fire door that is automatic closing by smoke detection, the maximum delay before the fire door starts to close is _____ seconds after the smoke detector is actuated.

    a. 3   b. 5

    c. 10   d. 15

Reference_____

13. Where automatic-closing fire doors are installed, which one of the following locations does not require closing upon actuation of smoke detectors?

    a. cross-corridor doors in an office building

    b. doors in a fire barrier separating control areas

    c. a pair of doors installed in a fire wall

    d. a single door installed in a smoke partition

Reference_____

14. What is the maximum size of a wired glass panel that is permitted in a fire window assembly?

    a. 100 square inches   b. 1,296 square inches

    c. 54 inches by 54 inches   d. unlimited

Reference_____

15. What is the minimum required fire-protection rating for a fire window assembly located in a 2-hour fire barrier?

    a. $^3/_4$ hour   b. 1 hour

    c. $1^1/_2$ hours   d. fire windows are not permitted

Reference _____

16. The total area of interior fire window assemblies located in fire partitions and fire barriers is limited to a maximum of _____ percent of the area of the common wall within any room.

    a. 10   b. 20

    c. 25   d. 50

Reference_____

17. What is the minimum fire damper rating for a damper located in a 1-hour fire barrier?

    a. 20 minutes   b. 45 minutes

    c. 1 hour   d. $1^1/_2$ hour

Reference_____

18. Where a spot-type detector is used to actuate a smoke damper installed within an unducted opening in a smoke barrier wall, the detector shall be located a maximum of _____ from the damper.

    a. 12 inches vertically   b. 30 inches vertically

    c. 3 feet horizontally   d. 5 feet horizontally

Reference_____

19. Where a duct passes through a fire wall, which of the following dampers is/are required?

    a. fire damper only

    b. smoke damper only

    c. both a fire damper and a smoke damper

    d. neither a fire damper nor a smoke damper

Reference_____

20. In general, which of the following dampers is/are required where an air transfer opening penetrates a shaft enclosure?

    a. fire damper only

    b. smoke damper only

    c. both fire and smoke dampers

    d. neither a fire damper nor a smoke damper

Reference_____

21. Which of the following materials is not specifically listed by the code as a fireblocking material?

    a. 2-inch nominal lumber

    b. two thicknesses of 1-inch nominal lumber with broken lap joints

    c. one thickness of 0.719-inch wood structural panel with joints backed

    d. one thickness of $^1/_2$-inch particleboard with joints backed

Reference_____

22. The maximum spacing of fireblocking within concealed spaces of exterior architectural elements erected with combustible framing shall be _____ feet.

    a. 10   b. 20

    c. 60   d. 100

Reference_____

23. Which of the following materials is not specifically listed by the code as an acceptable draftstopping material?

    a. $^3/_8$-inch gypsum board   b. $^3/_8$-inch wood structural panels

    c. $^3/_8$-inch particleboard   d. glass fiber batts

Reference_____

24. In a nonsprinklered Type V office building, the maximum attic area permitted between draftstops is _____.

    a. 1,000 square feet   b. 3,000 square feet

    c. 100 feet in any direction   d. unlimited

Reference_____

25. Other than for cellulose loose-fill insulation that is not spray applied, the maximum flame spread index of insulating materials concealed within a Type II building shall be _____.

    a. 25   b. 50

    c. 75   d. unlimited

Reference_____

26. A through-penetration firestop system, where used to protect a membrane penetration in a fire barrier wall, shall at minimum have a(n) _____ rating of not less than the rating of the wall penetrated.

    a. F   b. L

    c. T   d. F and T

Reference_____

27. A firestop system may not be required for the protection of unlisted steel electrical boxes in a 1-hour fire partition, provided the boxes have a maximum individual size of _____ square inches.

    a. 16   b. 25

    c. 100   d. 144

Reference_____

28. What is the minimum required fire protection rating for a fire window located in a $^1/_2$-hour fire-resistance-rated fire partition?

    a. 0, no rating is required   b. 20 minutes

    c. 30 minutes   d. 45 minutes

Reference_____

29. Where wood sleepers are installed for the installation of wood flooring in a church sanctuary, the maximum size of any open spaces shall be _____ square feet unless the entire underfloor space is filled with an approved material.

    a. 10   b. 20

    c. 100   d. 144

Reference_____

30. Where plaster is used for fire-resistance purposes, it shall be provided with an additional layer of approved lath where its minimum thickness exceeds _____ inch.

    a. $^3/_8$ b. $^1/_2$

    c. $^3/_4$ d. 1

Reference_____

31. Where the fire protective covering of a structural member is subject to impact damage that is due to the handling of merchandise, it shall be adequately protected from damage to a minimum height of _____ feet.

    a. 4   b. 5

    c. 6   d. 8

Reference _____

32. Fire protection is not required at the bottom flange of lintels with a maximum span of _____ feet where the lintel is a part of the structural frame.

    a. 3   b. 4

    c. 5   d. 6

Reference _____

33. A required smoke damper shall have a minimum leakage rating of Class _____.

    a. 0   b. I

    c. II   d. III

Reference _____

34. Fireblocking shall be installed within concealed spaces of exterior wall elements erected with combustible framing so that any open spaces will be a maximum of _____ square feet in area.

    a. 100   b. 200

    c. 1,000   d. 3,000

Reference _____

35. For fire-resistance purposes, $^1/_2$ inch of unsanded gypsum plaster is deemed to be equivalent to _____ inch of portland cement sand plaster.

    a. $^1/_4$ b. $^3/_8$

    c. $^3/_4$ d. 1

Reference _____

2006 IBC Chapter 9
Fire Protection Systems

OBJECTIVE: To obtain an understanding of the design and installation of fire protection systems, including automatic sprinkler systems, standpipe systems, fire alarm and detection systems, smoke control systems, and smoke and heat vents.

REFERENCE: Chapter 9, 2006 *International Building Code*

KEY POINTS:
- What are the quality standards for fire protection systems?
- How shall fire alarm systems be monitored?
- What are the different classifications for standpipe systems? For other types of systems?
- For Group A-1, A-3 and A-4 occupancies, what size fire area requires the installation of an automatic sprinkler system? How many occupants? What location in the building?
- When must a restaurant or café be sprinklered?
- Which conditions would allow Group E occupancies to be nonsprinklered?
- A sprinkler system is required in Groups F-1, M and S-1 occupancies when the floor area within any fire area exceeds what size?
- Which high-hazard occupancies require installation of a sprinkler system? Institutional occupancies?
- Which residential occupancies require the installation of an automatic sprinkler system? In which portions of the building must the sprinkler system be located?
- Under what conditions must sprinkler protection be provided for exterior balconies and ground-floor patios of dwelling units?
- For the purpose of fire department access, what is considered a "windowless" building?
- In "windowless" buildings, how are basements regulated as compared to above-grade levels?
- When must buildings be sprinklered due to their height?
- Which locations or uses are exempt from the sprinkler requirements?
- What are the different classes of standpipe systems? How do they differ?
- When are standpipes required?
- When a standpipe system is mandated, where must the hose connections be located?

KEY POINTS:
(Cont'd)

- In which occupancies are manual fire alarm systems required? Automatic fire alarm systems?
- Where are smoke alarms mandated?
- Where fire alarms are required, where must visible appliances be installed?
- What is the scope and purpose of a smoke-control system?
- What are the three different methods of mechanical smoke control? Which of the three is the primary means of controlling smoke?
- What are the critical elements for the acceptance of a smoke-control system?
- In which occupancies are smoke and heat vents required?
- When smoke and heat vents are provided, what are their minimum size and spacing requirements?
- What is the minimum distance between curtain boards? What is the maximum permitted size of the curtained area?
- Under which conditions can a mechanical smoke exhaust system be substituted for smoke and heat vents?
- What are the requirements for a fire command center?

Code Text: *The provisions of Chapter 9 shall specify where fire protection systems are required and shall apply to the design, installation and operation of fire protection systems. A fire protection system is approved devices, equipment and systems or combinations of systems used to detect a fire, activate an alarm, extinguish or control a fire, control or manage smoke and products of a fire or any combination thereof. Fire protection systems shall be installed, repaired, operated and maintained in accordance with* the International Building Code and the International Fire Code.

Discussion and Commentary: The code provides requirements for three distinct systems considered vital to a safe building environment. The first system is intended to control and limit fire spread and to provide building occupants and fire personnel with the means of fighting a fire. The second system provides for detection of a fire condition and a means of notification. The third system is intended to control smoke migration.

General requirements for fire protection systems:

- Systems to be installed, repaired, operated and maintained in accordance with the *International Building Code* and *International Fire Code.*
- Systems not required by the IBC are permitted to be installed for partial or complete protection, provided such systems meet the requirements of the IBC.
- Any system for which an exception to, or reduction in, the provisions of the IBC has been granted shall be considered a required system.
- No person shall remove or modify any system installed or maintained under the provisions of either code without approval of the building official.
- All systems shall be tested in accordance with the requirements of the IBC and IFC, in the presence of the building official and at the expense of the owner or owner's representative.
- It is unlawful to occupy portions of a structure until the required fire protection systems within that portion have been tested and approved.

Unless specifically excepted, approved supervising stations are mandated for automatic sprinkler systems, fire alarm systems, and Group H occupancy manual alarm, automatic fire-extinguishing and emergency alarm systems.

Code Text: *A fire area is the aggregate floor area enclosed and bounded by fire walls, fire barriers, exterior walls or fire-resistance-rated horizontal assemblies of a building. Where buildings, or portions thereof, are divided into fire areas so as not to exceed the limits established for requiring a fire protection system in accordance with Chapter 9, such fire areas shall be separated by fire barriers having a fire-resistance rating of not less than that determined in accordance with Section 706.3.9.*

Discussion and Commentary: The concept behind fire areas is that of compartmentalization. As a building is subdivided into smaller spaces through the use of fire-resistance-rated elements, the potential hazards tend to be confined to each compartment. Therefore, as the level of hazards decreases, the need for protection diminishes. In the IBC, this reduced level of protection is reflected in the fact that an automatic sprinkler system may not be required. The primary purpose for the creation of fire areas is to address the requirements of Section 903.2 for sprinkler protection.

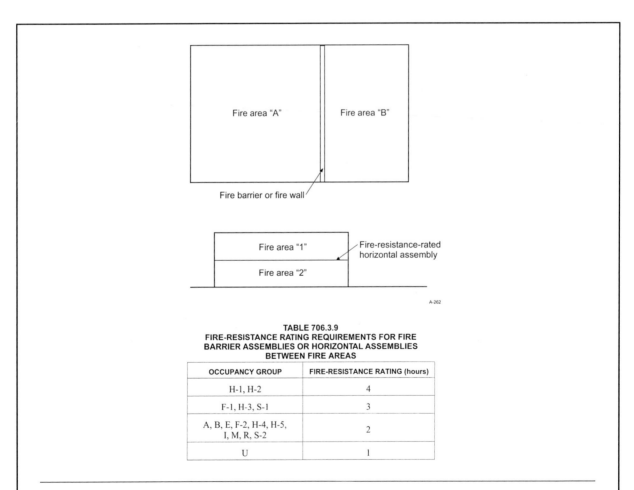

TABLE 706.3.9
FIRE-RESISTANCE RATING REQUIREMENTS FOR FIRE
BARRIER ASSEMBLIES OR HORIZONTAL ASSEMBLIES
BETWEEN FIRE AREAS

OCCUPANCY GROUP	FIRE-RESISTANCE RATING (hours)
H-1, H-2	4
F-1, H-3, S-1	3
A, B, E, F-2, H-4, H-5, I, M, R, S-2	2
U	1

To determine the appropriate level of fire resistance for fire barriers used to create one or more fire areas, refer to Table 706.3.9. This table mandates the minimum hourly rating for fire barriers separating one or more occupancies into different fire areas.

Code Text: *An automatic sprinkler system shall be provided throughout buildings and portions thereof used as Group A occupancies as provided in Section 903.2.1. For Group A-1, A-2, A-3 and A-4 occupancies, the automatic sprinkler system shall be provided throughout the floor area where the Group A-1, A-2, A-3 or A-4 occupancy is located, and in all floors between the Group A occupancy and the level of exit discharge. For Group A-5 occupancies, the automatic sprinkler system shall be provided in the spaces indicated in Section 903.2.1.5.*

Discussion and Commentary: Although most Group A occupancies lack the combustible loading that creates a high degree of fire severity, protection provided by an automatic sprinkler system is deemed necessary due to the hazards of having large numbers of people in concentrated areas. Based on varying thresholds, a sprinkler system may be required in any of the five assembly occupancies.

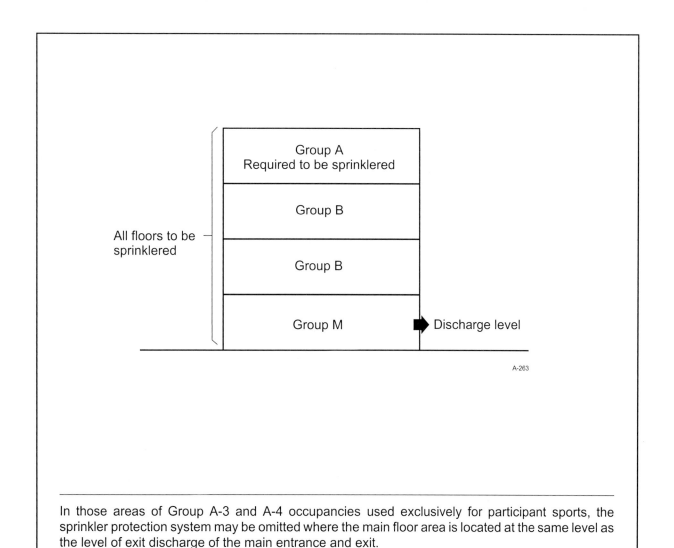

In those areas of Group A-3 and A-4 occupancies used exclusively for participant sports, the sprinkler protection system may be omitted where the main floor area is located at the same level as the level of exit discharge of the main entrance and exit.

Code Text: *An automatic sprinkler system shall be provided for Group A-1 occupancies where one of the following conditions exists: 1) the fire area exceeds 12,000 square feet (1115 m²), 2) the fire area has an occupant load of 300 or more, or 3) the fire area is located on a floor other than the level of exit discharge. The same criteria apply to Group A-3 and A-4 occupancies. An automatic sprinkler system shall be provided for Group A-2 occupancies where one of the following conditions exist: 1) the fire area exceeds 5,000 square feet (465 m²), 2) the fire area has an occupant load of 100 or more, or 3) the fire area is located on a floor other than the level of exit discharge. An automatic sprinkler system shall be provided for Group A-5 occupancies in the following areas: concession stands, retail areas, press boxes and other accessory use areas in excess of 1,000 square feet (93 m²).*

Discussion and Commentary: The thresholds at which Group A occupancies are required to be sprinklered vary based upon past fire records and the potential hazards associated with the different types of assembly uses.

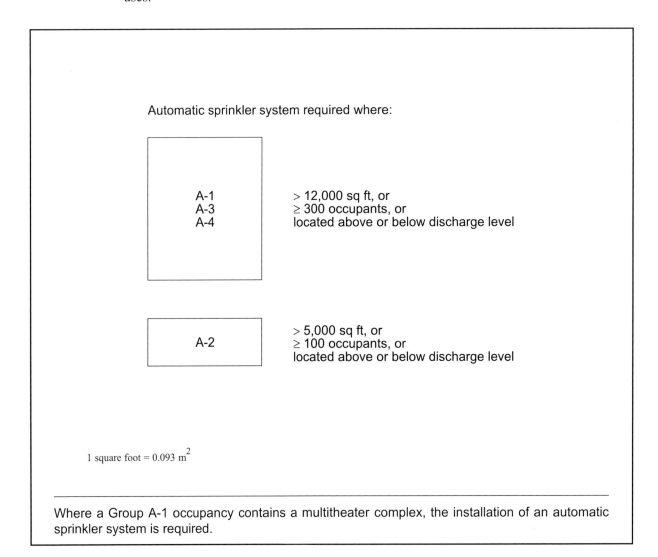

Automatic sprinkler system required where:

A-1
A-3
A-4

> 12,000 sq ft, or
≥ 300 occupants, or
located above or below discharge level

A-2

> 5,000 sq ft, or
≥ 100 occupants, or
located above or below discharge level

1 square foot = 0.093 m²

Where a Group A-1 occupancy contains a multitheater complex, the installation of an automatic sprinkler system is required.

Topic: Group E Occupancies

Category: Fire Protection Systems

Reference: IBC 903.2.2

Subject: Automatic Sprinkler Systems

Code Text: *An automatic sprinkler system shall be provided for Group E occupancies as follows: 1) throughout all Group E fire areas greater than 20,000 square feet (1858 m²) in area, and 2) throughout every portion of educational buildings below the level of exit discharge. See exception for buildings with direct egress at ground level from each classroom.*

Discussion and Commentary: As a group, educational occupancies tend to have a very good fire record. This stems from the ongoing supervision of activities in the building, as well as the rapid egress of students in response to emergencies. However, because of the amount of combustibles in Group E occupancies and the potentially high occupant load, it is typically necessary to provide sprinkler systems for those undivided floor areas exceeding 20,000 square feet or in basements.

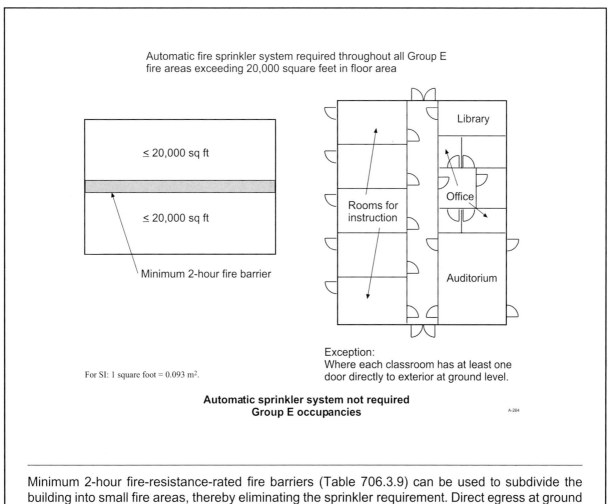

Automatic fire sprinkler system required throughout all Group E fire areas exceeding 20,000 square feet in floor area

≤ 20,000 sq ft

≤ 20,000 sq ft

Minimum 2-hour fire barrier

For SI: 1 square foot = 0.093 m².

Library

Rooms for instruction

Office

Auditorium

Exception:
Where each classroom has at least one door directly to exterior at ground level.

A-264

Automatic sprinkler system not required
Group E occupancies

Minimum 2-hour fire-resistance-rated fire barriers (Table 706.3.9) can be used to subdivide the building into small fire areas, thereby eliminating the sprinkler requirement. Direct egress at ground level from each classroom is also deemed to be an equivalent method of ensuring occupant safety.

Code Text: *An automatic sprinkler system shall be provided throughout all buildings contain a Group F-1 occupancy where one of the following conditions exists: 1) where a Group F-1 fire area exceeds 12,000 square feet (1115 m², 2) where a Group F-1 fire area is located more than three stories above grade plane, or 3) where the combined area of all Group F-1 fire areas on all floors, including any mezzanines, exceeds 24,000 square feet (2230 m²). Same criteria applies to Group M and S-1 occupancies.*

Discussion and Commentary: Because of the potential presence of high levels of combustible materials in factories, sales buildings and warehouses, the IBC limits the size and location of fire areas not protected by an automatic sprinkler system. If any fire area in the building exceeds the threshold, the sprinkler system must be provided throughout the entire building, not just in the fire area that exceeds the area or height limitations.

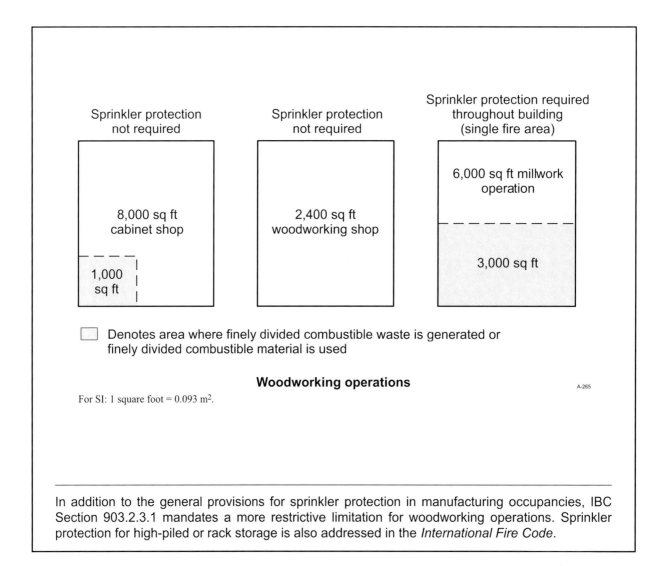

Sprinkler protection not required

8,000 sq ft cabinet shop

1,000 sq ft

Sprinkler protection not required

2,400 sq ft woodworking shop

Sprinkler protection required throughout building (single fire area)

6,000 sq ft millwork operation

3,000 sq ft

Denotes area where finely divided combustible waste is generated or finely divided combustible material is used

Woodworking operations

A-265

For SI: 1 square foot = 0.093 m².

In addition to the general provisions for sprinkler protection in manufacturing occupancies, IBC Section 903.2.3.1 mandates a more restrictive limitation for woodworking operations. Sprinkler protection for high-piled or rack storage is also addressed in the *International Fire Code*.

Topic: Groups H and I Occupancies   **Category:** Fire Protection Systems
Reference: IBC 903.2.4.1, 903.2.5   **Subject:** Automatic Sprinkler Systems

Code Text: *An automatic sprinkler system shall be installed in Group H occupancies. An automatic sprinkler system shall be provided throughout buildings with a Group I fire area. See exception for allowance of NFPA 13R or 13D systems in Group I-1 facilities.*

Discussion and Commentary: Hazardous occupancies require automatic sprinkler systems to protect not only the building's occupants and contents, but also the surrounding property. The sprinkler system only need be provided in the portion of the building classified as Group H. Buildings containing institutional uses must be protected throughout due to the lack of mobility of the occupants. The sprinkler system is intended to limit the size and the spread of a fire, thereby allowing extra time for moving occupants of the institutional building into an adjoining smoke compartment or through a horizontal exit.

[F] TABLE 903.2.4.2
GROUP H-5 SPRINKLER DESIGN CRITERIA

LOCATION	OCCUPANCY HAZARD CLASSIFICATION
Fabrication areas	Ordinary Hazard Group 2
Service corridors	Ordinary Hazard Group 2
Storage rooms without dispensing	Ordinary Hazard Group 2
Storage rooms with dispensing	Extra Hazard Group 2
Corridors	Ordinary Hazard Group 2

In a semiconductor fabrication facility classified as a Group H-5 occupancy, the sprinkler system must be installed throughout the entire building. For sprinkler design criteria, the code identifies the occupancy hazard classifications based on the various areas and locations.

Code Text: *An automatic sprinkler system installed in accordance with Section 903.3 shall be provided throughout all buildings with a Group R fire area.*

Discussion and Commentary: Statistics bear out that the majority of fire deaths and injuries occur in residential occupancies. It has also been statistically shown that buildings provided with sprinkler systems perform quite well under fire conditions. This mandate for the installation of automatic sprinkler systems in all buildings containing any Group R occupancy is based upon the desire to reduce such fire deaths and injuries in all residential buildings regulated by the *International Building Code.* This provision, like most requirements found in Chapter 9, is also found in the *International Fire Code.*

The scope of the IBC, Section 101.2, defers certain residential occupancies to the construction regulations of the *International Residential Code.* As such, this sprinkler requirement applies only to those residential structures constructed under the requirements of the *International Building Code.*

Code Text: *An automatic sprinkler system shall be installed throughout every story or basement of all buildings where the floor area exceeds 1,500 square feet (139.4 m²) and where there are not complying exterior wall openings.*

Discussion and Commentary: The IBC considers those structures with inadequate exterior openings for fire department access and/or rescue to be "windowless buildings," which require the installation of an automatic sprinkler system. Two methods of providing appropriate openings are set forth; one method is for openings below grade, and the other is for openings entirely above adjoining ground level. In all cases, at least one side of the building must be provided with complying openings in each 50 lineal feet of exterior wall. Basements are more highly regulated than floors above grade.

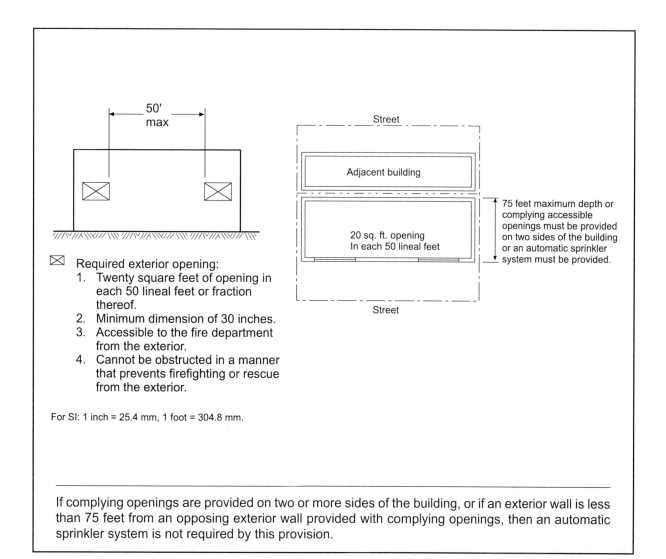

50'
max

☒ Required exterior opening:
  1. Twenty square feet of opening in each 50 lineal feet or fraction thereof.
  2. Minimum dimension of 30 inches.
  3. Accessible to the fire department from the exterior.
  4. Cannot be obstructed in a manner that prevents firefighting or rescue from the exterior.

For SI: 1 inch = 25.4 mm, 1 foot = 304.8 mm.

Street

Adjacent building

20 sq. ft. opening
In each 50 lineal feet

75 feet maximum depth or complying accessible openings must be provided on two sides of the building or an automatic sprinkler system must be provided.

Street

If complying openings are provided on two or more sides of the building, or if an exterior wall is less than 75 feet from an opposing exterior wall provided with complying openings, then an automatic sprinkler system is not required by this provision.

Code Text: *An automatic sprinkler system shall be installed throughout buildings with a floor level having an occupant load of 30 or more that is located 55 feet (16 764 mm) or more above the lowest level of fire department vehicle access.* See exceptions that exempt airport control towers, open parking structures and Group F-2 occupancies.

Discussion and Commentary: Because of difficulties associated with manual suppression of a fire in buildings constructed a substantial height above the fire department's point of attack, an automatic sprinkler system is required throughout the building, regardless of occupancy. Note that buildings that qualify for a sprinkler system by this provision, often termed "mid-rise" buildings, are not necessarily high-rise buildings as defined in Section 403.1.

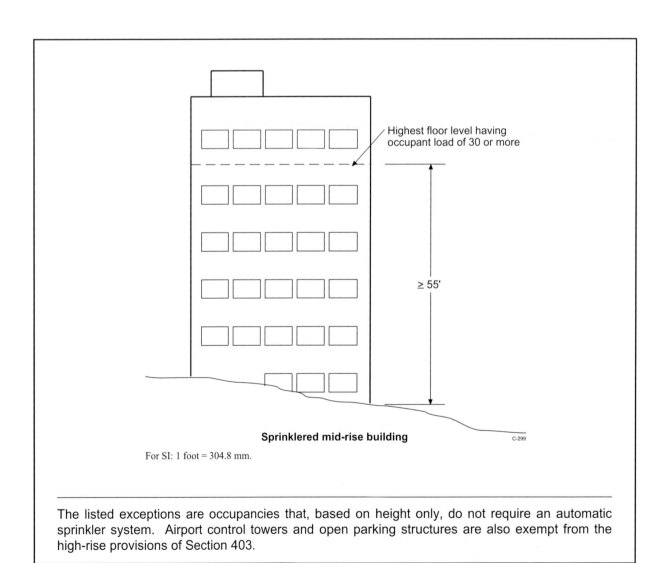

Highest floor level having occupant load of 30 or more

≥ 55'

Sprinklered mid-rise building

C-299

For SI: 1 foot = 304.8 mm.

The listed exceptions are occupancies that, based on height only, do not require an automatic sprinkler system. Airport control towers and open parking structures are also exempt from the high-rise provisions of Section 403.

Code Text: *Standpipe systems shall be installed where required by Sections 905.3.1 through 905.3.7 and in the locations indicated in Sections 905.4, 905.5 and 905.6.*

Discussion and Commentary: Installed exclusively for the fighting of fires, a standpipe system is a wet or dry system composed of piping, valves, outlets and related equipment designed to provide water at specified pressures. Standpipe systems are permitted to be combined with automatic sprinkler systems. Divided into Classes I, II and III, standpipe systems are generally required in structures of substantial height. Connections for Class I standpipes, which are solely for use by the fire department, shall be located in protected areas to allow for staging operations. Exit enclosures are typical locations for Class I connections.

REQUIRED STANDPIPE INSTALLATIONS

LOCATION OR USE	NONSPRINKLERED BUILDING	SPRINKLERED BUILDING
Building with highest story located at more than 30 feet above lowest level of fire department vehicle access	Class III [1,2,3]	Class I
Building with lowest story located at more than 30 feet below highest level of fire department vehicle access	Class III [1,2,3]	Class I
Group A occupancies with occupant load exceeding 1,000	Class I [4]	No requirement
Covered mall buildings	—	Class I
Stages more than 1,000 square feet	Class III	Class III [5]
Underground buildings	—	Class I

1 Class I standpipes permitted in basements equipped with automatic sprinkler system
2 Class I manual dry standpipes permitted in open parking garages subject to freezing temperatures, provided hose connections located as for Class II systems
3 Class I manual standpipes permitted in open parking garages where highest floor is less than 150 feet above the lowest level of fire department vehicle access
4 Not required in open-air seating spaces without enclosed spaces
5 Hose connections permitted to be supplied by sprinkler system

Fire hose cabinets in which hoses are attached to outlets on Class II standpipes (as well as the use of portable fire extinguishers) are provided as a means by which the building occupants can control the fire prior to either sprinkler activation or fire personnel arrival.

Code Text: *An approved manual, automatic or manual and automatic fire alarm system installed in accordance with the provisions of the IBC and NFPA 72 shall be provided in new buildings and structures in accordance with Sections 907.2.1 through 907.2.23 and provide occupant notification in accordance with Section 907.9, unless other requirements are provided by another section of the IBC. The automatic fire detectors shall be smoke detectors. Where ambient conditions prohibit installation of automatic smoke detection, other automatic fire detection shall be allowed.*

Discussion and Commentary: For many of the occupancies identified by the IBC, it is necessary to provide some level of notification to the building occupants and/or a supervised location reserved for a fire emergency. The threshold at which an alarm and/or detection system is required varies according to the occupancy classification.

OCCUPANCY	CONDITIONS	SYSTEM TYPE	EXCEPTIONS
A	Occupant load ≥ 300	Manual fire alarm system	1
A	Occupant load ≥ 1,000	Voice/alarm	1
B	Occupant load ≥ 500, or > 100 above or below discharge level	Manual fire alarm system	1
E	Occupant load ≥ 50	Manual fire alarm system	1, 3
F	Two or more stories, and ≥ 500 above or below discharge level	Manual fire alarm system	1
H	H-5 and where organic coatings are manufactured	Manual fire alarm system	None
H	Highly toxic gases, organic peroxides and oxidizers	Automatic smoke detection system	None
I	All Group I occupancies	Manual fire alarm system	4
I	Corridors and habitable spaces in Group I-1	Automatic smoke detection system	12
I	Corridors in Group I-2 nursing homes, detoxification facilities and spaces open to corridors	Automatic fire detection system	5
I	Group I-3 occupancies	Manual and automatic fire alarm system	6
I	Group I-3 occupancies	Smoke detectors	7
M	Occupant load ≥ 500, or 100 above or below discharge level	Manual fire alarm system	1, 8
R	Group R-1	Manual fire alarm system	9, 10
R	Group R-1	Automatic fire detection system	11
R	Group R-2 with: 1) Any unit ≥ three stories above lowest discharge level, or 2) Any unit > one story below highest discharge level, or 3) > 16 dwelling units	Manual fire alarm system	9, 10, 11

1. Manual fire alarm boxes not required where building is sprinklered and notification appliance activated upon sprinkler flow.
2. When approved, prerecorded announcement may be momentarily deactivated to allow for live voice announcements.
3. Manual fire alarm boxes not required where six specific conditions are met.
4. Manual fire alarm boxes in patient sleeping areas of Groups I-1 and I-2 not required when nurses' control stations are visible and continuously accessible.
5. Corridor smoke detection not required where patient room detectors comply with UL 268 or where integral detectors are installed on automatic door-closing devices.
6. Manual fire alarm boxes not required other than at staff-attended locations with direct supervision over areas where boxes have been omitted.
7. Smoke detectors not required in conditions II and III, in sprinklered smoke compartments with four occupants or where other approved methods are used.
8. Signal may be activated only at a constantly attended location.
9. Manual fire alarm system not required in low-height buildings with 1-hour separation of units.
10. Manual fire alarm boxes not required in sprinklered buildings where notification appliances activate upon sprinkler flow.
11. System not required in sprinklered buildings that do not have interior corridors serving the units.
12. Smoke detection not required in habitable spaces in sprinklered facility.

Audible alarm notification appliances are to be provided and shall create a distinctive sound that is used for no other purpose. Visual alarm notification appliances are also required, to varying degrees, in public and common areas, employee work areas, and Group I-1, R-1 and R-2 occupancies.

Code Text: *Manual fire alarm boxes shall be located not more than 5 feet (1524 mm) from the entrance to each exit. Additional manual fire alarm boxes shall be located so that travel distance to the nearest box does not exceed 200 feet (60 960 mm). The height of the manual fire alarm boxes shall be a minimum of 42 inches (1067 mm) and a maximum of 48 inches (1219 mm), measured vertically, from the floor level to the activating handle or lever of the box. Manual fire alarm boxes shall be red in color.*

Discussion and Commentary: The required location of fire alarm boxes adjacent to exit doors provides an opportunity for the alarm to be transmitted in a timely manner. In multistory buildings, such locations also encourage the actuation of a manual fire alarm box on the fire floor prior to entering the stair enclosure, resulting in the alarm being received from the actual fire floor and not another floor along the path of egress.

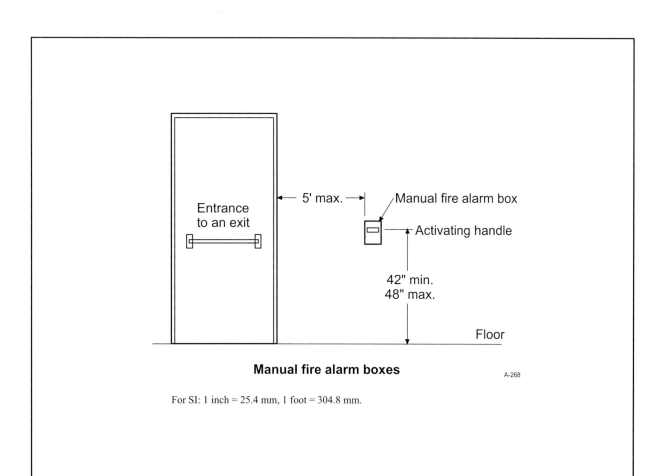

Manual fire alarm boxes

A-268

For SI: 1 inch = 25.4 mm, 1 foot = 304.8 mm.

Where a manual fire alarm system is required, manual fire alarm boxes (pull stations) must be installed. However, in some occupancies the code permits the elimination of such boxes if water flow in an automatic sprinkler system installed throughout the building activates the notification appliances.

Code Text: *Section 909 applies to mechanical or passive smoke control systems when they are required by some other provision of the IBC. Smoke control systems regulated by Section 909 serve a different purpose than the smoke- and heat-venting provisions found in Section 910.*

Discussion and Commentary: The provisions of this section do not apply unless specifically mandated for a special use, such as an atrium. It is the intent that none of the requirements apply unless directed by other provisions of the code. Where a smoke control system is provided, it may be either passive or mechanical, or a combination of the two systems. A mechanical system is an engineered system that uses mechanical fans either to produce pressure differences across smoke barriers or to establish airflows to limit and direct smoke movement. A passive system is a system of smoke barriers arranged to limit the migration of smoke.

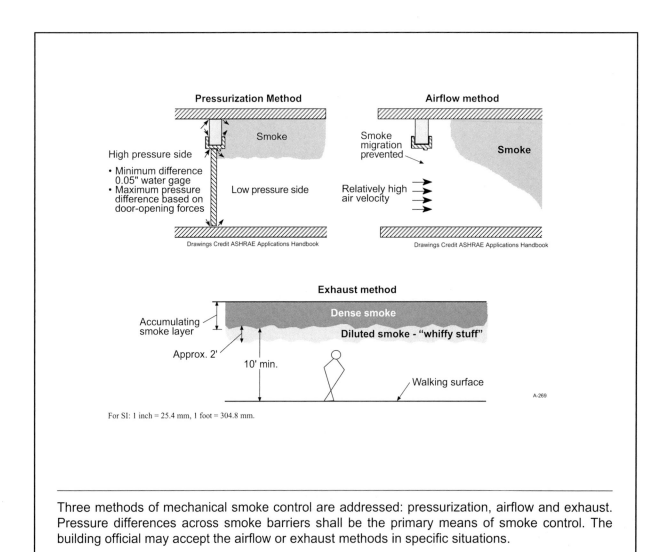

For SI: 1 inch = 25.4 mm, 1 foot = 304.8 mm.

Three methods of mechanical smoke control are addressed: pressurization, airflow and exhaust. Pressure differences across smoke barriers shall be the primary means of smoke control. The building official may accept the airflow or exhaust methods in specific situations.

Topic: Where Required

Category: Fire Protection Systems

Reference: IBC 910.2

Subject: Smoke and Heat Vents

Code Text: *Smoke and heat vents shall be installed in the roofs of one-story buildings or portions thereof 1) used as Group F-1 or S-1 occupancies having more than 50,000 square feet (4645 m²) in undivided area, 2) containing high-piled combustible stock or rack storage in any occupancy group in accordance with Section 413 and the IFC, or 3) used as a Group F-1 or S-1 occupancy where the maximum exit access travel distance is increased in accordance with Section 1016.2.*

Discussion and Commentary: Smoke and heat vents shall be installed in conjunction with draft curtains, which confine the smoke and hot gases so that they are not diluted. Draft curtains thus increase the effectiveness of automatic vents. Those areas of buildings that are equipped with early suppression fast response (ESFR) sprinklers are not required to be provided with automatic smoke and heat vents.

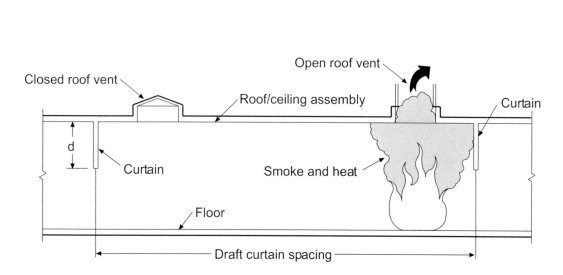

Note: In general, several small vents are more effective than a larger vent of equal area.

Roof vents and curtain boards

A-272

The code permits the building official to approve the use of a mechanical smoke exhaust system to ventilate the building as an alternative to smoke and heat vents. The exhaust fans in such a system are regulated for size, location, operation, wiring, control, supply air and interlocks.

Quiz

Study Session 8

IBC Chapter 9

1. Automatic sprinkler systems shall be monitored by an approved supervising station where the minimum number of sprinklers is _____.

    a. 10
    b. 20
    c. 50
    d. 100

    Reference_____

2. Which one of the following classes of standpipe systems is intended primarily for the use of building occupants or the fire department during initial response?

    a. Class I
    b. Class II
    c. Class III
    d. Class IV

    Reference_____

3. Which one of the following types of standpipe systems requires water from a fire department pumper to be pumped into the system in order to supply the system demand?

    a. automatic dry
    b. automatic wet
    c. manual wet
    d. semiautomatic dry

    Reference_____

4.	In a Group A-2 occupancy, an automatic sprinkler system shall be provided throughout any fire area having a minimum occupant load of _____.

　　　a.	50　　　　　　　　　　　　b.	60

　　　c.	100　　　　　　　　　　　d.	3000

Reference_____

5.	A stadium press box in a Group A-5 occupancy having a maximum floor area of _____ square feet need not be sprinklered.

　　　a.	400　　　　　　　　　　　b.	1,000

　　　c.	5,000　　　　　　　　　　d.	12,000

Reference_____

6.	Unless each classroom has at least one exterior door at grade level, a sprinkler system is required for all Group E fire areas exceeding _____ square feet.

　　　a.	2,500　　　　　　　　　　b.	5,000

　　　c.	12,000　　　　　　　　　d.	20,000

Reference_____

7.	Where woodworking operations in a Group F-1 occupancy generate finely divided combustible waste, an automatic sprinkler system is required for a minimum size fire area of _____ square feet.

　　　a.	1,000　　　　　　　　　　b.	2,500

　　　c.	5,000　　　　　　　　　　d.	12,000

Reference_____

8.	In a Group H-5 occupancy, which of the following locations requires an occupancy hazard sprinkler classification of Extra Hazard Group 2?

　　　a.	fabrication areas　　　　　b.	service corridors

　　　c.	egress corridors　　　　　d.	storage rooms with dispensing

Reference_____

9. An automatic sprinkler system shall be provided throughout a building with a Group M fire area located a minimum of _____ stories above grade plane.

    a. 2   b. 3

    c. 4   d. 6

Reference_____

10. Buildings containing which of the following residential occupancies must be sprinklered under all conditions?

    a. Groups R-1 and R-2 only   b. Groups R-1, R-2 and R-4 only

    c. Groups R-2 and R-4 only   d. All Group R occupancies

Reference_____

11. A building shall be fully sprinklered where the combined area of all Group S-1 fire areas exceeds _____ square feet.

    a. 2,500   b. 5,000

    c. 12,000   d. 24,000

Reference_____

12. A single-story Group S-1 repair garage need not be fully sprinklered where the fire area contains a maximum of _____ square feet.

    a. 2,500   b. 10,000

    c. 12,000   d. 24,000

Reference_____

13. A Group S-2 parking garage used to store commercial buses need not be sprinklered where the fire area has a maximum size of _____ square feet.

    a. 2,500   b. 5,000

    c. 10,000   d. 12,000

Reference_____

14. Where any portion of a basement exceeds a minimum of _____ feet from complying exterior openings, the basement shall be provided with an automatic sprinkler system.

    a. 50   b. 75

    c. 100   d. 150

Reference_____

15. Which of the following NFPA automatic sprinkler systems is designed for installation in one- and two-family dwellings?

    a. 13   b. 13D

    c. 13R   d. 17A

Reference_____

16. Which of the following spaces requiring sprinkler system protection is not required to use quick-response or residential sprinklers?

    a. light-hazard occupancies

    b. sleeping units in Group R-1

    c. patient sleeping rooms in Group I-2

    d. sales rooms of Group M

Reference_____

17. A minimum of _____ shall be maintained between automatic sprinklers and the top of piles of combustible fibers.

    a. 12 inches   b. 18 inches

    c. 3 feet   d. 5 feet

Reference_____

18. A secondary water supply, where required in high-rise buildings, shall have a minimum duration of _____ minutes.

    a. 15   b. 30

    c. 60   d. 90

Reference_____

19. The manual actuation device for a fire-extinguishing system for a commercial cooking system shall be located a minimum of _____ feet and a maximum of _____ feet from the kitchen exhaust system.

    a. 3, 6   b. 5, 10

    c. 6, 12   d. 10, 20

Reference_____

20. Connections for Class II standpipe systems shall be located so that all portions of the building are within _____ feet of a nozzle attached to _____ feet of hose.

    a. 20, 50   b. 30, 100

    c. 40, 125   d. 40, 150

Reference_____

21. In a nonsprinklered Group B occupancy, a manual fire alarm system shall be installed where there are a minimum of _____ occupants above or below the level of exit discharge.

    a. 101   b. 201

    c. 301   d. 501

Reference_____

22. A manual fire alarm system is not required in a Group E occupancy with a maximum occupant load of _____ persons.

    a. 49   b. 99

    c. 299   d. 499

Reference_____

23. Manual fire alarm boxes shall be located a maximum of _____ feet from the entrance to each exit.

    a. 5   b. 10

    c. 12   d. 20

Reference_____

24. In a Group R-1 hotel providing 220 sleeping units, a minimum of _____ such units shall be provided with visible alarm notification devices.

    a. 3   b. 11

    c. 17   d. 22

    Reference_____

25. The minimum sound pressure level for audible alarm notification appliances in an office building shall be _____ decibels.

    a. 60   b. 70

    c. 90   d. 105

    Reference_____

26. The manual actuation device for an automatic fire-extinguishing system serving a commercial cooking system shall be installed a minimum of _____ inches and a maximum of _____ inches above the floor..

    a. 38, 42   b. 34, 48

    c. 38, 42   d. 42, 48

    Reference_____

27. A Class III wet standpipe is not required for stages having a maximum size of _____ square feet.

    a. 100   b. 400

    c. 500   d. 1,000

    Reference_____

28. Where natural ventilation is utilized for venting a smokeproof enclosure, each vestibule shall be provided with a minimum _____-square-foot opening in the exterior wall.

    a. 9   b. 16

    c. 24   d. 35

    Reference_____

29. Other than for an aircraft repair hangar, a one-story Group S-1 occupancy shall be provided with smoke and heat vents where it exceeds _____ square feet in undivided area.

    a. 8,000　　　　　　　　　b. 10,000

    c. 15,000　　　　　　　　 d. 50,000

    Reference_____

30. Where required in a Group F-1 occupancy, smoke and heat vents shall be spaced at maximum intervals of _____ feet when measured center to center.

    a. 60　　　　　　　　　　b. 75

    c. 100　　　　　　　　　 d. 120

    Reference_____

31. Sprinkler protection is required for exterior balconies, decks and ground floor patios of dwelling units where the building is of _____ construction.

    a. combustible　　　　　　b. Type III

    c. Type IV　　　　　　　　d. Type V

    Reference _____

32. A manual fire alarm system shall be installed in all multistory Group F occupancies having a minimum occupant load of _____ above or below the lowest level of exit discharge.

    a. 50　　　　　　　　　　b. 100

    c. 300　　　　　　　　　 d. 500

    Reference _____

33. The vestibule space in a smokeproof enclosure shall have a minimum width of _____ inches and a minimum length in the direction of egress travel of _____ inches.

    a. 44, 60　　　　　　　　 b. 44, 72

    c. 60, 60　　　　　　　　 d. 60, 84

    Reference _____

34. Where automatic smoke and heat vents contain heat-sensitive glazing designed to shrink and drop out of the vent opening when exposed to fire, the vents shall open a maximum of _____ minute(s) after the established fire exposure.

   a. 1   b. 2

   c. 5   d. 10

   Reference _____

35. Where a fire command center is required by other provisions of the code, the room shall be a minimum of _____ square feet in area.

   a. 96   b. 100

   c. 120   d. 144

   Reference _____

9

2006 IBC Sections 1001 – 1007, 1011 and 1013
Means of Egress I

OBJECTIVE: To obtain an understanding of the general system design requirements of a means of egress system, including the determination of occupant load, the required width of egress components, means of egress identification and illumination, accessible means of egress and the provisions regulating guards.

REFERENCE: Sections 1001 – 1007, 1011 and 1013, 2006 *International Building Code*

KEY POINTS:
- What is the definition of a means of egress system? What are its three distinct elements?
- What is the minimum ceiling height permitted along a means of egress?
- What limitations are placed on protruding objects extending below the minimum ceiling height? Projecting horizontally over a walking surface?
- For an elevation change along the egress path, at what point is a ramp required rather than a step or stairway?
- In areas without fixed seats, what is the correct method of determining occupant load?
- What method shall be used to calculate the occupant load in areas with fixed seating, such as benches, pews or booths?
- How may the design occupant load be increased over what is calculated?
- In which types of rooms or spaces must the maximum occupant load be posted?
- How is exiting addressed where exits serve more than one floor?
- How is exiting from a mezzanine regulated?
- Are yards, patios and courts regulated in the same manner as interior areas?
- How shall the minimum width of different egress components be determined?
- What is the maximum allowable encroachment of a door into the required egress width?
- When must the means of egress be illuminated? What is the minimum required illumination at the floor level?
- Which locations must be provided with emergency power for egress illumination?
- For what duration is an emergency power system for means of egress illumination required to provide power?
- What is considered an accessible means of egress? How many are required in a building?
- When is an area of refuge needed? How are they to be constructed?

- How must an accessible means of egress be identified?
- What is the purpose of an exterior area for rescue assistance? How must this area be separated from the interior of the building it serves?
- Where are exit signs required? When must they be illuminated? What level of illumination is required? Which types of power sources are necessary?
- What is the definition of a guard?
- When are guards required? What is the minimum height requirement from the walking surface to the top of a guard?
- How must guards be constructed to limit passage through the protective barrier?

Code Text: *Buildings or portions thereof shall be provided with a means of egress system as required by Chapter 10. The provisions of Chapter 10 shall control the design, construction and arrangement of means of egress components required to provide an approved means of egress from structures and portions thereof.*

Discussion and Commentary: The *International Building Code* regulates the design, construction and maintenance of an exiting system through two general categories—system design and egress components. Any building elements that are a part of the system must be reviewed for compliance with the criteria for the number, location, width or capacity, height, continuity and arrangement of egress components, and all other applicable provisions.

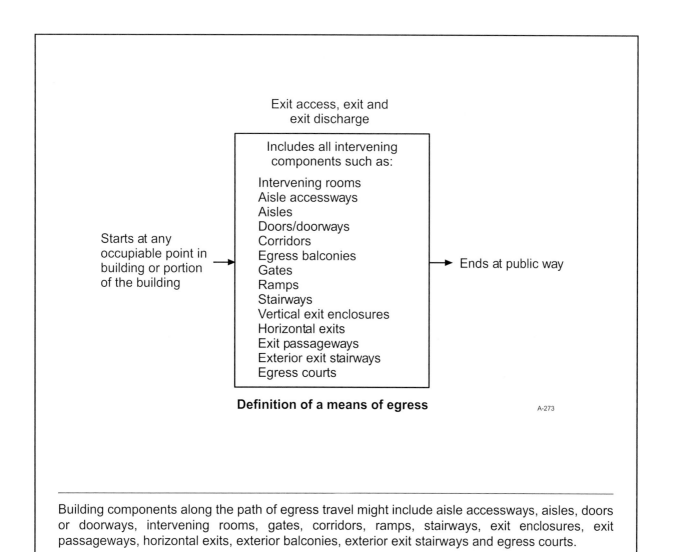

Exit access, exit and exit discharge

Includes all intervening components such as:

Intervening rooms
Aisle accessways
Aisles
Doors/doorways
Corridors
Egress balconies
Gates
Ramps
Stairways
Vertical exit enclosures
Horizontal exits
Exit passageways
Exterior exit stairways
Egress courts

Starts at any occupiable point in building or portion of the building

Ends at public way

Definition of a means of egress

A-273

Building components along the path of egress travel might include aisle accessways, aisles, doors or doorways, intervening rooms, gates, corridors, ramps, stairways, exit enclosures, exit passageways, horizontal exits, exterior balconies, exterior exit stairways and egress courts.

Code Text: *A means of egress is a continuous and unobstructed path of vertical and horizontal egress travel from any occupied portion of a building or structure to a public way. A means of egress consists of three separate and distinct parts: the exit access, the exit, and the exit discharge.*

Discussion and Commentary: The exit access begins at any occupied location within the building and does not end until it reaches the door to an exit enclosure, a horizontal exit or exit passageway, an exterior exit stairway or ramp, or an exterior door at ground level. Travel distance is regulated throughout the exit access, and the path of travel is seldom a fire-protected environment. At the exit discharge, which begins where the exit ends, egress remains regulated until the public way is reached.

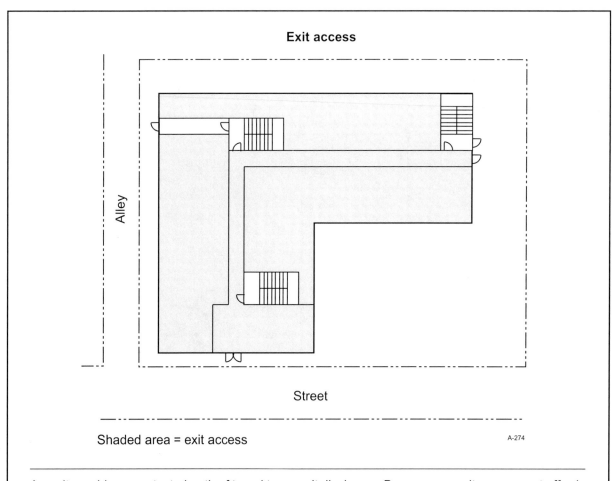

Exit access

Alley

Street

Shaded area = exit access

A-274

An exit provides a protected path of travel to an exit discharge. Because an exit component affords protection, travel distance is no longer a concern once the exit has been reached. Exit discharge travel distance to the public way is also unlimited.

Code Text: *The means of egress shall have a ceiling height of not less than 7 feet 6 inches (2286 mm).* See exceptions for sloped ceilings, ceilings of dwelling units and sleeping units, allowable projections, stair headroom and door height.

Discussion and Commentary: In addition to providing a travel path of adequate width, the code requires that the clear height of the means of egress be maintained at least $7^1/_2$ feet above the walking surface. There are several exceptions to this general requirement that permit limited reductions in the mandated height. Under most conditions, the vertical clearance at a stairway or doorway may be reduced to 80 inches. Protruding objects, such as sprinklers and light fixtures, are also permitted to extend below the minimum required ceiling height for up to 50 percent of ceiling area of the means of egress, provided such objects maintain a headroom clearance of at least 80 inches. Special provisions are applicable to sloped ceilings.

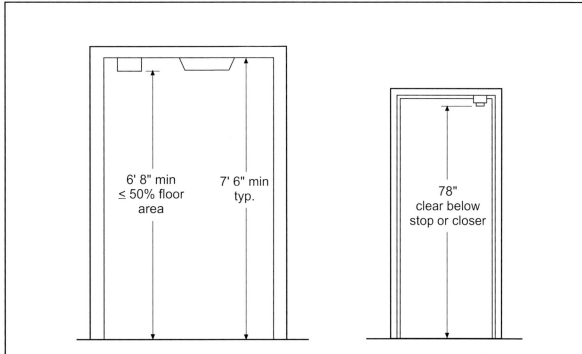

6' 8" min
≤ 50% floor
area

7' 6" min
typ.

78"
clear below
stop or closer

Corridor, aisle, passageway or any walking surface along egress of path travel

For SI: 1 inch = 25.4 mm, 1 foot = 304.8 mm.

A-277

The minimum ceiling heights established for environmental concerns are addressed in Section 1208.2. Habitable spaces, such as bedrooms and living rooms in residential occupancies, occupiable spaces and corridors must be at least 7 feet 6 inches in height. In other areas, reduced headroom is permitted.

Topic: Elevation Change

Category: Means of Egress

Reference: IBC 1003.5

Subject: General Egress

Code Text: *Where changes in elevation of less than 12 inches (305 mm) exist in the means of egress, sloped surfaces shall be used. Where the slope is greater than 1 unit vertical in 20 units horizontal (5-percent slope), ramps complying with Section 1010 shall be used. Where the difference in elevation is 6 inches (152 mm) or less, the ramp shall be equipped with either handrails or floor finish materials that contrast with adjacent floor finish materials. See exceptions for 1) a single 7-inch step in Groups F, H, R-2, R-3, S and U; 2) a stair with one or two risers with a handrail provided; and 3) a step in aisles serving seating.*

Discussion and Commentary: Along the egress path, there is a concern about slight changes in elevation that are not readily apparent to persons seeking to exit under emergency conditions. Therefore, a single riser or a pair of shallow risers is not permitted. Steps used to achieve minor differences in elevation frequently go unnoticed, and as such, can cause accidents.

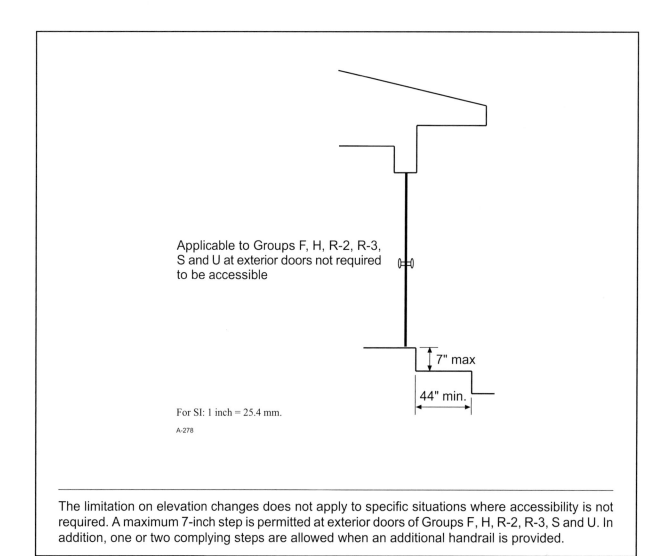

Applicable to Groups F, H, R-2, R-3, S and U at exterior doors not required to be accessible

7" max

44" min.

For SI: 1 inch = 25.4 mm.

A-278

The limitation on elevation changes does not apply to specific situations where accessibility is not required. A maximum 7-inch step is permitted at exterior doors of Groups F, H, R-2, R-3, S and U. In addition, one or two complying steps are allowed when an additional handrail is provided.

Code Text: *In determining means of egress requirements, the number of occupants for whom means of egress facilities shall be provided shall be determined in accordance with Section 1004. The number of occupants shall be computed at the rate of one occupant per unit of area as prescribed in Table 1004.1.1 . For areas without fixed seating, the occupant load shall not be less than that number determined by dividing the floor area under consideration by the occupant per unit of area factor assigned to the occupancy as set forth in Table 1004.1.1. See exception where building official is authorized to reduce occupant load below that calculated.*

Discussion and Commentary: For occupant load determination, it must be assumed that under normal conditions all portions of a building are fully occupied at the same time. The density characteristics of the various uses identified in Table 1004.1.1 are considered "occupant load factors." For most occupancies, the gross floor area is to be considered. However, a few of the occupant load factors are based on net floor area, which allows the deduction of areas such as corridors, stairways, toilet rooms, equipment rooms and closets.

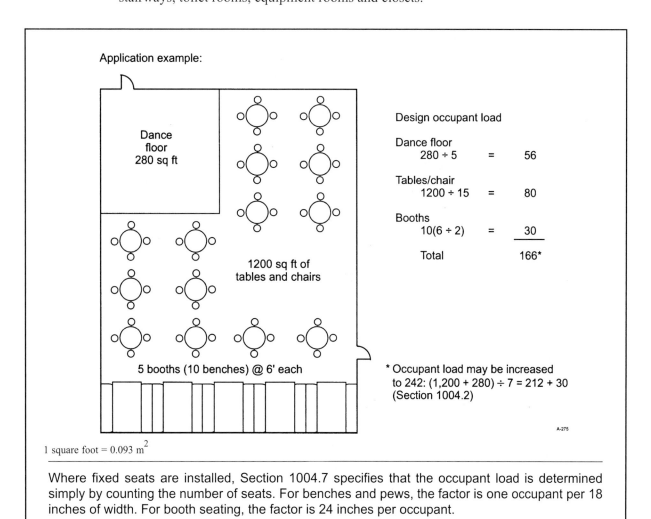

Application example:

Dance floor 280 sq ft

1200 sq ft of tables and chairs

5 booths (10 benches) @ 6' each

Design occupant load

Dance floor
280 ÷ 5   =   56

Tables/chair
1200 ÷ 15   =   80

Booths
10(6 ÷ 2)   =   30

Total   166*

* Occupant load may be increased to 242: (1,200 + 280) ÷ 7 = 212 + 30 (Section 1004.2)

A-275

1 square foot = 0.093 m^2

Where fixed seats are installed, Section 1004.7 specifies that the occupant load is determined simply by counting the number of seats. For benches and pews, the factor is one occupant per 18 inches of width. For booth seating, the factor is 24 inches per occupant.

Topic: Outdoor Areas

Category: Means of Egress

Reference: IBC 1004.8

Subject: Occupant Load

Code Text: *Yards, patios, courts and similar outdoor areas accessible to and usable by the building occupants shall be provided with means of egress as required by Chapter 10. Where outdoor areas are to be used by persons in addition to the occupants of the building, and the path of egress travel from the outdoor areas passes through the building, means of egress requirements shall be based on the sum of the occupant loads of the building plus the outdoor areas.* See exceptions for service areas and dwellings.

Discussion and Commentary: Although not limited in application, this provision primarily addresses the use of outdoor areas for dining and/or drinking in restaurants and similar establishments. The building official is authorized to establish an occupant load for the outdoor space in accordance with its anticipated use and to apply all means of egress provisions that would be appropriate.

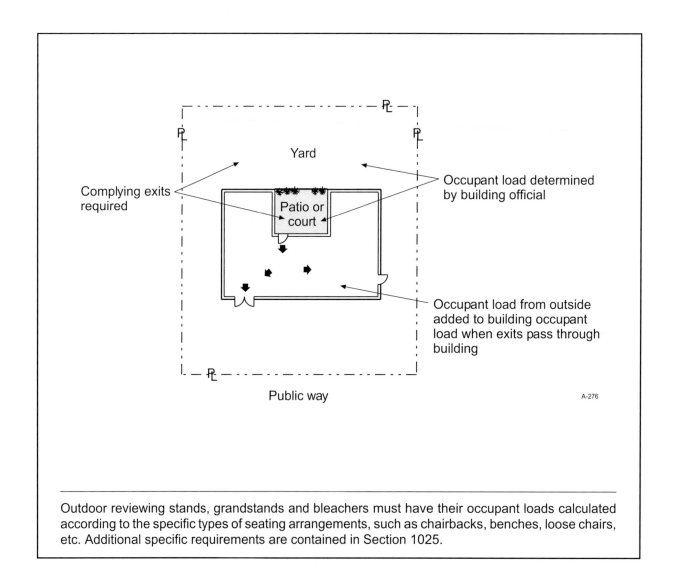

Outdoor reviewing stands, grandstands and bleachers must have their occupant loads calculated according to the specific types of seating arrangements, such as chairbacks, benches, loose chairs, etc. Additional specific requirements are contained in Section 1025.

Topic: Exiting from Multiple Levels	Category: Means of Egress
Reference: IBC 1004.4, 1004.5	Subject: Occupant Load

Code Text: *Where exits serve more than one floor, only the occupant load of each floor considered individually shall be used in computing the required capacity of the exits at that floor, provided that the exit capacity shall not decrease in the direction of egress travel. Where means of egress from floors above and below converge at an intermediate level, the capacity of the means of egress from the point of convergence shall not be less than the sum of the two floors.*

Discussion and Commentary: It is not necessary to add occupants together for width calculations as they travel vertically from one floor level to the next. Only the capacity for each individual floor is utilized in establishing the minimum required stairway width at each flight. However, once a minimum required width has been established along the stair path, it cannot be reduced.

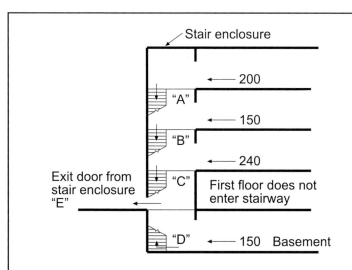

Given:
- A nonsprinklered four-story office building with basement
- Occupant load exiting into stair enclosure at each level as indicated
- First floor occupants exit to exterior without entering the stair enclosure

Exit element	Occupant load served	Required width
Stair-point "A"	200	60"
Stair-point "B"	150	60"[1]
Stair-point "C"	240	72"
Stair-point "D"	150	45"
Stair-point "E"	390[2]	78"

[1]Required width from above must be maintained
[2]Door at point "E" serves occupants from "C" and "D" egress convergence per Section 1004.5

There is an allowance for adding the occupant loads of floors above and below an intermediate level together where they converge along the exit path. The aggregate occupant load of such converging floors is to be used in determining the minimum required exit width.

Code Text: *The means of egress width shall not be less than required by* Section 1005. *The total width of means of egress in inches (mm) shall not be less than the total occupant load served by the means of egress multiplied by the factors in Table 1005.1 and not less than specified elsewhere in* the IBC. *See Section 1025 for determining the required width in assembly occupancies. Doors opening into the path of egress travel shall not reduce the required width to less than one-half during the course of the swing. When fully open, the door shall not project more than 7 inches (178 mm) into the required width.* See exception for doors within units of Groups R-2 and R-3.

Discussion and Commentary: In a given means of egress system, different components will afford different capacities. The most restrictive component will establish the capacity of the overall system. Doorways, aisles, stairways and corridors also have minimum established widths that must be provided.

$$\begin{array}{ccccc} \text{Occupant} & & \text{Factor from} & & \text{Minimum} \\ \text{Load} & \chi & \text{Table} & \leq & \text{Available} \\ \text{Served} & & \text{1005.1} & & \text{Width} \end{array}$$

TABLE 1005.1
EGRESS WIDTH PER OCCUPANT SERVED

OCCUPANCY	WITHOUT SPRINKLER SYSTEM		WITH SPRINKLER SYSTEM[a]	
	Stairways (inches per occupant)	Other egress components (inches per occupant)	Stairways (inches per occupant)	Other egress components (inches per occupant)
Occupancies other than those listed below	0.3	0.2	0.2	0.15
Hazardous: H-1, H-2, H-3 and H-4	0.7	0.4	0.3	0.2
Institutional: I-2	NA	NA	0.3	0.2

For SI: 1 inch = 25.4 mm. NA = Not applicable.
a. Buildings equipped throughout with an automatic sprinkler system in accordance with Section 903.3.1.1 or 903.3.1.2.

Width, in terms of a means of egress system or component, is the clear, unobstructed usable width afforded along the exit path by the individual components. Unless the code provides for a permitted projection, the minimum clear width may not be reduced throughout the travel path.

Code Text: *Multiple means of egress shall be sized such that the loss of any one means of egress shall not reduce the available capacity to less than 50 percent of the required capacity. The maximum capacity required from any story of a building shall be maintained to the termination of the means of egress.*

Discussion and Commentary: Where two complying means of egress are provided, the occupant load is to be distributed evenly between the two means of egress. However, where three or more means of egress are available, it is permissible to size one of the egress points for up to 50 percent of the occupant load, while distributing the remaining occupant load among the other means of egress. This distribution is not required to be equally applied; however, a dramatic imbalance of egress component capacities relative to occupant load distribution should be avoided.

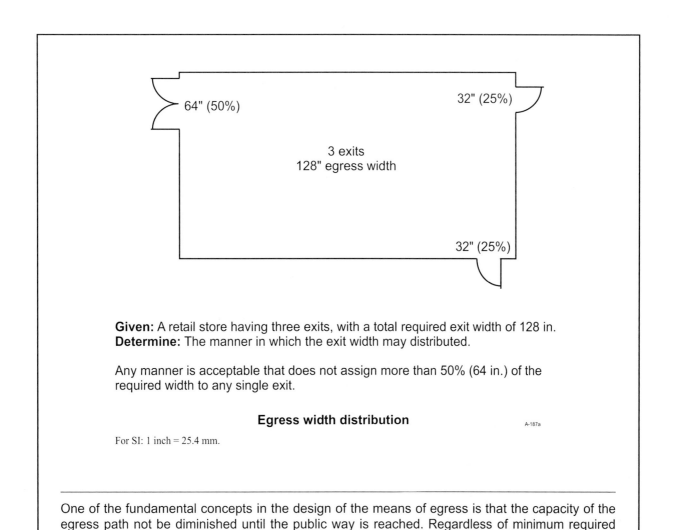

Given: A retail store having three exits, with a total required exit width of 128 in.
Determine: The manner in which the exit width may distributed.

Any manner is acceptable that does not assign more than 50% (64 in.) of the required width to any single exit.

Egress width distribution   A-187a

For SI: 1 inch = 25.4 mm.

One of the fundamental concepts in the design of the means of egress is that the capacity of the egress path not be diminished until the public way is reached. Regardless of minimum required component width, the calculated width based on the occupant load served must be maintained.

Code Text: *The means of egress, including the exit discharge, shall be illuminated at all times the building space served by the means of egress is occupied.* See exceptions for 1) Group U occupancies; 2) aisle accessways in Group A; 3) dwelling and sleeping units in Groups R-1, R-2 and R-3; and 4) sleeping units of Group I. *The power supply for means of egress illumination shall normally be provided by the premises electrical supply. In the event of power supply failure, an emergency electrical system shall automatically illuminate aisles and unenclosed egress stairways in rooms and spaces which require two or more means of egress; and corridors, exit enclosures and exit passageways, exterior egress components above grade, interior discharge elements, and exterior landings for exit discharge doorways in buildings required to have two or more exits.*

Discussion and Commentary: Often identified as *emergency lighting*, a completely separate source of power from the premise's wiring system is required when the life-safety risk in a building becomes sufficiently great. This threshold is recognized as the point at which the occupant load of the room, area or building is high enough so that two means of egress are required.

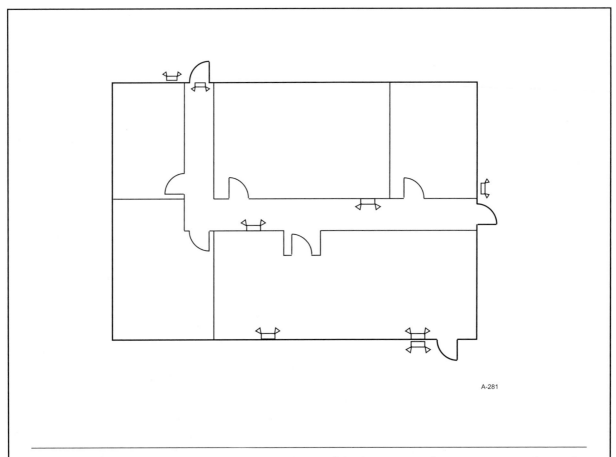

A-281

For the building occupant to be able to negotiate safely the means of egress system, the entire system must be illuminated any time the building is occupied. The illumination must provide an intensity of at least one foot-candle at the floor level.

Code Text: *Accessible spaces shall be provided with not less than one accessible means of egress. Where more than one means of egress is required by Sections 1015.1 or 1019.1 from any accessible space, each accessible portion of the space shall be served by not less than two accessible means of egress.* See exceptions for 1) alterations to existing buildings, 2) accessible mezzanines, and 3) assembly spaces with sloped floors. *Each required accessible means of egress shall be continuous to a public way and shall consist of one or more of the following components: accessible routes, stairways within exit enclosures, exterior exit stairways, elevators, platform lifts, horizontal exits, ramps and areas of refuge.* See exceptions for use of exterior areas for assisted rescue.

Discussion and Commentary: An accessible means of egress is *a continuous and unobstructed way of egress travel, from any accessible point in a building or facility to a public way.*

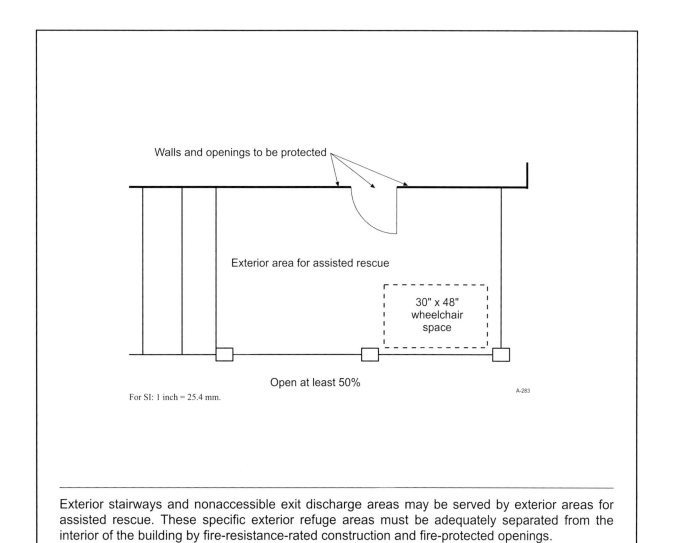

Exterior stairways and nonaccessible exit discharge areas may be served by exterior areas for assisted rescue. These specific exterior refuge areas must be adequately separated from the interior of the building by fire-resistance-rated construction and fire-protected openings.

Topic: Areas of Refuge

Reference: IBC 1007.6

Category: Means of Egress

Subject: Accessible Means of Egress

Code Text: *Every required area of refuge shall be accessible from the space it serves by an accessible means of egress. Every required area of refuge shall have direct access to an enclosed stairway complying with Section 1007.3 and 1020.1 or an elevator complying with Section 1007.4.*

Discussion and Commentary: An area of refuge is defined as *an area where persons unable to use stairways can remain temporarily to await instructions or assistance during emergency evacuation.* An area of refuge needs to be separated from the remainder of the story by a smoke barrier or horizontal exit unless the refuge area is located within a stairway enclosure. A two-way communication system with appropriate instructions must be provided in each area of refuge and must also be identified by complying signs.

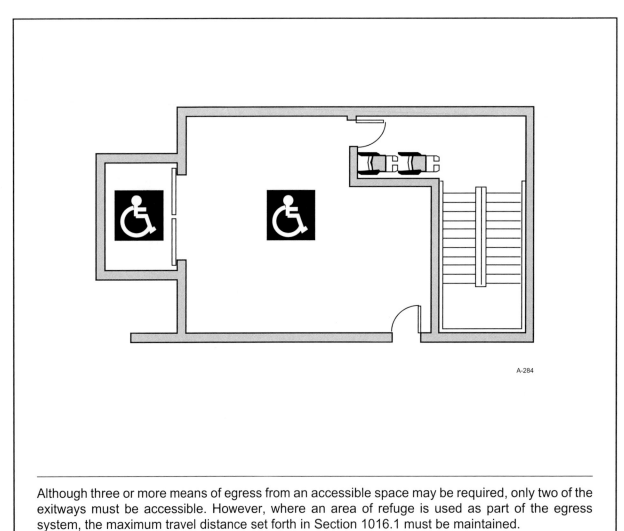

A-284

Although three or more means of egress from an accessible space may be required, only two of the exitways must be accessible. However, where an area of refuge is used as part of the egress system, the maximum travel distance set forth in Section 1016.1 must be maintained.

Code Text: *Exit and exit access doors shall be marked by an approved exit sign readily visible from any direction of egress travel. Access to exits shall be marked by readily visible exit signs in cases where the exit or the path of egress travel is not immediately visible to the occupants.* See exceptions for uses or conditions where exit signs are not required.

Discussion and Commentary: Exit signs are only mandated when the room or area under consideration is required to have multiple exits or exit access doors. Other locations are also specified where the presence of an exit sign is deemed unnecessary, such as clearly identifiable main exterior doors. Although the appropriate locations for exit signs should be identified during the plan review phase of a project, the true evaluation of their effectiveness should be done just prior to occupancy, when the correct location and orientation of the signs can be checked.

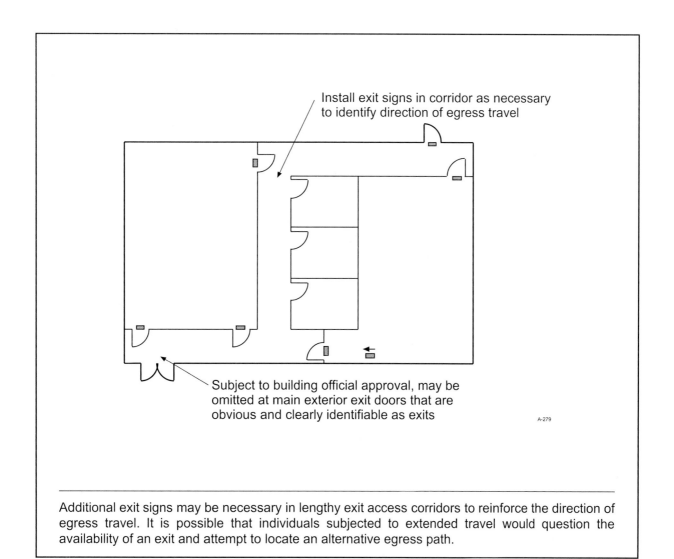

Install exit signs in corridor as necessary to identify direction of egress travel

Subject to building official approval, may be omitted at main exterior exit doors that are obvious and clearly identifiable as exits

A-279

Additional exit signs may be necessary in lengthy exit access corridors to reinforce the direction of egress travel. It is possible that individuals subjected to extended travel would question the availability of an exit and attempt to locate an alternative egress path.

Code Text: *Exit signs shall be internally or externally illuminated.* See exception for tactile signs. *To ensure continued illumination for a duration of not less than 90 minutes in case of primary power loss, the sign illumination means shall be connected to an emergency system provided from storage batteries, unit equipment or an on-site generator.*

Discussion and Commentary: To ensure visibility under all conditions, required exit signs must always be illuminated. The building official may approve alternative types of signs or lighting systems to those specified, provided that equivalent light levels can be achieved. A separate source of power shall be provided to all required exit signs, regardless of the occupant load served, much in the same manner as it is required for path of travel illumination.

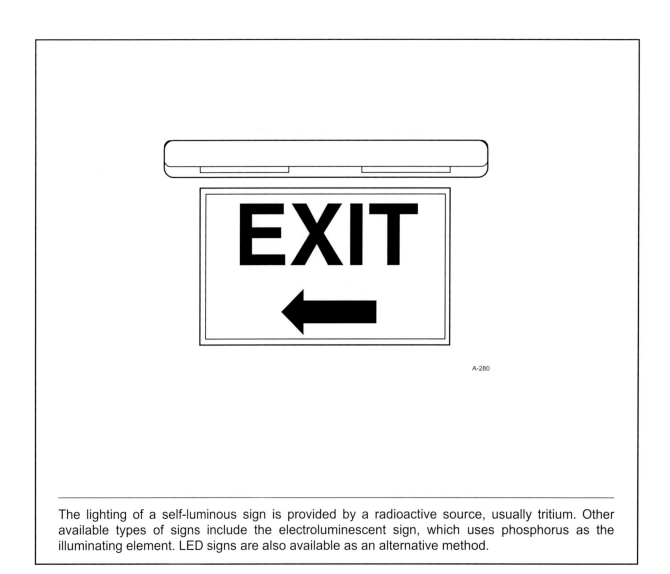

A-280

The lighting of a self-luminous sign is provided by a radioactive source, usually tritium. Other available types of signs include the electroluminescent sign, which uses phosphorus as the illuminating element. LED signs are also available as an alternative method.

Code Text: *Guards shall be located along open-sided walking surfaces, mezzanines, industrial equipment platforms, stairways, ramps and landings that are located more than 30 inches (762 mm) above the floor or grade below. Guards shall form a protective barrier not less than 42 inches (1067 mm) high. Open guards shall have balusters or ornamental patterns such that a 4-inch-diameter (102 mm) sphere cannot pass through.* See multiple exceptions.

Discussion and Commentary: Guards must be of adequate height and structural stability to prevent an individual from accidentally falling from the protected area. They must be designed also to prevent small children from intentionally crawling through the barrier. In certain industrial-type areas, the degree of protection is reduced because of the nonpublic uses involved. In addition, guards are not mandated in specific applications relating to loading docks, stages, platforms and vehicle service pits.

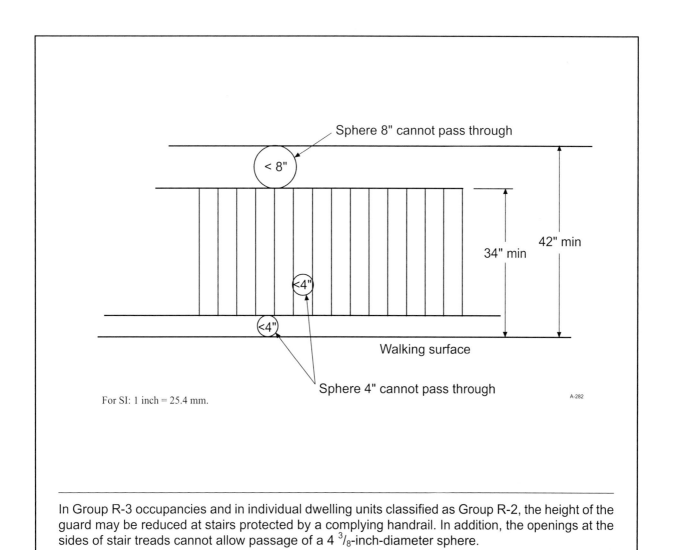

Sphere 8" cannot pass through

< 8"

42" min

34" min

Walking surface

Sphere 4" cannot pass through

For SI: 1 inch = 25.4 mm.

A-282

In Group R-3 occupancies and in individual dwelling units classified as Group R-2, the height of the guard may be reduced at stairs protected by a complying handrail. In addition, the openings at the sides of stair treads cannot allow passage of a 4 3/8-inch-diameter sphere.

Topic: Rooftop Access and Equipment   **Category:** Means of Egress
Reference: IBC 1013.5   **Subject:** Guards

Code Text: *Guards shall be provided where appliances, equipment, fans, roof hatch openings or other components that require service are located within 10 feet (3048 mm) of a roof edge or open side of a walking surface and such edge or open side is located more than 30 inches (762 mm) above the floor, roof or grade below. The guard shall be constructed so as to prevent the passage of a 21-inch-diameter (533 mm) sphere. The guard shall extend not less than 30 inches (762 mm) beyond each end of such appliance, equipment, fan or component.*

Discussion and Commentary: The requirement for guards primarily addresses the hazard created when service personnel are working on rooftop equipment. Where such activity occurs close to a roof edge, it is critical that guards be provided to prevent an accidental fall.

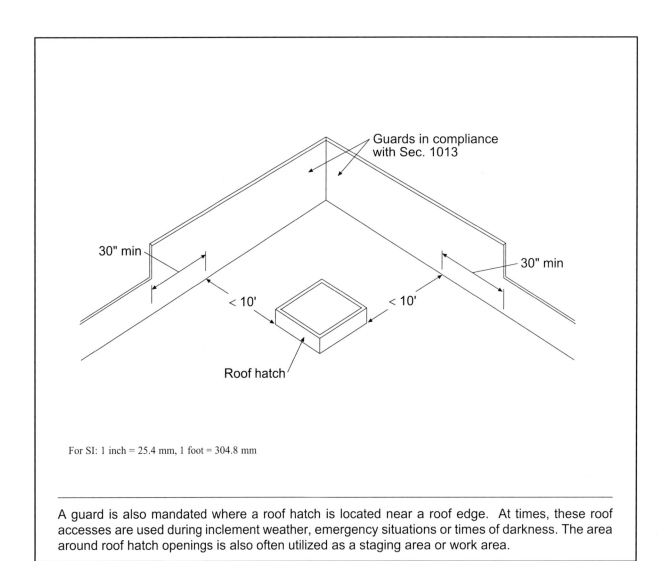

For SI: 1 inch = 25.4 mm, 1 foot = 304.8 mm

A guard is also mandated where a roof hatch is located near a roof edge. At times, these roof accesses are used during inclement weather, emergency situations or times of darkness. The area around roof hatch openings is also often utilized as a staging area or work area.

Quiz

Study Session 9
IBC Sections 1001 – 1007, 1011 and 1013

1. A court or yard that provides access to a public way for one or more exits is considered a(n) _____.

   a. exit accessway   b. egress court

   c. public way   d. horizontal exit

   Reference_____

2. Which of the following elements is not a distinct and separate part of the means of egress?

   a. exit discharge   b. exit access

   c. exit   d. exit convergence

   Reference_____

3. That portion of the exit access that occupants must traverse before two separate and distinct paths of egress travel to two exits are available is defined as a _____.

   a. means of egress   b. single egress path

   c. common path of egress travel d. limited egress travel distance

   Reference_____

4. Panic hardware that is listed for use on fire door assemblies is considered to be _____ hardware.

    a. fire egress   b. fire exit

    c. panic   d. panic and fire

Reference_____

5. An alternating tread device has a series of steps that are positioned a minimum of _____ degrees and maximum of _____ degrees from horizontal.

    a. 30, 45   b. 45, 60

    c. 50, 70   d. 60, 75

Reference_____

6. In a dining room without fixed seating, the calculated occupant load is determined by dividing the floor area by a factor of one occupant per _____ square feet.

    a. 7 net   b. 15 net

    c. 15 gross   d. 20 net

Reference_____

7. A 1,500-square-foot (net) woodworking shop classroom in a high school is considered to have a design occupant load of _____ persons.

    a. 25   b. 30

    c. 75   d. 100

Reference_____

8. For areas having fixed seats and aisles, the occupant load for bench seating without armrests is based on one occupant for each _____ inches of seating length.

    a. 15   b. 18

    c. 24   d. 30

Reference_____

9. In a fully-sprinklered Group B office building having an occupant load of 3,200 occupants, the minimum total calculated means of egress width for egress other than stairways shall be _____ inches.

    a. 960   b. 640

    c. 480   d. 320

Reference_____

10. A stairway serving 400 occupants in a fully-sprinklered Group I-2 hospital shall be a minimum of _____ inches in width.

    a. 60   b. 80

    c. 120   d. 160

Reference_____

11. Multiple means of egress shall be sized so that the loss of any one means of egress shall not reduce the available capacity to less than _____ of the required capacity.

    a. 10 percent   b. 25 percent

    c. $33^1/_3$ percent   d. 50 percent

Reference_____

12. When fully open, a door is permitted to project into the required width of the path of egress travel a maximum of _____.

    a. one-half the required width   b. one-half the actual width

    c. $3^1/_2$ inches   d. 7 inches

Reference_____

13. Up to 50 percent of the ceiling area of a means of egress may have a minimum ceiling height of _____ where reduced by protruding objects.

    a. 78 inches   b. 80 inches

    c. 84 inches   d. 90 inches

Reference_____

14. At a doorway, the minimum headroom clearance below any door closer or stop shall be _____ inches.

    a. 76   b. 78

    c. 80   d. 84

Reference_____

15. Other than handrails serving stairs and ramps, the maximum projection into the travel path of a horizontal projection located 48 inches above the walking surface shall be _____ inches.

    a. $1^1/_2$ b. $3^1/_2$

    c. 4   d. $4^1/_2$

Reference_____

16. At an exterior door not required to be accessible in a Group F-1 occupancy, what is the maximum permitted elevation change?

    a. $^1/_2$ inch   b. 1 inch

    c. 7 inches   d. 8 inches

Reference_____

17. Exit signs shall be located so that the maximum distance from any point in an exit access corridor to the nearest visible exit sign is _____ feet, or the listed viewing distance for the sign, whichever is less.

    a. 50   b. 75

    c. 100   d. 150

Reference_____

18. Externally-illuminated exit signs shall have a minimum intensity at the face of the sign of _____ foot-candles.

    a. 1   b. 5

    c. 10   d. 12

Reference_____

19. Emergency lighting facilities for means of egress illumination shall initially provide_____ along the path of egress at floor level.

    a. at least 1 foot-candle

    b. an average of 1 foot-candle

    c. at least 5 foot-candles

    d. an average of 0.2 foot-candle

    Reference_____

20. A guard need not be located along an open-sided walking surface located a maximum of _____ inches above the floor below.

    a. 15

    b. 30

    c. 36

    d. 42

    Reference_____

21. In a mercantile occupancy, a required guard shall form a protective barrier a minimum of _____ inches in height.

    a. 36

    b. 38

    c. 42

    d. 44

    Reference_____

22. In areas of a Group S-1 occupancy not open to the public, horizontal intermediate rails in a required guard shall be constructed so that a _____ sphere cannot pass through any opening.

    a. 4-inch

    b. 6-inch

    c. 12-inch

    d. 21-inch

    Reference_____

23. Where a roof-top HVAC unit requiring occasional service and maintenance is located a maximum of _____ from the roof edge, a complying guard shall be provided.

    a. 10 feet

    b. 5 feet

    c. 3 feet

    d. 30 inches

    Reference_____

24. In a nonsprinklered building, an enclosed stairway utilized as an accessible means of egress shall be a minimum of _____ inches in clear width between handrails.

    a. 36   b. 44

    c. 48   d. 60

Reference_____

25. An area of refuge serving 450 occupants shall be provided with a minimum of _____ wheelchair space(s).

    a. 1   b. 2

    c. 3   d. 5

Reference_____

26. Where a barrier is installed below a protruding object having a vertical clearance of less than 80 inches, the maximum height of the barrier shall be _____ inches above the floor.

    a. 27   b. 30

    c. 36   d. 42

Reference_____

27. In all cases, the occupant load in a room or building shall not be increased beyond a maximum of one occupant per _____ square feet.

    a. 3   b. 5

    c. 6   d. 7

Reference_____

28. The means of egress in which of the following areas is required to be illuminated when the space is occupied?

    a. aisle accessways in Group A   b. sleeping units in Group I

    c. dwelling units in Group R-2   d. aisles in Groups F and S

Reference_____

29. The exterior wall adjacent to an exterior area for assisted rescue does not require a fire-resistance rating where the area is located a minimum of _____ feet horizontally from the wall.

    a. 5   b. 10

    c. 15   d. 20

Reference_____

30. A tactile sign stating EXIT shall be provided adjacent to the door of all of the following means of egress components, except for _____.

    a. an egress stairway   b. the exit discharge

    c. a horizontal exit   d. an exit passageway

Reference_____

31. Where seating booths are provided, the occupant load shall be based on one person for each _____ inches of booth length.

    a. 18   b. 20

    c. 21   d. 24

Reference _____

32. During the course of the door swing, a door is permitted to project into the required width of the path of egress travel a maximum of _____.

    a. one-half the required width   b. one-half the actual width

    c. 4 inches   d. $4^{1}/_{2}$ inches

Reference _____

33. Under general conditions, the means of egress illumination level shall not be less than _____ foot-candle(s) at the walking surface level.

    a. 0.1   b. 1

    c. 5   d. 10

Reference _____

34. From a height of 34 inches to 42 inches in a required guard, a sphere with a maximum diameter of _____ inches shall not pass through.

    a. 4
            b. $4^3/_8$

    c. 6
            d. 8

Reference _____

35. Within an individual dwelling unit of a Group R-2 occupancy, openings in required guards on the sides of stair treads shall not allow the passage of a maximum _____-inch-diameter sphere.

    a. 4
            b. $4^3/_8$

    c. 6
            d. 8

Reference _____

2006 IBC Sections 1008 – 1010, and 1012
Means of Egress II

OBJECTIVE: To obtain an understanding of the general component requirements of a means of egress system, including those regulating doors, gates, stairways, ramps and turnstiles located along the egress path.

REFERENCE: Sections 1008 – 1010, and 1012, 2006 *International Building Code*

KEY POINTS:
- What is the minimum height and width of an egress door?
- What are the limitations on projections into the required clear door width?
- When must doors swing in the direction of egress travel?
- What is the maximum opening force permitted for an interior side-swinging door without a closer? Other side-swinging doors? Sliding and folding doors?
- What are "special" doors? How are they regulated differently than other doors?
- At a door, what change in elevation is permitted for a landing or floor surface?
- How must landings at doors be sized? What is the maximum allowable amount that doors may encroach into the required landing size?
- What is the maximum height of a threshold at a doorway?
- Which types of locks and latches are required on egress doors?
- When is the unlatching of an egress door permitted to take more than a single operation?
- Why do turnstiles create special egress concerns? What are the limitations on their use?
- When is panic hardware required?
- How are gates regulated differently than doors?
- How is the minimum required width of a stairway determined?
- At a stairway and its landings, what is the minimum headroom clearance?
- What is the minimum rise of a stair riser? Maximum rise? Minimum tread run?
- What degree of tolerance is permitted between the largest and the smallest tread run within a flight of stairs? Between the greatest and the smallest riser height?
- How are stairway landings regulated for size?
- What is the maximum vertical rise permitted between stairway landings?
- Which special limitations apply to winders? Curved or spiral stairways? Alternate tread devices?

KEY POINTS: • When is a stairway required to provide access to a roof?

(Cont'd) • What is the maximum permissible ramp slope? The maximum permissible rise for a ramp?

• How are ramp landings regulated for width? Length? Construction? Edge protection?

• When are handrails required for ramps?

• What must be the minimum height of a handrail located above the nosing of stairway treads and landings? Maximum height?

• How are handrails to be regulated for graspability?

• What are the exceptions to the general provision that handrails be continuous and without interruption?

• To what extent must handrails extend beyond the top and bottom risers of a stair flight?

• How much clear space is needed between a handrail and a wall or other surface?

• Under which conditions are intermediate handrails required?

Code Text: *Doors provided for egress purposes in numbers greater than required by the IBC shall meet the requirements of Section 1008. Means of egress doors shall be readily distinguishable from the adjacent construction and finishes such that the doors are easily recognizable as means of egress doors. Mirrors or similar reflecting materials shall not be used on means of egress doors. Means of egress doors shall not be concealed by curtains, drapes, decorations or similar materials.*

Discussion and Commentary: During a fire or other incident, occupants will attempt to exit through those doors that they believe will eventually lead to the exterior. Accordingly, any doors that would suggest an egress path must meet all of the door requirements. In addition, means of egress doors must be obvious and available for immediate use by the building occupants.

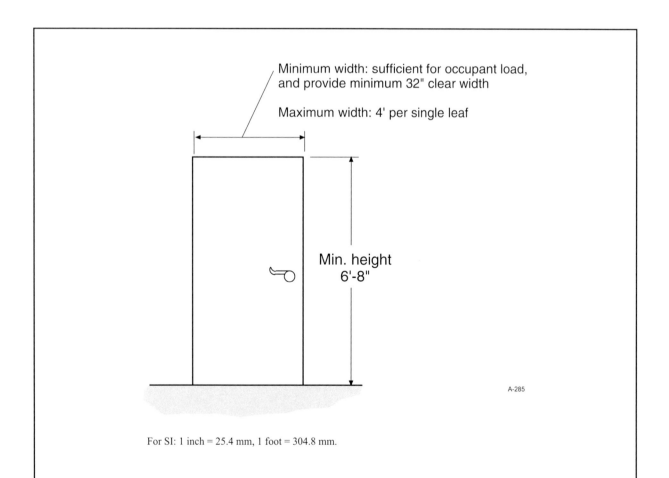

Minimum width: sufficient for occupant load, and provide minimum 32" clear width

Maximum width: 4' per single leaf

Min. height 6'-8"

A-285

For SI: 1 inch = 25.4 mm, 1 foot = 304.8 mm.

In accordance with Section 1018.2, any building or structure used for human occupancy must have at least one exterior door opening that complies with the minimum width (32 inches) and height (80 inches) requirements of Section 1008.1.1.

Code Text: *The minimum width of each door opening shall be sufficient for the occupant load thereof and shall provide a clear width of not less than 32 inches (813 mm). Clear openings of doorways with swinging doors shall be measured between the face of the door and the stop, with the door open 90 degrees (1.57 rad). The height of doors shall not be less than 80 inches (2032 mm).* See multiple exceptions for both width and height.

Discussion and Commentary: A clear width of 32 inches is required only to a height of 34 inches above the floor or ground. Beyond this point, projections up to 4 inches into the required width are permitted. Although a single doorway is expected to be used for the egress of one individual at a time, it must also be adequate for wheelchair users.

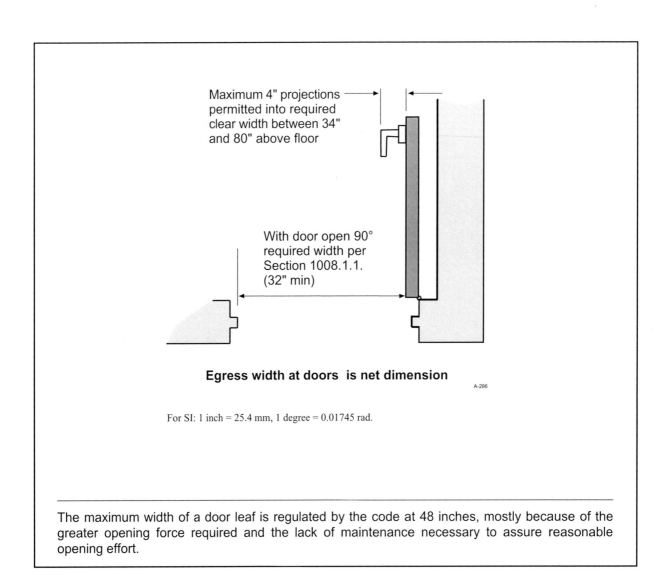

Maximum 4" projections permitted into required clear width between 34" and 80" above floor

With door open 90° required width per Section 1008.1.1. (32" min)

Egress width at doors is net dimension

A-286

For SI: 1 inch = 25.4 mm, 1 degree = 0.01745 rad.

The maximum width of a door leaf is regulated by the code at 48 inches, mostly because of the greater opening force required and the lack of maintenance necessary to assure reasonable opening effort.

Topic: Door Swing

Category: Means of Egress

Reference: IBC 1008.1.2

Subject: Doors, Gates and Turnstiles

Code Text: *Egress doors shall be side-hinged swinging.* See multiple exceptions addressing special conditions. *Doors shall swing in the direction of egress travel where serving an occupant load of 50 or more persons or a Group H occupancy.*

Discussion and Commentary: Numerous fire deaths in buildings have been attributed to improper exit doors, but no single incident is more infamous than the 1942 Coconut Grove fire in Boston. Inward-swinging exterior exit doors were a significant factor in the loss of 492 lives. As a result, doors serving sizable occupant loads or Group H occupancies must swing in the direction of exit flow. For assembly and educational occupancies, the use of panic hardware increases the likelihood that egress doors can be opened easily.

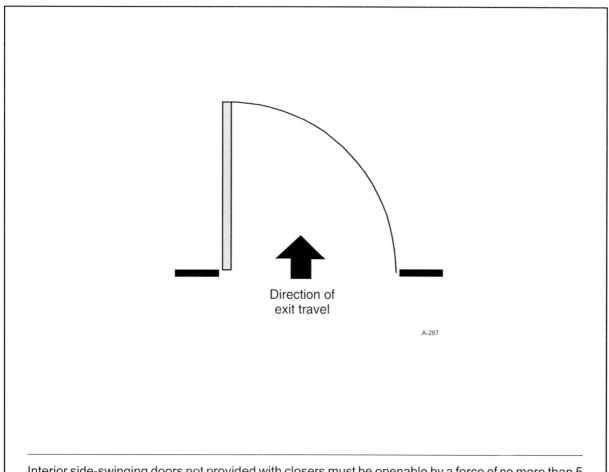

Direction of
exit travel

A-287

Interior side-swinging doors not provided with closers must be openable by a force of no more than 5 pounds. Where a closer is installed, such as for fire doors or exterior doors, the opening force limitation is increased to 15 pounds to ensure that the device can bring the door to a fully closed position.

Code Text: *Special doors and security grilles shall comply with the requirements of Sections 1008.1.3.1 through 1008.1.3.5.*

Discussion and Commentary: In general, doors in means of egress systems must be of the pivoted or side-hinged swinging type. Other doors, identified as special doors, are also addressed in the code. Such doors include revolving doors, power-operated doors, horizontal sliding doors, access-controlled egress doors and security grilles. These types of doors are specifically limited in their use because the difficult or unusual operation of such doors increases the likelihood of obstructed travel in an emergency.

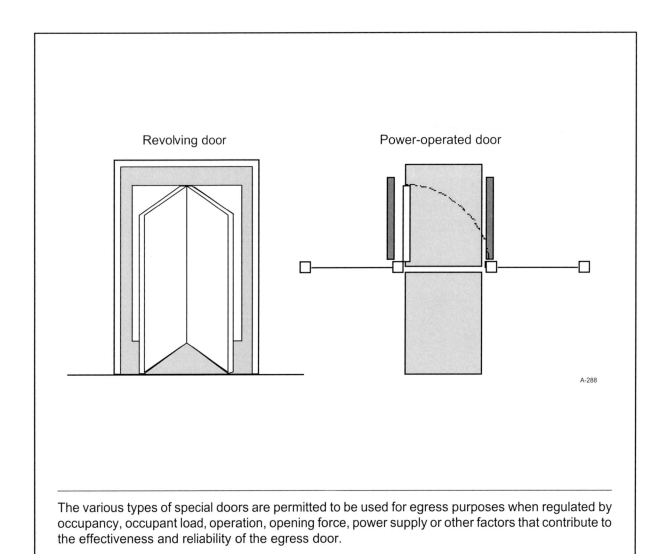

Revolving door

Power-operated door

A-288

The various types of special doors are permitted to be used for egress purposes when regulated by occupancy, occupant load, operation, opening force, power supply or other factors that contribute to the effectiveness and reliability of the egress door.

Topic: Floor Elevation

Category: Means of Egress

Reference: IBC 1008.1.4

Subject: Doors, Gates and Turnstiles

Code Text: *There shall be a floor or landing on each side of a door. Such floor or landing shall be at the same elevation on each side of the door.* See exceptions.

Discussion and Commentary: To avoid a surprise change in the elevation of a walking surface as it passes through a doorway, limitations have been placed on the height differential. The IBC generally requires that no elevation change occur at a door. In many cases, however, such a change in elevation takes place due to a variation in the type or thickness of floor finish materials. Where this occurs, a difference of no more than $^1/_2$ inch is permitted. Otherwise, only those exceptions that apply to certain dwelling units or to exterior doors not on an accessible route may have an elevation change at a doorway.

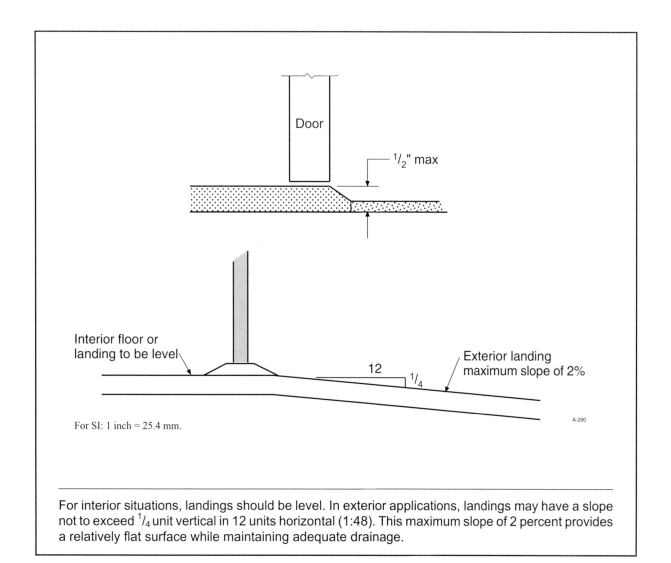

For SI: 1 inch = 25.4 mm.

For interior situations, landings should be level. In exterior applications, landings may have a slope not to exceed $^1/_4$ unit vertical in 12 units horizontal (1:48). This maximum slope of 2 percent provides a relatively flat surface while maintaining adequate drainage.

Topic: Landings at Doors

Category: Means of Egress

Reference: IBC 1008.1.5

Subject: Doors, Gates and Turnstiles

Code Text: *Landings shall have a width not less than the width of the stairway or the door, whichever is greater. Doors in the fully open position shall not reduce a required dimension by more than 7 inches (178 mm). When a landing serves an occupant load of 50 or more, doors in any position shall not reduce the landing to less than one-half its required width.*

Discussion and Commentary: This provision, which allows a door to project into the path of exit travel on a stairway's landing, comprises two issues. The first issue is that a door is not a fixed obstruction; it swings across the landing when it is used by occupants of the building. The second issue is that a door in any position is allowed to obstruct only one-half of the required width of the landing. The expectation is that the additional width of the landing will be provided for the occupants using the stairway as the door swings toward its fully open or fully closed position.

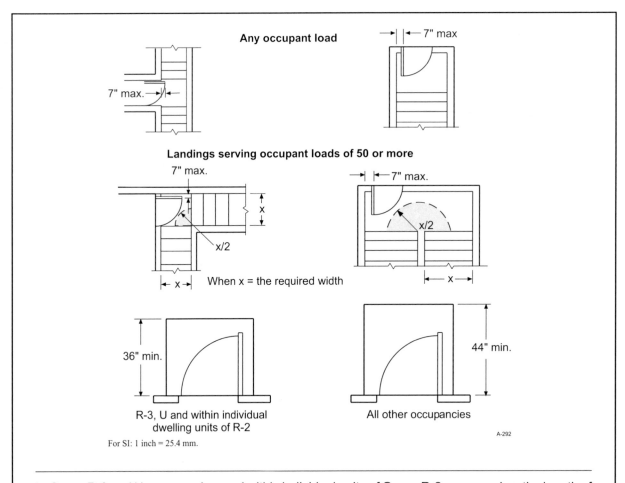

In Group R-3 and U occupancies, and within individual units of Group R-2 occupancies, the length of a landing can be no less than 36 inches. In all other occupancy groups, the minimum landing length is 44 inches, measured in the direction of travel.

Topic: Door Arrangement

Reference: IBC 1008.1.7

Category: Means of Egress

Subject: Doors, Gates and Turnstiles

Code Text: *Space between two doors in series shall be 48 inches (1219 mm) minimum plus the width of a door swinging into the space. Doors in series shall swing either in the same direction or away from the space between doors.* See exceptions for dwelling units and horizontal sliding power-operated doors.

Discussion and Commentary: Where two doors are installed in a manner to create a vestibule or similar space, they must be located so to allow building occupants effective and efficient movement through one door prior to continuing through the second door. This is especially true where the person opening the door has limited mobility and is required to make special effort in opening the door and passing through the doorway. The IBC recognizes these concerns by mandating an adequate spatial separation between doors provided in a series.

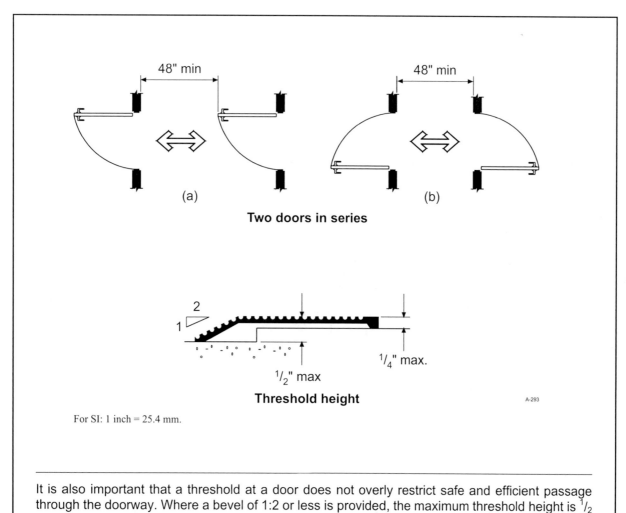

(a)   (b)

Two doors in series

Threshold height

A-293

For SI: 1 inch = 25.4 mm.

It is also important that a threshold at a door does not overly restrict safe and efficient passage through the doorway. Where a bevel of 1:2 or less is provided, the maximum threshold height is $1/2$ inch. Otherwise, an abrupt change in elevation is limited to $1/4$ inch.

Code Text: *Except as specifically permitted by Section 1008.1.8, egress doors shall be readily openable from the egress side without the use of a key or special knowledge or effort. Door handles, pulls, latches, locks and other operating devices shall be installed 34 inches (864 mm) minimum and 48 inches (1219 mm) maximum above the finished floor. Locks and latches shall be permitted to prevent operation of doors where any of four conditions exists. Manually operated flush bolts or surface bolts are not permitted.* See two exceptions. *The unlatching of any leaf shall not require more than one operation.* See exceptions identifying four locations that allow for multiple operations.

Discussion and Commentary: Every element along the path of exit travel through a means of egress system, particularly doors, must be under the control of, and operable by, the person seeking egress. The intent is that the hardware installed be of a type familiar to most users, readily recognizable and usable under any emergency conditions.

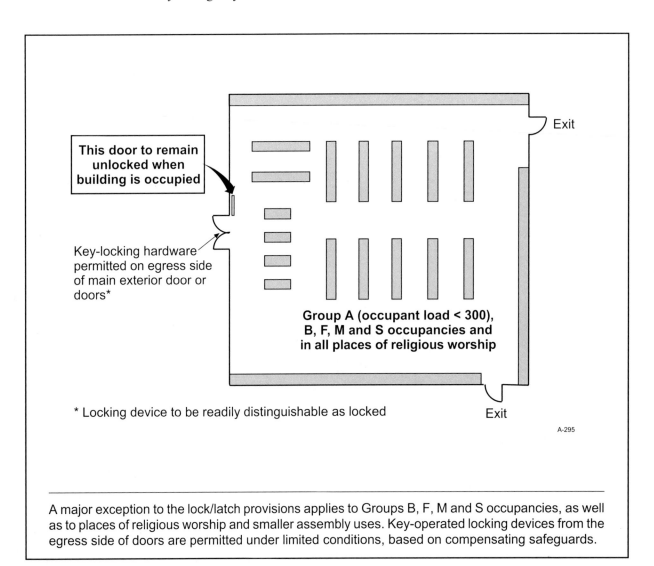

This door to remain unlocked when building is occupied

Key-locking hardware permitted on egress side of main exterior door or doors*

Group A (occupant load < 300), B, F, M and S occupancies and in all places of religious worship

Exit

* Locking device to be readily distinguishable as locked

Exit

A-295

A major exception to the lock/latch provisions applies to Groups B, F, M and S occupancies, as well as to places of religious worship and smaller assembly uses. Key-operated locking devices from the egress side of doors are permitted under limited conditions, based on compensating safeguards.

Topic: Panic and Fire Exit Hardware

Category: Means of Egress

Reference: IBC 1008.1.9

Subject: Doors, Ramps and Turnstiles

Code Text: *Each door in a means of egress from an occupancy of Group A or E having an occupant load of 50 or more and any Group H occupancy shall not be provided with a latch or lock unless it is panic hardware or fire exit hardware.*

Discussion and Commentary: Panic hardware is a door-latching assembly incorporating a device that releases the latch when force is applied in the direction of exit travel. It is utilized in assembly occupancies because of the hazard that occurs when a large number of occupants reach an exit door at the same instance. In educational occupancies, the same concern is present, along with the need for children to be able to operate the latch of an exit door easily during an emergency. Fire exit hardware is merely panic hardware listed for use on a fire door assembly.

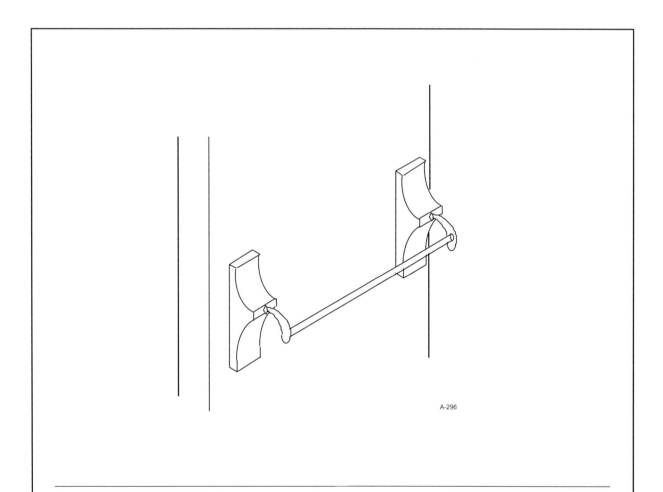

A-296

To ensure that contact with the door actuates the releasing device, the code requires that the actuating portion extend for at least one half of the door width. Where balanced or pivoted doors are used, the device width is again limited to one-half of the door width for leverage purposes.

Code Text: *Turnstiles or similar devices that restrict travel to one direction shall not be placed so as to obstruct any required means of egress.* See exception that allows installation of turnstiles serving as a portion of the means of egress if each turnstile serves a maximum occupant load of 50 persons and meets four additional provisions.

Discussion and Commentary: Turnstiles may serve as an exit component under very specific conditions. Where the listed conditions are met, each turnstile can be assigned a maximum exit capacity of 50 persons. However, permanent turnstiles cannot be considered as providing any of the required egress capacity when serving an occupant load over 300, no matter how many turnstiles are installed. In such cases, additional side-hinged swinging doors must be installed within 50 feet of the permanent turnstiles.

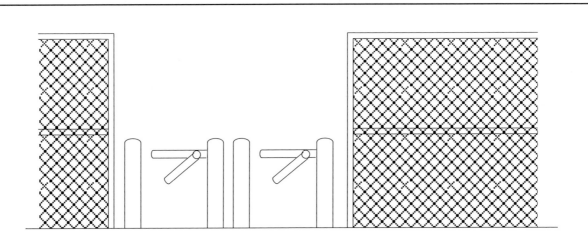

Each turnstile credited for up to 50-person capacity for egress where each turnstile:

- Will turn freely in direction of egress when power is lost, and upon manual release by employee in area

- Only given credit for 50% of required egress capacity (egress other than by turnstiles required)

- Limited to 39 inches in height

- Has minimum of 16.5 inches clear width at and below height of 39 inches

- Has minimum of 22 inches clear width at height above 39 inches

Where the turnstile has a height exceeding 39 inches, the restriction to egress is much like that of a revolving door. Therefore, the egress limitations for revolving doors as set forth in Section 1008.1.3.1 apply to this higher type of turnstile.

Topic: Stairway Width

Reference: IBC 1009.1, 1012.7

Category: Means of Egress

Subject: Stairways and Handrails

Code Text: *The width of stairways shall be determined as specified in Section 1005.1 (calculated based on occupant load) but such width shall not be less than 44 inches (1118 mm).* See exceptions for small occupant loads, spiral stairways, aisle stairs, and where a stairway lift is installed. *Projections into the required width at each handrail shall not exceed 4.5 inches (114 mm) at or below the handrail height. Projections into the required width shall not be limited above the minimum headroom height required in Section 1009.2.*

Discussion and Commentary: A stairway is considered one or more flights of stairs, each made up of one or more risers and any connecting landings. Any change of elevation along a travel path, unless accomplished by a ramp, must include a stair or stairway. Although stairways are generally required to be at least 44 inches in width, a 36-inch-wide stairway is permitted where serving an occupant load of 50 or less.

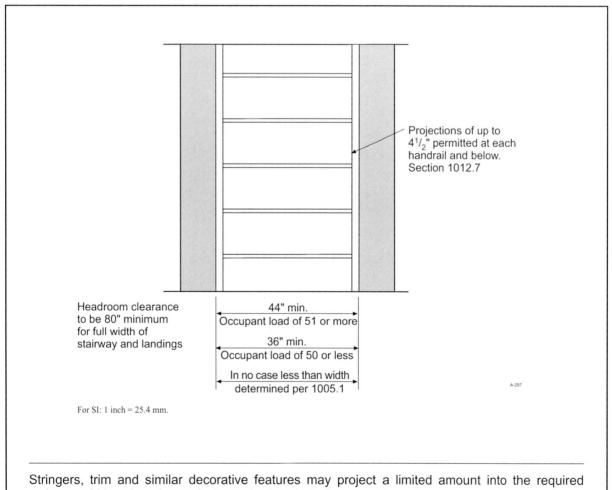

Projections of up to 4¹/₂" permitted at each handrail and below. Section 1012.7

Headroom clearance to be 80" minimum for full width of stairway and landings

44" min.
Occupant load of 51 or more

36" min.
Occupant load of 50 or less

In no case less than width determined per 1005.1

A-297

For SI: 1 inch = 25.4 mm.

Stringers, trim and similar decorative features may project a limited amount into the required stairway width unless located above the handrail. Between the rail and the required headroom height of 80 inches, no projection is permitted.

Code Text: *Stair riser heights shall be 7 inches (178 mm) maximum and 4 inches (102 mm) minimum. Stair tread depths shall be 11 inches (279 mm) minimum.* See exceptions, including the allowance for greater riser heights (7.75 inches) and shallower tread depths (10 inches) in Group R-3 and associated Group U occupancies, and within individual dwelling units of Group R-2.

Discussion and Commentary: The stairway 7-11 rule is the result of much research and discussion on stairway design. In addition to the proportional criteria that has been developed, the uniformity of the treads and risers in a flight of stairs is critical. The maximum variation between the highest and lowest risers and between the shallowest and deepest treads is limited to $^3/_8$ inch within any flight, which is intended as a permissible construction tolerance.

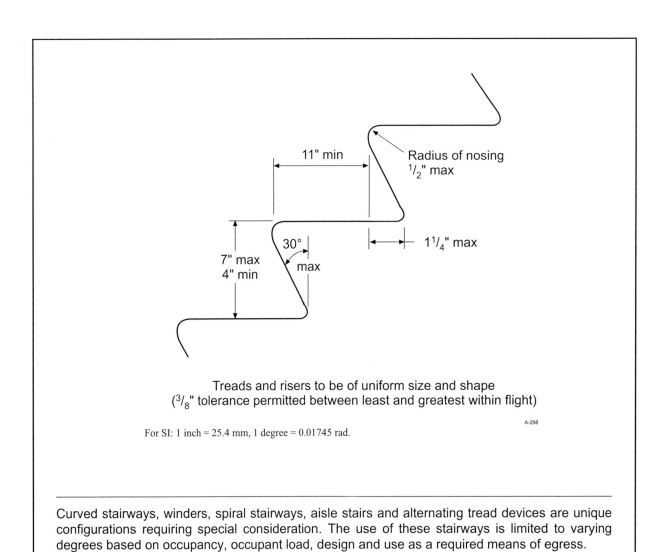

Treads and risers to be of uniform size and shape
($^3/_8$" tolerance permitted between least and greatest within flight)

A-298

For SI: 1 inch = 25.4 mm, 1 degree = 0.01745 rad.

Curved stairways, winders, spiral stairways, aisle stairs and alternating tread devices are unique configurations requiring special consideration. The use of these stairways is limited to varying degrees based on occupancy, occupant load, design and use as a required means of egress.

Code Text: *There shall be a floor or landing at the top and bottom of each stairway. The width of landings shall be not less than the width of stairways they serve. Every landing shall have a minimum dimension measured in the direction of travel equal to the width of the stairway. Such dimension need not exceed 48 inches (1219 mm) where the stairway has a straight run.* See exceptions for aisles stairs and doors opening onto a stairway landing.

Discussion and Commentary: A landing that serves a stairway is required to have a length equal to or greater than the stairway width unless the stairway has a straight run. This measurement is based on the actual width of the stairway, not the required width. It is important to ensure that the capacity of the stairway is not reduced as occupants travel between stairway flights. A length of 48 inches is acceptable for straight stairway travel, insofar as the capacity is not reduced for travel through the landing.

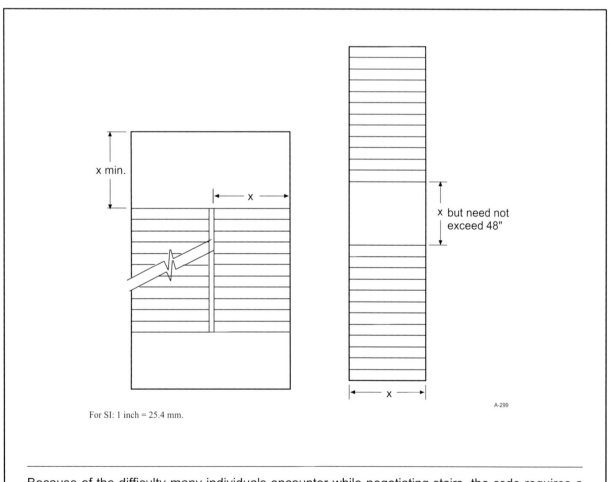

For SI: 1 inch = 25.4 mm.

Because of the difficulty many individuals encounter while negotiating stairs, the code requires a maximum vertical rise between landings of 12 feet. When placed at limited intervals, landings can be used as a resting place for the stair user and can also make stair travel less intimidating.

Topic: Handrail Locations

Category: Means of Egress

Reference: IBC 1009.10, 1012.8

Subject: Stairways and Handrails

Code Text: *Stairways shall have handrails on each side and shall comply with Section 1012. See five exceptions where a single handrail or no handrail is required. Stairways shall have intermediate handrails located in such a manner so that all portions of the stairway width required for egress capacity are within 30 inches (762 mm) of a handrail. On monumental stairs, handrails shall be located along the most direct path of egress travel.*

Discussion and Commentary: The handrail, a very important safety element of a stairway, must be located within relatively easy reach of every stair user. Therefore, in most applications, a rail must be provided on both sides of the stairway. In the case of extremely wide stairways, such as monumental stairs, the requirement for additional rails located throughout the width of the stairway is based on the required stairway width, not the actual width.

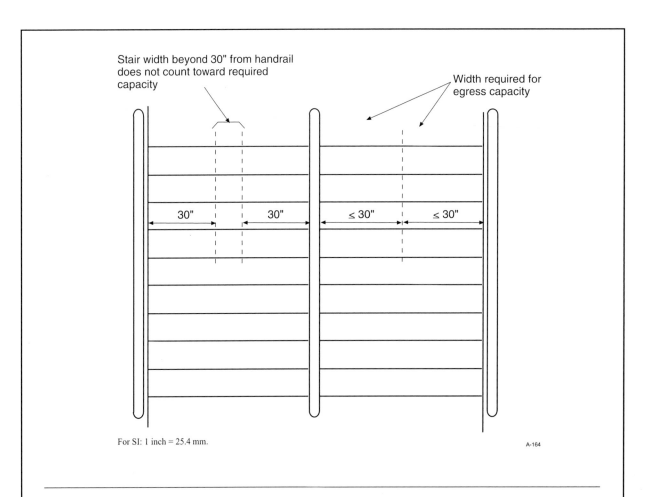

Stair width beyond 30" from handrail does not count toward required capacity

Width required for egress capacity

30" 30" ≤ 30" ≤ 30"

For SI: 1 inch = 25.4 mm.

A-164

Various exceptions permit the use of a single handrail, and in some cases no rail, within a dwelling unit. In addition, aisle stairs with a center handrail, or those serving seating on only one side, are permitted to use a single handrail.

Code Text: *Ramps used as part of a means of egress shall have a running slope not steeper than one unit vertical in 12 units horizontal (8-percent slope). The slope of other pedestrian ramps shall not be steeper than one unit vertical in eight units horizontal (12.5-percent slope). The rise for any ramp run shall be 30 inches (762 mm) maximum. The minimum width of a means of egress ramp shall not be less than that required for corridors by Section 1017.2. The clear width of a ramp and the clear width between handrails, if provided, shall be 36 inches (914 mm). Ramps with a rise greater than 6 inches (152 mm) shall have handrails on both sides complying with Section 1012.*

Discussion and Commentary: Although many of the governing ramp provisions are designed for accessibility purposes, egress capabilities must also be considered. Unlike allowances for corridors, there are no permitted projections into the required height or width of ramps.

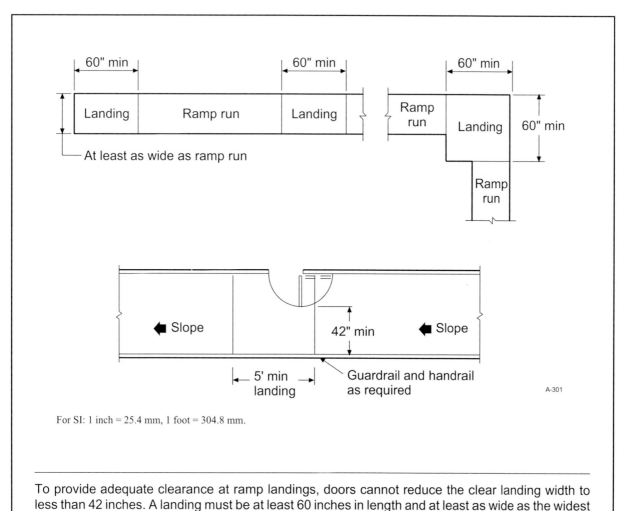

For SI: 1 inch = 25.4 mm, 1 foot = 304.8 mm.

To provide adequate clearance at ramp landings, doors cannot reduce the clear landing width to less than 42 inches. A landing must be at least 60 inches in length and at least as wide as the widest ramp run adjoining the landing.

Code Text: *Edge protection complying with Sections 1010.9.1* (curb, rail, wall or barrier) *or 1010.9.2* (extended floor or ground surface) *shall be provided on each side of ramp runs and at each side of ramp landings.* See exceptions for curb ramps and ramp landings. *A curb, rail, wall or barrier shall be provided that prevents the passage of a 4-inch-diameter (102 mm) sphere, where any portion of the sphere is within 4 inches (102 mm) of the floor or ground surface, or the floor or ground surface of the ramp run or landing shall extend 12 inches (305 mm) minimum beyond the inside face of a handrail complying with Section 1012.*

Discussion and Commentary: Edge protection at ramps and ramp landings is necessary to prevent the wheels of a wheelchair from leaving the ramp or landing surface, or becoming lodged between the edge of the ramp and any adjacent construction. The protection is also beneficial to those individuals who utilize various forms of walking aids, including canes and crutches.

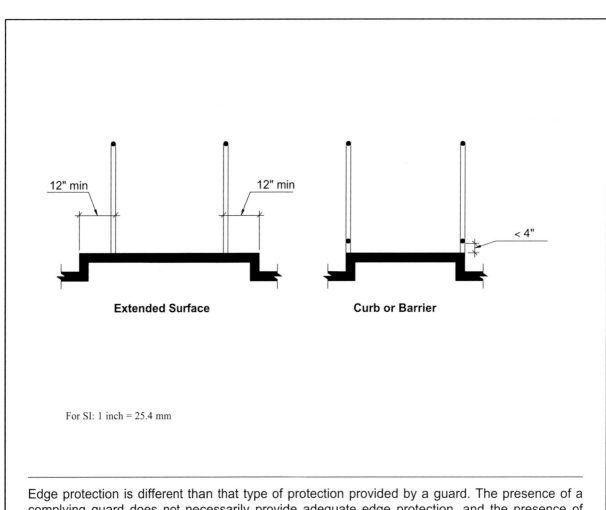

Edge protection is different than that type of protection provided by a guard. The presence of a complying guard does not necessarily provide adequate edge protection, and the presence of adequate edge protection does not satisfy the requirements for a guard.

Topic: Handrail Dimensions

Category: Means of Egress

Reference: IBC 1012.2, 1012.3

Subject: Stairways and Handrails

Code Text: *Handrail height, measured above stair tread nosings, or finish surface of ramp slope, shall be uniform, not less than 34 inches (864 mm) and not more than 38 inches (965 mm). Handrails with a circular cross section shall have an outside diameter of at least 1.25 inches (32 mm) and not greater than 2 inches ((51 mm)or shall provide equivalent graspability. If the handrail is not circular, it shall have a perimeter dimension of at least 4 inches (102 mm) and not greater than 6.25 inches (160 mm) with a maximum cross-section dimension of 2.25 inches (57 mm).*

Discussion and Commentary: Handrail height shall be measured from the nosing of the treads to the top of the rail. Handrails located above or below this height range are not easily reached by most individuals. The shape of the rail should allow for easy grasping by most users.

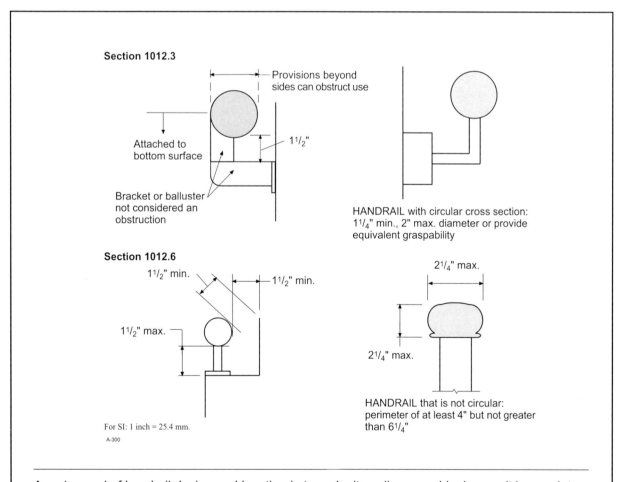

Section 1012.3

Provisions beyond sides can obstruct use

Attached to bottom surface

1¹/₂"

Bracket or baluster not considered an obstruction

HANDRAIL with circular cross section: 1¹/₄" min., 2" max. diameter or provide equivalent graspability

Section 1012.6

1¹/₂" min.

1¹/₂" min.

1¹/₂" max.

2¹/₄" max.

2¹/₄" max.

HANDRAIL that is not circular: perimeter of at least 4" but not greater than 6¹/₄"

For SI: 1 inch = 25.4 mm.

A-300

A major goal of handrail design and location is to make it easily graspable; hence, it is mandatory that the rail be placed at least 1¹/₂ inches from any abutting elements, such as a wall. However, the projection of the rail into the required width is limited to no more than 4¹/₂ inches.

Quiz

Study Session 10
IBC Sections 1008 – 1010, and 1012

1. In general, a door opening shall provide a minimum clear width of _____ inches.

    a. 30 b. 32

    c. 34 d. 36

    Reference_____

2. In a Group I-2 occupancy, means of egress doors used for the movement of beds shall have a minimum width of _____ inches.

    a. 32 b. 36

    c. $41^1/_2$ d. 44

    Reference_____

3. The minimum required width for door openings does not apply to storage closets less than _____ square feet in floor area.

    a. 10 b. 25

    c. 50 d. 100

    Reference_____

4. At a door opening, the maximum permitted projection into the required clear width shall be _____ inch(es) at any point between 34 inches and 80 inches above the floor.

   a. 0, no projections are permitted

   b. $^{1}/_{2}$

   c. 1

   d. 4

   Reference_____

5. In which of the following uses must an egress door be a side-hinged swinging door where serving an occupant load of 10 or less?

   a. office   b. storage

   c. manufacturing   d. retail sales

   Reference_____

6. In a Group F-2 occupancy, egress doors shall swing in the direction of egress travel where serving a minimum occupant load of _____ persons.

   a. 10   b. 30

   c. 50   d. 100

   Reference_____

7. For a side-swinging fire door provided with a self-closing device, the egress door shall swing to a full-open position when subjected to a maximum _____ force.

   a. 5-pound   b. 15-pound

   c. 30-pound   d. 50-pound

   Reference_____

8. A revolving door shall be provided with a side-hinged swinging door located in the same wall and within _____ feet of the revolving door.

   a. 5   b. 10

   c. 20   d. 50

   Reference_____

9. Entrance doors in a means of egress in a Group B office building may be equipped with an approved entrance and egress access control system under limited conditions, including that any manual unlocking device be located a maximum of _____ feet from the secured doors.

    a. 3   b. 4

    c. 5   d. 10

    Reference_____

10. Exterior landings at doors shall have a maximum slope of _____ unit vertical in 12 units horizontal.

    a. $^1/_8$ b. $^1/_4$

    c. $^1/_2$ d. 1

    Reference_____

11. Other than at sliding doors serving dwelling units, the maximum height for thresholds at doorways is _____ inch.

    a. $^1/_8$ b. $^1/_4$

    c. $^1/_2$ d. 1

    Reference_____

12. Approved, listed delayed egress locks are permitted under specific conditions in all but which one of the following occupancies?

    a. Group A   b. Group I

    c. Group F   d. Group S

    Reference_____

13. In general, door handles, pulls, latches, locks and other operating devices shall be installed a minimum of _____ inches and a maximum of _____ inches above the finished floor.

    a. 32, 48   b. 34, 42

    c. 34, 48   d. 36, 54

    Reference_____

14. In a Group E occupancy, panic hardware shall be provided on exterior egress doors when serving a minimum occupant load of _____ persons.

    a. 1, panic hardware is always required

    b. 10

    c. 50

    d. 100

    Reference_____

15. Stairways, other than spiral stairways, shall have a minimum headroom clearance of _____ inches.

    a. 76   b. 78

    c. 80   d. 84

    Reference_____

16. Within an individual dwelling unit of a Group R-2 apartment building, a stair shall have a maximum riser height of _____ inches and a minimum tread depth of _____ inches.

    a. 7, 11   b. $7^3/_4$, 10

    c. 8, 9   d. $8^1/_4$, 9

    Reference_____

17. For a 60-inch-wide stairway having a straight run, any intermediate landing shall be a minimum of _____ inches in length.

    a. 36   b. 44

    c. 48   d. 60

    Reference_____

18. For winder treads, the minimum permitted tread depth at the narrow edge shall be _____ inches.

    a. 4   b. 6

    c. 8   d. 10

    Reference_____

19. Where serving a maximum of five occupants, a spiral stairway may be used as a means of egress component from a space having a maximum floor area of _____ square feet.

    a. 200   b. 250

    c. 500   d. 1,000

Reference_____

20. Stairway handrails shall be located a minimum of _____ inches and a maximum of _____ inches above stair tread nosings.

    a. 30, 34   b. 30, 38

    c. 32, 34   d. 34, 38

Reference_____

21. Where handrails are not continuous between stair flights, they shall continue to slope for _____ beyond the bottom riser.

    a. the depth of one tread

    b. a minimum of 12 inches

    c. 12 inches plus one tread depth

    d. one-half the length of the landing

Reference_____

22. Projections into the required width at each stairway handrail shall be limited to a maximum of _____ inches at any point at and below the handrail height.

    a. 0, no projections permitted   b. $1^{1}/_{2}$

    c. $3^{1}/_{2}$ d. $4^{1}/_{2}$

Reference_____

23. The maximum permitted rise of any ramp shall be _____ inches.

    a. 30   b. 44

    c. 48   d. 60

Reference_____

24. Only those ramps having a maximum rise of _____ inches are permitted without complying handrails.

    a. 6   b. 12

    c. 24   d. 30

Reference_____

25. Unless having a maximum height of _____ inches, turnstiles shall be regulated as for revolving doors.

    a. 34   b. 36

    c. 38   d. 39

Reference_____

26. Where a power-operated door must be opened manually, the maximum force to set the door in motion shall be _____ pounds.

    a. 5   b. 15

    c. 30   d. 50

Reference_____

27. The space required between two doors in a series shall be a minimum of _____ inches plus the width of a door swinging into the space.

    a. 30   b. 44

    c. 48   d. 60

Reference_____

28. A stairway serving an occupant load of 35 in an office suite shall have a minimum width of _____ inches.

    a. 30   b. 36

    c. 42   d. 44

Reference_____

29. Stairway handrails having a circular cross section shall have a minimum outside diameter of _____ inches and a maximum diameter of _____ inches.

    a. $1^1/_4$, 2   b. $1^1/_4$, $2^5/_8$

    c. $1^1/_2$, 2   d. $1^1/_2$, $2^5/_8$

Reference_____

30. Where a railing is used as edge protection along the side of a ramp run, it must have a rail mounted to prevent the passage of a maximum _____ -inch sphere, where any portion of the sphere is within 4 inches of the ground.

    a. 4   b. 6

    c. 8   d. 12

Reference_____

31. Doors from electrical rooms with equipment rated at a minimum of _____ amperes that contain overcurrent, switching or control devices must be provided with panic hardware.

    a. 600   b. 1,000

    c. 1,200   d. 1,500

Reference _____

32. In general, the minimum required width for a stairway in a Group B occupancy shall be _____ inches.

    a. 36   b. 42

    c. 44   d. 48

Reference _____

33. Where a ramp is not a part of an accessible route, the length of the landing shall be a minimum of _____ inches measured in the direction of travel.

    a. 44   b. 48

    c. 54   d. 60

Reference _____

34. At ramps where handrails are not continuous between runs, the handrails shall extend horizontally a minimum of _____ inches beyond the top and bottom of the ramp.

    a. 6   b. 8

    c. 12   d. 18

Reference _____

35. Stairways shall have intermediate handrails located so that all portions of the stairway width required for egress capacity are within _____ inches of a handrail.

    a. 30   b. 32

    c. 35   d. 44

Reference _____

2006 IBC Sections 1014 – 1017
Means of Egress III

OBJECTIVE: To obtain an understanding of the system design requirements for the exit access, including number of exits, separation of egress doorways and maximum travel distances, as well as the requirements for the exit access components, including aisles, corridors and egress balconies.

REFERENCE: Section 1014 through 1017, 2006 *International Building Code*

KEY POINTS:
- What portion of the means of egress system is the exit access?
- Is egress permitted through an adjoining or intervening room? If so, under what conditions?
- How must egress be provided where more than one tenant occupies any single floor of a building or structure?
- What is a common path of travel?
- Why is the common path of travel so limiting?
- Where must complying aisles be provided?
- What is the minimum permitted aisle width in public areas of Groups B and M? In nonpublic areas?
- What is an aisle accessway? What is the minimum aisle accessway width where the aisle serves seating arrangements at tables or counters? Merchandise pads?
- How do the exiting provisions for egress balconies compare with those for corridors?
- When is a fire-resistance-rated separation mandated between the building and an egress balcony?
- When are at least two exits or exit access doorways required from a room?
- At what point is access required to three or more exits?
- Why are multiple exit paths required to be separated at a specified minimum distance from each other?
- At what minimum distance must two exits or exit access doorways be separated? Three or more exits or exit access doorways?
- What benefit is derived from exit separation in a sprinklered building?
- Why is travel distance regulated?
- How is the maximum travel distance measured? Where does it start? Where does it end?

KEY POINTS:
(Cont'd)

- How is travel distance measured where travel involves stairways?
- Which occupancies permit the least travel distance? The most?
- What are the travel distance limitations for low-hazard storage and manufacturing buildings? Which special conditions must be met?
- To what amount can the travel distance be increased on an exterior egress balcony?
- What is the purpose of a corridor? When must a corridor be of fire-resistance-rated construction?
- What is the minimum required width of a corridor?
- What is a dead-end condition? When are dead ends limited in length?
- How is the use of corridors for supply, return, exhaust or ventilation air regulated?

Code Text: *Egress from a room or space shall not pass through adjoining or intervening rooms or areas, except where such adjoining rooms or areas are accessory to the area served; are not a high-hazard occupancy; and provide a discernible path of egress travel to an exit.* See exception for Group H, S and F occupancies.

Discussion and Commentary: A workable means of egress system must be direct, obvious and unobstructed. Therefore, egress may only travel through an intervening room, space or area where the exit path is discernable. Travel must be such that it is readily apparent which direction the occupant must go to continue toward the exit. In addition, access through a high-hazard space is prohibited unless traveling from another high-hazard space. It is expected that any intervening room used for egress will be accessory to the room from which egress begins.

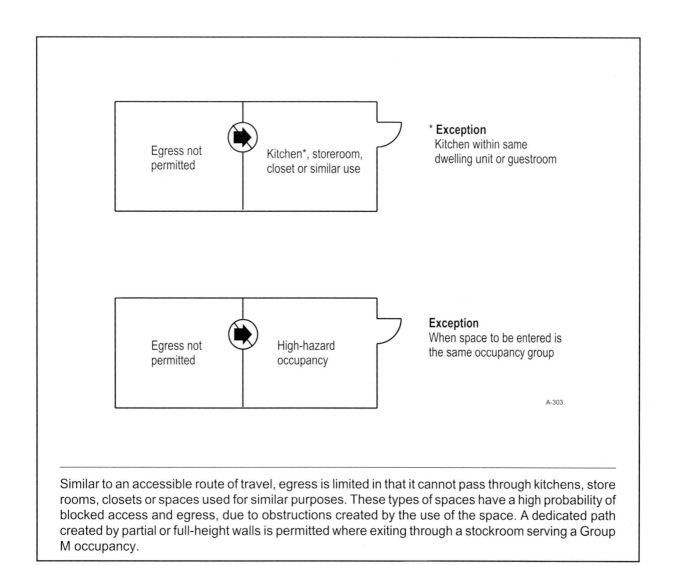

Similar to an accessible route of travel, egress is limited in that it cannot pass through kitchens, store rooms, closets or spaces used for similar purposes. These types of spaces have a high probability of blocked access and egress, due to obstructions created by the use of the space. A dedicated path created by partial or full-height walls is permitted where exiting through a stockroom serving a Group M occupancy.

Code Text: *In occupancies other than Groups H-1, H-2 and H-3, the common path of egress travel shall not exceed 75 feet (22 860 mm). In occupancies in Groups H-1, H-2 and H-3, the common path of egress travel shall not exceed 25 feet (7620 mm).* See three exceptions where a 100-foot common path is acceptable and an exception permitting a 125-foot common path in a Group R-2 occupancy.

Discussion and Commentary: A common path of egress travel is defined as that portion of exit access which the occupants are required to traverse before two separate and distinct paths of egress travel to two exits are available. The concept of limiting the common path of egress travel addresses the concern that multiple egress options must be available to occupants where the expected egress travel distance becomes excessive. Although the overall travel distance in a building may be of considerable length, such travel is greatly limited where only one egress path is available.

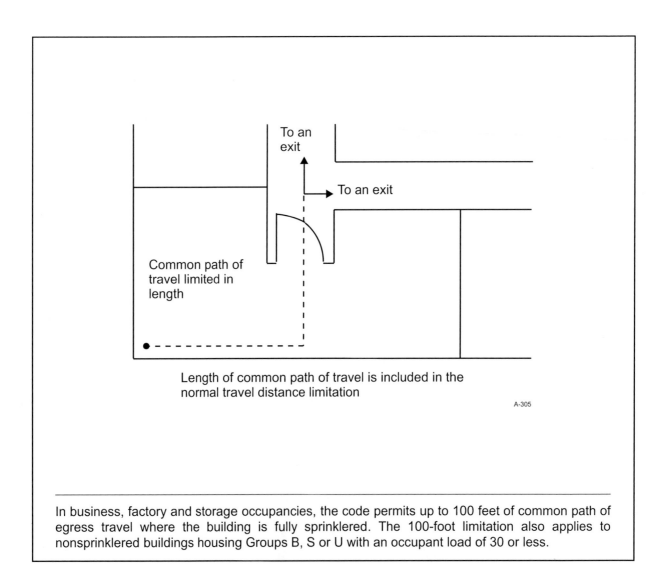

In business, factory and storage occupancies, the code permits up to 100 feet of common path of egress travel where the building is fully sprinklered. The 100-foot limitation also applies to nonsprinklered buildings housing Groups B, S or U with an occupant load of 30 or less.

Category: Means of Egress
Subject: Exit Access

Code Text: *Aisles shall be provided from all occupied portions of the exit access which contain seats, tables, furnishings, displays and similar fixtures or equipment. Aisles serving assembly areas, other than seating at tables, shall comply with Section 1025. Aisles serving reviewing stands, grandstands and bleachers shall also comply with Section 1025. The required width of aisles shall be unobstructed.* See exception for doors, handrails, trim and decorative features. *In Group B and M occupancies, the minimum clear aisle width shall be determined by Section 1005.1 for the occupant load served, but shall not be less than 36 inches (914 mm).* See exception for nonpublic aisles.

Discussion and Commentary: Well-defined aisles must be provided throughout office spaces, retail stores and similar facilities. The mandated clear width varies based on the presence of obstructions, such as chairs, clothes racks or other items that can easily interrupt the egress flow to an exit.

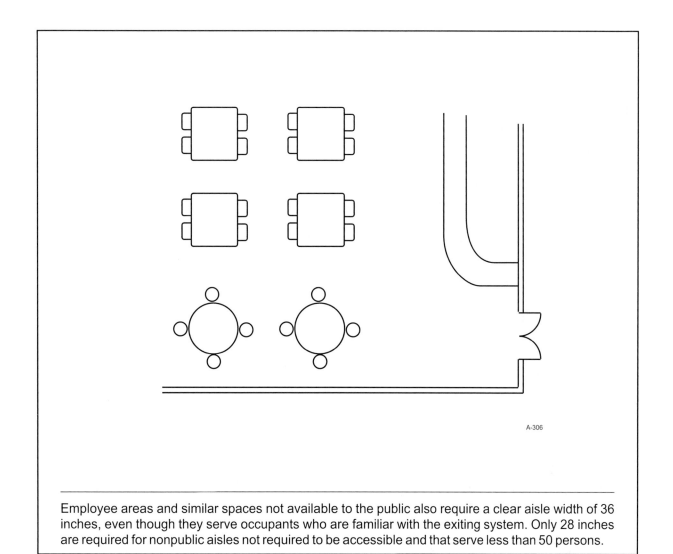

A-306

Employee areas and similar spaces not available to the public also require a clear aisle width of 36 inches, even though they serve occupants who are familiar with the exiting system. Only 28 inches are required for nonpublic aisles not required to be accessible and that serve less than 50 persons.

Code Text: *An aisle accessway shall be provided on at least one side of each element within the merchandise pad. The minimum clear width for an aisle accessway not required to be accessible shall be 30 inches (762 mm). The required clear width of the aisle accessway shall be measured perpendicular to the elements and merchandise within the merchandise pad. The 30-inch (762 mm) minimum clear width shall be maintained to provide a path to an adjacent aisle or aisle accessway.*

Discussion and Commentary: A merchandise pad is defined as the merchandise display area that contains multiple counters, shelves, racks and other movable fixtures. Bounded by aisles, permanent fixtures and walls, the merchandise pad also includes aisle accessways utilized to provide both access to an aisle and circulation throughout the pad area.

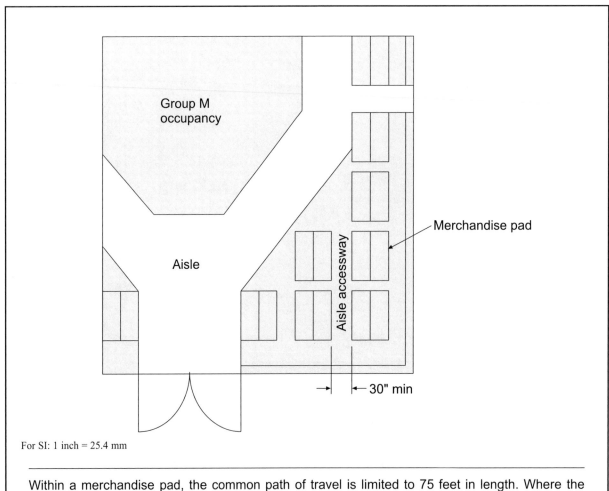

For SI: 1 inch = 25.4 mm

Within a merchandise pad, the common path of travel is limited to 75 feet in length. Where the occupant load of the area served by the common path exceeds 50 persons, the common path cannot exceed 30 feet in length from any point in the merchandise pad.

| **Topic:** Seating at Tables | **Category:** Means of Egress |
| **Reference:** IBC 1014.4.3 | **Subject:** Exit Access |

Code Text: *Aisle accessways serving arrangements of seating at tables or counters shall have sufficient clear width to conform to the capacity requirements of Section 1005.1. Aisle accessways shall provide a minimum of 12 inches (305 mm) plus 0.5 inch (12.7 mm) of width for each additional 1 foot (305 mm), or fraction thereof, beyond 12 feet (3658 mm) of aisle accessway length measured from the center of the seat farthest from an aisle.* See exception for aisle accessways of limited lengths and occupant loads. *The length of travel along the aisle accessway shall not exceed 30 feet (9144 mm) from any seat to the point where a person has a choice of two or more paths of egress travel to separate exits.*

Discussion and Commentary: To facilitate progress toward an established aisle, it is important that a minimum degree of egress width be established within areas furnished with tables and chairs. An aisle accessway, defined as *that portion of an exit access that leads to an aisle,* is thus regulated.

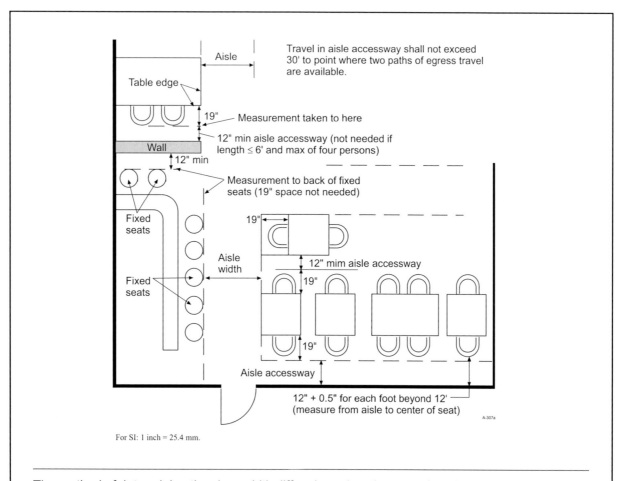

For SI: 1 inch = 25.4 mm.

The method of determining the clear width differs based on the type of seating that is provided. For fixed seats, the measurement is made from the back of the seats. Otherwise, the clear width is measured to a line 19 inches from the edge of the table or counter.

Topic: Egress Balconies

Category: Means of Egress

Reference: IBC 1014.5, 1014.5.1

Subject: Exit Access

Code Text: *Balconies used for egress purposes shall conform to the same requirements as corridors for width, headroom, dead ends and projections. Exterior egress balconies shall be separated from the interior of the building by walls and opening protectives as required for corridors. See exception for elimination of separation.*

Discussion and Commentary: Although the openness of exterior balconies provides some degree of protection from smoke and toxic gases created by a fire, travel along such balconies usually places the occupants at considerable risk. Therefore, the IBC regulates egress balcony travel in a manner consistent with unprotected travel inside the structure. An increase of 100 feet in maximum allowable travel distance is permitted by Section 1016.3 for egress balcony travel.

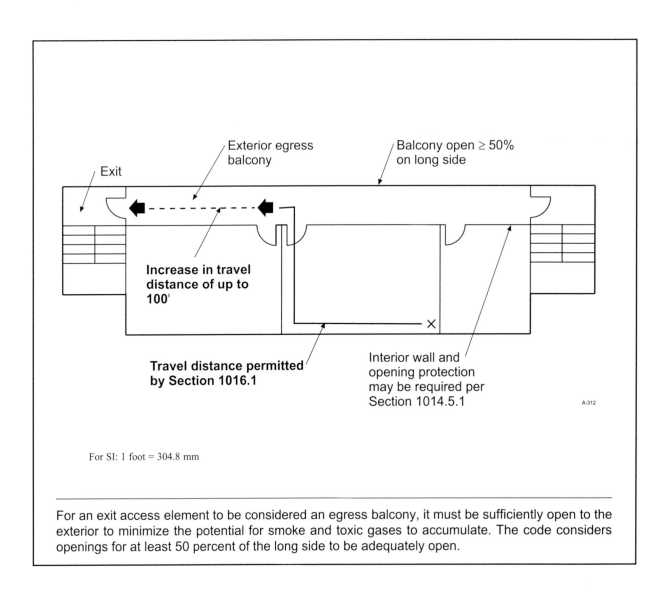

For SI: 1 foot = 304.8 mm

For an exit access element to be considered an egress balcony, it must be sufficiently open to the exterior to minimize the potential for smoke and toxic gases to accumulate. The code considers openings for at least 50 percent of the long side to be adequately open.

Topic: Doorways Required

Reference: IBC 1015.1

Category: Means of Egress

Subject: Exit and Exit Access Doorways

Code Text: *Two exits or exit access doorways from any space shall be provided where 1) the occupant load of the space exceeds the values in Table 1015.1, or 2) the common path of egress travel exceeds the limitations of Section 1014.3, or (3) required by Sections 1015.3, 1015.4 and 1015.5. See exceptions for Group I-2 occupancies.*

Discussion and Commentary: Two basic criteria establish the point at which it is necessary to provide at least two paths of egress travel from a portion of a building. The first is a function of both the occupancy involved and the occupant load of the space. The second is related to the maximum travel distance an occupant encounters prior to reaching a point where a choice of egress paths is available. Based on the occupancy involved, where the established occupant load or the common travel distance is exceeded, the space must have multiple egress doorways.

TABLE 1015.1
SPACES WITH ONE MEANS OF EGRESS

OCCUPANCY	MAXIMUM OCCUPANT LOAD
A, B, Eª, F, M, U	49
H-1, H-2, H-3	3
H-4, H-5, I-1, I-3, I-4, R	10
S	29

a. Day care maximum occupant load is 10.

Requirements for three or four exits are found in Section 1019.1 and are based solely on occupant load. The same criteria applies for access to exits, where a very high occupant load is assigned to a room or space. Multiple exit access doorways must then be provided.

Code Text: *Required exits shall be located in a manner that makes their availability obvious. Where two exits or exit access doorways are required from any portion of the exit access, the exit doors or exit access doorways shall be placed a distance apart equal to not less than one-half of the length of the maximum overall diagonal dimension of the building or area to be served measured in a straight line between exit doors or exit access doorways. Interlocking or scissor stairs shall be counted as one exit stairway.* See exceptions for sprinklered buildings and where a rated corridor connects two exit enclosures.

Discussion and Commentary: One of the fundamental concepts of exiting is that a single fire incident should not render all means of egress unusable. In this regard, egress doorways are required to be located so as to minimize the probability of such an occurrence. The required separation in sprinklered buildings is reduced to a distance of one-third of the overall diagonal.

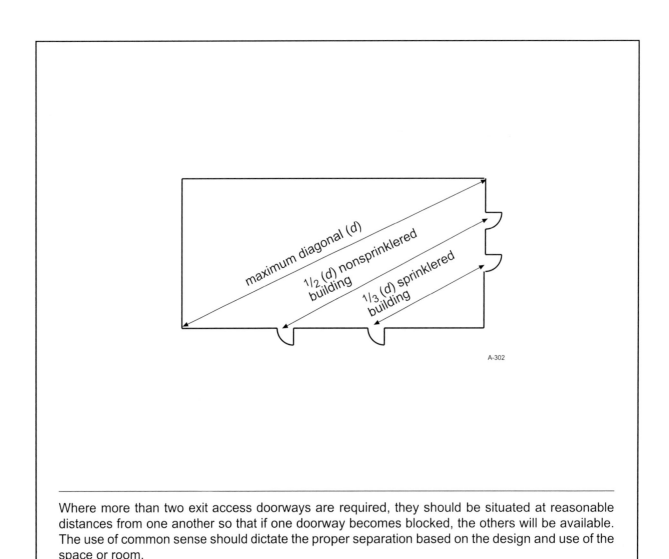

Where more than two exit access doorways are required, they should be situated at reasonable distances from one another so that if one doorway becomes blocked, the others will be available. The use of common sense should dictate the proper separation based on the design and use of the space or room.

Topic: Travel Distance Limitations

Category: Means of Egress

Reference: IBC 1016.1

Subject: Exit Access Travel Distance

Code Text: *Exits shall be so located on each story that the maximum length of exit access travel, measured from the most remote point within a story to the entrance to an exit along the natural and unobstructed path of egress travel, shall not exceed the distances given in Table 1016.1.*

Discussion and Commentary: Travel distance is considered the portion of egress travel between any occupiable location in a building and the nearest door of an exit. Because quick evacuation from a building is the foremost method of protecting the occupants in many fire incidents, the length of travel is limited until the occupant reaches one of the "protected components." Travel should be measured around any obstruction that is considered fixed or permanent, including low-height office partitions, retail shelving, storage racks, fixed seating, etc.

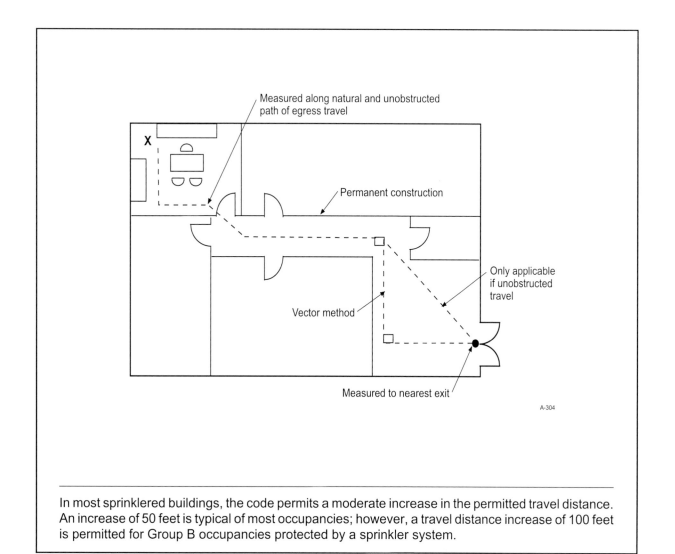

In most sprinklered buildings, the code permits a moderate increase in the permitted travel distance. An increase of 50 feet is typical of most occupancies; however, a travel distance increase of 100 feet is permitted for Group B occupancies protected by a sprinkler system.

Code Text: *Where the path of exit access includes unenclosed stairways or ramps within the exit access or includes unenclosed exit ramps or stairways as permitted in Section 1020.1, the distance of travel on such means of egress components shall also be included in the travel distance measurement. The measurement along stairways shall be made on a plane parallel and tangent to the stair tread nosings in the center of the stairway.* See exceptions for 1) open parking garages, 2) outdoor facilities, and 3) stairs permitted to be unenclosed per Exceptions 8 and 9 of Section 1020.1.

Discussion and Commentary: Exit travel distance may include travel on an exit stairway if the stairway is not constructed as a vertical exit enclosure. In such cases, travel up or down the unenclosed (open) stairway would need to be included in the travel distance determination.

TABLE 1016.1
EXIT ACCESS TRAVEL DISTANCE[a]

OCCUPANCY	WITHOUT SPRINKLER SYSTEM (feet)	WITH SPRINKLER SYSTEM (feet)
A, E, F-1, I-1, M, R, S-1	200	250[b]
B	200	300[c]
F-2, S-2, U	300	400[c]
H-1	Not Permitted	75[c]
H-2	Not Permitted	100[c]
H-3	Not Permitted	150[c]
H-4	Not Permitted	175[c]
H-5	Not Permitted	200[c]
I-2, I-3, I-4	150	200[c]

For SI: 1 foot = 304.8 mm.

a. See the following sections for modifications to exit access travel distance requirements:
Section 402: For the distance limitation in malls.
Section 404: For the distance limitation through an atrium space.
Section 1016.2 For increased limitations in Groups F-1 and S-1.
Section 1025.7: For increased limitation in assembly seating.
Section 1025.7: For increased limitation for assembly open-air seating.
Section 1019.2: For buildings with one exit.
Chapter 31: For the limitation in temporary structures.

b. Buildings equipped throughout with an automatic sprinkler system in accordance with Section 903.3.1.1 or 903.3.1.2. See Section 903 for occupancies where automatic sprinkler systems in accordance with Section 903.3.1.2 are permitted.

c. Buildings equipped throughout with an automatic sprinkler system in accordance with Section 903.3.1.1.

As an example, where an unenclosed stairway is provided as a sole means of egress from a mezzanine, the travel distance would be measured from the most remote point on the mezzanine, down the stairway and continue until reaching the door of an exit.

Topic: Corridor Construction

Category: Means of Egress

Reference: IBC 1017.1

Subject: Corridors

Code Text: *Corridors shall be fire-resistance rated in accordance with Table 1017.1. The corridor walls required to be fire-resistance rated shall comply with Section 708 for fire partitions.* See four exceptions where a rating is not required.

Discussion and Commentary: A fire-resistance-rated corridor is intended to protect occupants of the corridor during egress travel from an incident in an enclosed space bordering the corridor. The construction of the corridor provides a minimum level of protection from fire and smoke through the use of fire-resistance-rated walls and ceilings, as well as fire-protected openings. Smoke infiltration is limited also by smoke- and draft-control door assemblies and smoke dampers. Occupancy group, occupant load and presence of a fire sprinkler system are the major factors in determining whether or not a corridor must have a fire-resistance rating.

TABLE 1017.1
CORRIDOR FIRE-RESISTANCE RATING

OCCUPANCY	OCCUPANT LOAD SERVED BY CORRIDOR	REQUIRED FIRE-RESISTANCE RATING (hours)	
		Without sprinkler system	With sprinkler system[c]
H-1, H-2, H-3	All	Not Permitted	1
H-4, H-5	Greater than 30	Not Permitted	1
A, B, E, F, M, S, U	Greater than 30	1	0
R	Greater than 10	Not Permitted	0.5
I-2[a], I-4	All	Not Permitted	0
I-1, I-3	All	Not Permitted	1[b]

a. For requirements for occupancies in Group I-2, see Section 407.3.
b. For a reduction in the fire-resistance rating for occupancies in Group I-3, see Section 408.7.
c. Buildings equipped throughout with an automatic sprinkler system in accordance with Section 903.3.1.1 or 903.3.1.2 where allowed.

Exceptions eliminate the need for a fire-resistance-rated corridor in certain Group E occupancies, in sleeping units or dwelling units of residential occupancies, in open parking garages and in Group B occupancies that are permitted a single means of egress by Section 1015.1.

Topic: Corridor Width

Category: Means of Egress

Reference: IBC 1017.2

Subject: Corridors

Code Text: *The minimum corridor width shall be as determined in Section 1005.1 (calculated width), but not less than 44 inches (1118 mm).* See exceptions requiring a minimum of 1) 24 inches for access to building systems or equipment, 2) 36 inches for occupant loads less than 50, 3) 36 inches with a dwelling unit, 4) 72 inches in a Group E corridor with an occupant load of 100 or more, 5) 72 inches in specified surgical Group I health care centers, and 6) 96 inches in Group I-2 areas where required for bed movement.

Discussion and Commentary: To allow for adequate circulation throughout a building and, more importantly, for egress purposes, a minimum width requirement is established. In addition, complying routes of travel to accessible spaces must be provided. Only in areas used for access to electrical, mechanical or plumbing systems or equipment is the width permitted to be reduced to less than 36 inches. A minimum 36-inch width is mandated for corridors within dwelling units, or for those corridors serving an occupant load of 50 or less. Only specific projections such as doors are permitted to encroach a limited distance into the required corridor width.

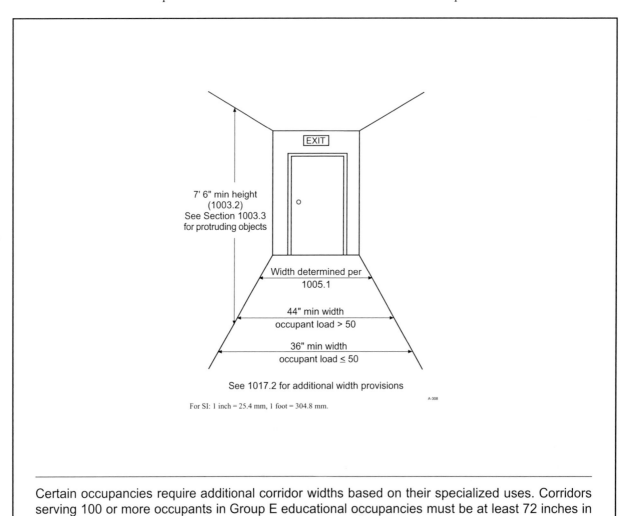

7' 6" min height
(1003.2)
See Section 1003.3
for protruding objects

Width determined per
1005.1

44" min width
occupant load > 50

36" min width
occupant load ≤ 50

See 1017.2 for additional width provisions

For SI: 1 inch = 25.4 mm, 1 foot = 304.8 mm.

Certain occupancies require additional corridor widths based on their specialized uses. Corridors serving 100 or more occupants in Group E educational occupancies must be at least 72 inches in width, and healthcare occupancies require increased widths for bed movement.

Code Text: *Where more than one exit or exit access doorway is required, the exit access shall be arranged such that there are no dead ends in corridors more than 20 feet (6096 mm) in length.* See exceptions for increased dead-end lengths.

Discussion and Commentary: Limitations on dead-end corridors are established where two or more exit or exit access doorways are required. The intent is to limit the distance building occupants must travel before they determine that there is no way out and that they must retrace steps in order to locate an exit or exit access doorway. Where only a single means of egress is permitted, a dead-end condition is not limited in length; however, the provisions of Section 1014.3 for common paths of travel must be considered.

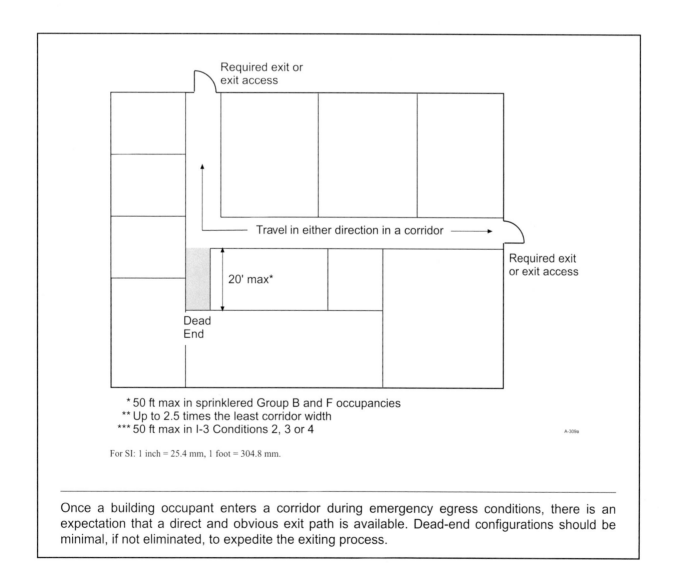

* 50 ft max in sprinklered Group B and F occupancies
** Up to 2.5 times the least corridor width
*** 50 ft max in I-3 Conditions 2, 3 or 4

A-309a

For SI: 1 inch = 25.4 mm, 1 foot = 304.8 mm.

Once a building occupant enters a corridor during emergency egress conditions, there is an expectation that a direct and obvious exit path is available. Dead-end configurations should be minimal, if not eliminated, to expedite the exiting process.

Code Text: *Corridors shall not serve as supply, return, exhaust, relief or ventilation air ducts or plenums. See three exceptions addressing return air and makeup air for exhaust systems. Use of the space between the corridor ceiling and the floor or roof structure above as a return air plenum is permitted for one or more of the following conditions:* See five conditions.

Discussion and Commentary: The use of corridors for the movement of air is strictly limited by the code. Because a corridor is intended to be a relatively safe environment for occupants exiting a building, it is not advisable to introduce air movement that might increase the potential for fire, smoke or toxic gases to enter the corridor. It is possible, under specific conditions, to use the space above a corridor ceiling as a return air plenum. For example, where the corridor is not required to be of fire-resistance-rated construction, or where the above-ceiling space is isolated from a rated corridor by fire-resistance-rated construction, the upper area may be used for return air.

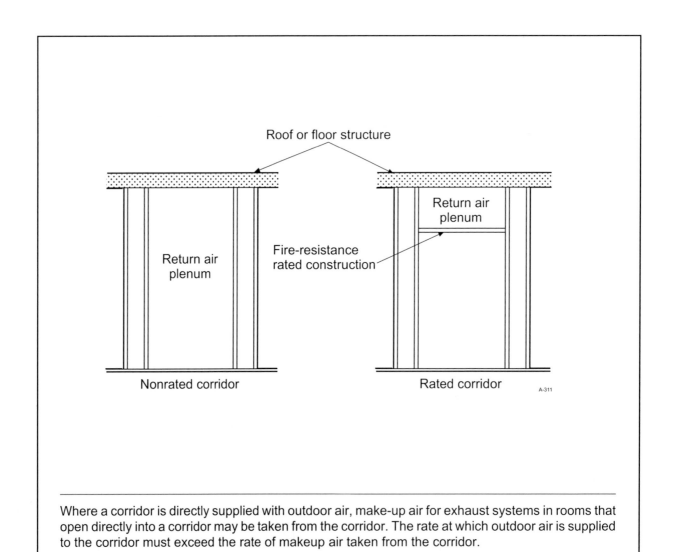

Where a corridor is directly supplied with outdoor air, make-up air for exhaust systems in rooms that open directly into a corridor may be taken from the corridor. The rate at which outdoor air is supplied to the corridor must exceed the rate of makeup air taken from the corridor.

Code Text: *Fire-resistance-rated corridors shall be continuous from the point of entry to an exit, and shall not be interrupted by intervening rooms.* See exceptions for travel through foyers, lobbies and reception rooms.

Discussion and Commentary: Once an occupant enters a corridor required to be of fire-resistance-rated construction, he or she expects that travel to an exit will be direct. The level of protection within the corridor must not be reduced at any point along the egress path. Where an intervening room or space interrupts the corridor, it is quite likely that the exitway will be obstructed or confusing. Where corridor travel includes or terminates at a lobby, foyer or reception room, the condition is considered acceptable, insofar as such spaces are usually an extension of the circulation and egress path.

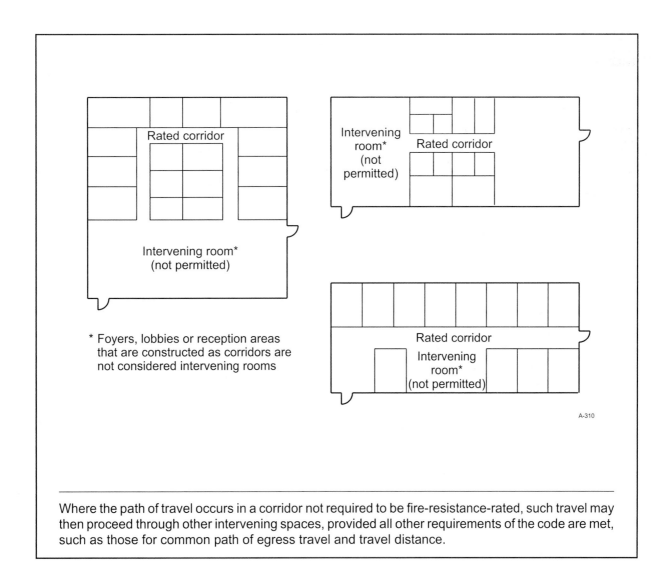

* Foyers, lobbies or reception areas that are constructed as corridors are not considered intervening rooms

Where the path of travel occurs in a corridor not required to be fire-resistance-rated, such travel may then proceed through other intervening spaces, provided all other requirements of the code are met, such as those for common path of egress travel and travel distance.

Quiz

Study Session 11
IBC Sections 1014 – 1017

1. Where the common path of travel is within the permitted limits, a Group B occupancy may have a single means of egress where the maximum occupant load is _____.

   a. 10
   b. 20
   c. 29
   d. 49

   Reference_____

2. In which one of the following occupancies having an occupant load of 40 must at least two exits or exit access doorways always be provided?

   a. A-2
   b. B
   c. F-1
   d. S-1

   Reference_____

3. What is the minimum number of exits required for a Group E occupancy having an occupant load of 1,200?

   a. 2
   b. 3
   c. 4
   d. 6

   Reference_____

4. Where two means of egress are required from a room in a fully sprinklered building, they shall be separated a minimum of _____ of the length of the maximum overall diagonal dimension of the area served.

    a. one-half   b. one-third

    c. one-fourth   d. one-sixth

    Reference_____

5. Which of the following conditions is not specifically required where egress from a room passes through an adjoining room?

    a. a discernable path of travel must be provided

    b. the adjoining room shall be accessory to the area served

    c. no more than one means of egress can pass through the adjoining room

    d. unless of the same occupancy, the adjoining room cannot be classified as high hazard

    Reference_____

6. In general, through which one of the following spaces is egress not specifically prohibited?

    a. private offices   b. store rooms

    c. kitchens   d. closets

    Reference_____

7. In a Group I-2 occupancy, patient sleeping rooms may egress through one intervening room where the intervening room is used as an exit access for a maximum of _____ patient beds.

    a. 2   b. 5

    c. 8   d. 10

    Reference_____

8. At least two exit access doors are required from a Group I-2 suite of rooms, other than patient sleeping rooms, where the minimum suite floor area is _____ square feet.

    a. 1,000   b. 2,500

    c. 5,000   d. 10,000

    Reference_____

9. In a Group B occupancy located in a fully-sprinklered building, the maximum permitted exit access travel distance is _____ feet.

    a. 200   b. 250

    c. 300   d. 400

Reference_____

10. In a nonsprinklered Group A-3 occupancy, what is the maximum permitted exit access travel distance?

    a. 150 feet   b. 200 feet

    c. 250 feet   d. 300 feet

Reference_____

11. Up to an additional 100 feet of travel distance is available where the last portion of exit access travel occurs _____.

    a. within a corridor   b. within a 1-hour-rated corridor

    c. on an exterior egress balcony   d. on an exterior exit stairway

Reference_____

12. In a Group H-3 occupancy, the common path of egress travel is limited to a maximum of _____ feet.

    a. 25   b. 75

    c. 100   d. 150

Reference_____

13. In an office tenant space having an occupant load of 24 persons, the maximum length of a common path of egress travel is _____ feet.

    a. 25   b. 75

    c. 100   d. 200

Reference_____

14. In the public areas of a Group M occupancy, aisles having fixtures or equipment on both sides shall be a minimum of _____ inches in width.

    a. 28   b. 36

    c. 42   d. 44

    Reference_____

15. What is the minimum required width of an aisle serving 15 persons in a nonpublic area of a Group M occupancy?

    a. no minimum width required   b. 28 inches

    c. 36 inches   d. 44 inches

    Reference_____

16. In areas of tables and seating, _____ feet is the maximum length of travel along an aisle accessway to the point where a person has a choice of two or more paths of egress travel to separate exits.

    a. 12   b. 20

    c. 30   d. 50

    Reference_____

17. A corridor serving 125 persons in a Group E occupancy shall be a minimum of _____ inches in width.

    a. 36   b. 44

    c. 60   d. 72

    Reference_____

18. A corridor serving an occupant load of 40 persons shall have a minimum width of _____ inches.

    a. 36   b. 44

    c. 48   d. 60

    Reference_____

19. In a nonsprinklered Group M retail sales building, a corridor serving a minimum occupant load of _____ persons shall be fire-resistance rated.

    a. any number of   b. 11

    c. 31   d. 51

Reference_____

20. In a fully-sprinklered building, a corridor serving 100 persons in which of the following occupancies must be fire-resistance rated?

    a. Group A-2   b. Group I-2

    c. Group M   d. Group R-1

Reference_____

21. In a Group I-1 occupancy, corridors requiring more than one exit or exit access doorway may have dead ends if limited to a maximum of _____ feet in length.

    a. 8   b. 20

    c. 30   d. 50

Reference_____

22. In a fully-sprinklered Group B office building, the maximum permitted length of a dead-end condition in a corridor requiring at least two means of egress is _____.

    a. 20 feet   b. 50 feet

    c. twice the corridor width   d. four times the corridor width

Reference_____

23. Utilization of corridors as return air plenums is permitted within tenant spaces having a maximum floor area of _____ square feet.

    a. 400   b. 1,000

    c. 3,000   d. 5,000

Reference_____

24. What is the minimum required width of an egress balcony serving an apartment building's occupant load of 55 persons?

    a. 36 inches

    b. 42 inches

    c. 44 inches

    d. 48 inches

    Reference_____

25. The long side of an egress balcony shall be a minimum of _____ percent open.

    a. 25

    b. 40

    c. 50

    d. $66^2/_3$

    Reference_____

26. In a Group A-2 dining room having a single exit access door, the common path of egress travel is limited to a maximum of _____ feet.

    a. 25

    b. 75

    c. 100

    d. 200

    Reference_____

27. At least two exits or exit access doors are required from a refrigeration machinery room where the room has a floor area exceeding _____ square feet.

    a. 0, two exits are always required

    b. 200

    c. 500

    d. 1,000

    Reference_____

28. A stairway serving as a means of egress from a catwalk serving a stage shall have a minimum width of _____ inches.

    a. 22

    b. 28

    c. 32

    d. 36

    Reference_____

29. What is the maximum permitted travel distance for a single-story Group F-1 factory provided with automatic heat and smoke roof vents and an automatic sprinkler system?

a. 200 feet

b. 250 feet

c. 300 feet

d. 400 feet

Reference_____

30. A 16-foot-wide dead-end corridor in a Group E high school is prohibited where the dead end is a minimum of _____ feet in length.

a. 20

b. 30

c. 32

d. 40

Reference_____

31. Where egress is permitted through a stockroom of a Group M occupancy, a minimum aisle width of _____ inches is required, defined by full or partial height walls.

a. 30

b. 36

c. 44

d. 48

Reference_____

32. Unless required to be accessible, an aisle accessway in a merchandise pad of a Group M occupancy shall be a minimum of _____ inches in width on at least one side of each merchandising element.

a. 30

b. 36

c. 44

d. 48

Reference_____

33. A minimum of two exit access doorways are required in boiler rooms with a floor area exceeding a minimum of _____ square feet and any fuel-fired equipment has an input capacity greater than 400,000 Btu.

a. 100

b. 400

c. 500

d. 1,000

Reference_____

34. In general, the means of egress from lighting and access catwalks, galleries and gridirons serving stages shall meet the requirements for Group _____ occupancies.

a. A-1

b. B

c. F-2

d. U

Reference _____

35. What is the minimum required corridor width in areas of Group I-2 occupancies where the movement of beds is required?

a. 44 inches

b. 60 inches

c. 72 inches

d. 96 inches

Reference _____

Study Session

12

2006 IBC Sections 1018 – 1026
Means of Egress IV

OBJECTIVE: To obtain an understanding of the provisions governing the exit and exit discharge portions of the means of egress, the special requirements applicable to egress from assembly occupancies, and the details for emergency escape and rescue openings.

REFERENCE: Sections 1018 through 1026, 2006 *International Building Code*

KEY POINTS:
- What is the definition of an exit? What elements of the building are considered exits?
- For which purposes are exits permitted to be used?
- How many exits from a building are required? From any floor level within a building?
- What degree of fire resistance is required for vertical exit enclosures?
- When is an enclosure not required for an exit stairway?
- How must exterior walls of a vertical exit enclosure be protected?
- How is travel in a vertical exit enclosure that extends beyond the discharge level addressed?
- Where are stairway floor number signs to be located?
- What is a smokeproof enclosure? When is a smokeproof enclosure required?
- What is an exit passageway?
- How is an exit passageway regulated for width?
- What level of fire-resistance-rated construction is mandated for an exit passageway?
- Which types of openings and penetrations are permitted in an exit passageway or a vertical exit enclosure? How are openings and penetrations to be protected?
- In what manner can an exit passageway be provided with ventilation?
- What is the function of a horizontal exit? How is it to be constructed?
- How is the capacity of a horizontal exit refuge area determined?
- What is the maximum height permitted for an exterior exit stairway?
- What is the minimum size exterior opening required for an exterior exit stairway?
- Where does the exit discharge begin? Where does it end?
- What is an egress court?

- What is the minimum size of an egress court? When are egress court walls required to be of fire-resistance-rated construction with protected openings?
- What is a safe dispersal area? What conditions must apply where such an area is utilized?
- What is the minimum capacity of the main exit in an assembly occupancy?
- What is smoke-protected assembly seating?
- What minimum aisle widths are required in assembly occupancies without smoke protection? With smoke protection?
- How is the guard height regulated in assembly areas where the guards interfere with occupant sightlines?
- Where are emergency escape and rescue openings required?
- What is the minimum height and width of an emergency escape and rescue opening? Minimum clear opening size? Maximum sill height from floor?

Code Text: An exit is *that portion of a means of egress system which is separated from other interior spaces of a building or structure by fire-resistance-rated construction and opening protectives as required to provide a protected path of egress travel between the exit access and the exit discharge. Exits include exterior exit doors at ground level, exit enclosures, exit passageways, exterior exit stairs, exterior exit ramps and horizontal exits.*

Discussion and Commentary: The path of travel through the exit access portion of the egress system must be designed to lead to one or more exits, which are locations where some degree of protection or safety from fire hazards is provided. Travel distance is no longer regulated once an exit is reached; therefore, travel within an exit component is virtually unlimited.

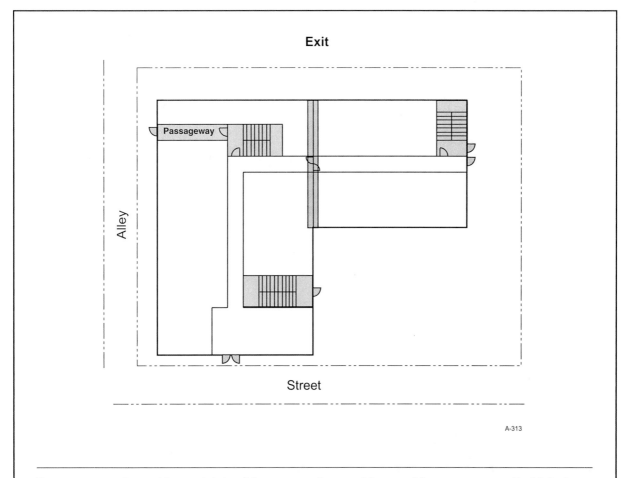

Because an exit must be maintained for egress, it cannot be used for any purpose that interferes with egress. In addition, once a mandated level of protection is provided for occupants reaching an exit, that level cannot be diminished prior to their reaching the exit discharge.

Code Text: *All rooms and spaces within each story shall be provided with and have access to the minimum number of approved independent exits required by Table 1019.1 based on the occupant load of the story, except as modified in Section 1015.1 or 1019.2. For the purposes of Chapter 10, occupied roofs shall be provided with exits as required for stories. The required number of exits from any story, basement or individual space shall be maintained until arrival at grade or the public way.*

Discussion and Commentary: The general provisions call for at least two unique and separate exits from any room, floor area or building. However, in many cases it has been determined that more than one exit provides little, if any, additional protection. Therefore, only one exit is permitted for egress from buildings of limited occupant loads, story heights and travel distances. In all cases, the exit must be continuous to the point where the exit discharge begins.

TABLE 1019.1
MINIMUM NUMBER OF EXITS FOR OCCUPANT LOAD

OCCUPANT LOAD (persons per story)	MINIMUM NUMBER OF EXITS (per story)
1-500	2
501-1,000	3
More than 1,000	4

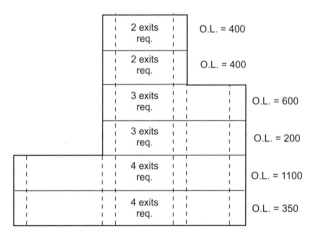

A-314

Until all exit paths reach the exterior at grade level, the minimum required number of exits must be maintained. Once exterior grade is reached, the required number of exit paths may be reduced, provided that the required exit width is maintained until arrival at the public way.

Code Text: *Only one exit shall be required in buildings as described below: 1) buildings described in Table 1019.2, provided that the building has not more than one level below the first story above grade plane, 2) buildings of Group R-3 occupancy, and 3) single-level buildings with the occupied space at the level of exit discharge provided that the story or space complies with Section 1015.1 as a space with one means of egress.*

Discussion and Commentary: Buildings with one exit are permitted where the configuration and occupancy meet certain characteristics which together do not present an unacceptable fire risk to the buildings' occupants. Those structures that are relatively small in size have a shorter travel distance and fewer occupants; thus, having access to a single exit does not significantly compromise the safety of the occupants.

TABLE 1019.2
BUILDINGS WITH ONE EXIT

OCCUPANCY	MAXIMUM HEIGHT OF BUILDING ABOVE GRADE PLANE	MAXIMUM OCCUPANTS (OR DWELLING UNITS) PER FLOOR AND TRAVEL DISTANCE
A, B[d], E[e], F, M, U	1 Story	49 occupants and 75 feet travel distance
H-2, H-3	1 Story	3 occupants and 25 feet travel distance
H-4, H-5, I, R	1 Story	10 occupants and 75 feet travel distance
S[a]	1 Story	29 occupants and 100 feet travel distance
B[b], F, M, S[a]	2 Stories	30 occupants and 75 feet travel distance
R-2	2 Stories[c]	4 dwelling units and 50 feet travel distance

For SI: 1 foot = 304.8 mm.

a. For the required number of exits for parking structures, see Section 1019.1.1.

b. For the required number of exits for air traffic control towers, see Section 412.1.

c. Buildings classified as Group R-2 equipped throughout with an automatic sprinkler system in accordance with Section 903.3.1.1 or 903.3.1.2 and provided with emergency escape and rescue openings in accordance with Section 1026 shall have a maximum height of three stories above grade plane.

d. Buildings equipped throughout with an automatic sprinkler system in accordance with Section 903.3.1.1 with an occupancy in Group B shall have a maximum travel distance of 100 feet.

e. Day care maximum occupant load is 10.

The restriction on the number of levels below the first story is intended to limit the vertical travel an occupant must accomplish to reach the exit discharge in a single-exit building. Above the first story, the limitation is more restrictive, with a single exit available to only five occupancies.

Code Text: *Interior exit stairways and interior exit ramps shall be enclosed with fire barriers constructed in accordance with Section 706 or horizontal assemblies constructed in accordance with Section 711, or both.. Exit enclosures shall have a fire-resistance rating of not less than 2 hours where connecting four stories or more and not less than 1 hour where connecting less than four stories. The number of stories connected by the exit enclosure shall include any basements but not any mezzanines.* See nine exceptions that identify where enclosures are not required.

Discussion and Commentary: Also addressed in Section 707, vertical openings created for stairways must typically be enclosed with fire-resistance-rated construction. Two commonly used exceptions, 8 and 9, permit unenclosed exit stairways between two interconnected floors, though they are limited to no more than 50 percent of the required number of exits. The exception is not applicable to Group H and I occupancies. In fully sprinklered buildings, all stairways serving only the first and second floors may be unenclosed.

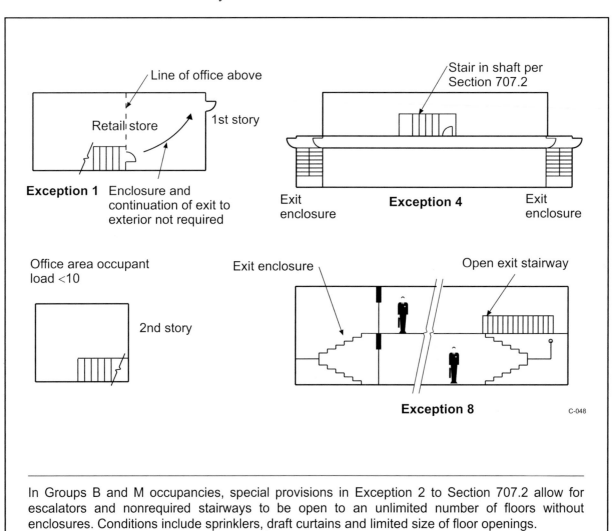

Exception 1 — Enclosure and continuation of exit to exterior not required

Line of office above

Retail store

1st story

Office area occupant load <10

2nd story

Stair in shaft per Section 707.2

Exit enclosure

Exception 4

Exit enclosure

Exit enclosure

Open exit stairway

Exception 8

C-048

In Groups B and M occupancies, special provisions in Exception 2 to Section 707.2 allow for escalators and nonrequired stairways to be open to an unlimited number of floors without enclosures. Conditions include sprinklers, draft curtains and limited size of floor openings.

Code Text: *Where nonrated walls or unprotected openings enclose the exterior of the stairway and the walls or openings are exposed to other parts of the building at an angle of less than 180 degrees (3.14 rad), the building exterior walls within 10 feet (3048 mm) horizontally of a nonrated wall or unprotected opening have a fire-resistance rating of not less than 1 hour. Openings within such exterior walls shall be protected by opening protectives having a fire protection rating of not less than $^3/_4$ hour. This construction shall extend vertically from the ground to a point 10 feet (3048 mm) above the topmost landing of the stairway or to the roof line, which ever is lower.*

Discussion and Commentary: Unless regulated according to fire separation distance or type of construction, the exterior wall of a vertical exit enclosure usually needs no fire-resistance rating. However, where exposure is possible from other exterior walls of the building in close proximity to the exit enclosure, a limited degree of fire separation is necessary.

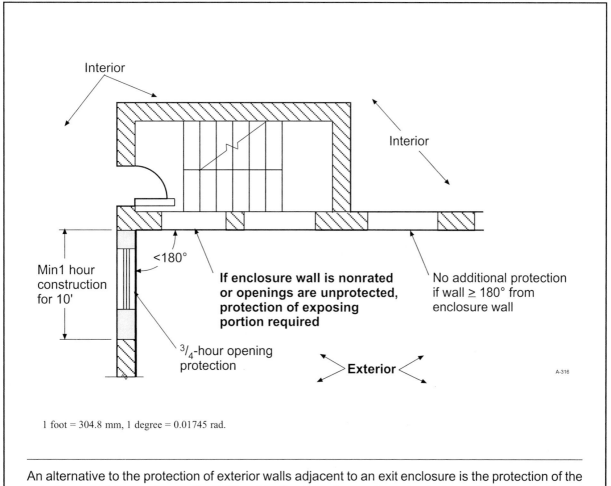

1 foot = 304.8 mm, 1 degree = 0.01745 rad.

An alternative to the protection of exterior walls adjacent to an exit enclosure is the protection of the exterior wall of the enclosure itself. Should a fire breach an adjacent exterior wall, its penetration of the exit enclosure would be halted for an acceptable time period.

Code Text: *In buildings required to comply with Section 403* (High-rise Buildings) *or 405* (Underground Buildings), *each of the exits of a building that serves stories where the floor surface is located more than 75 feet (22 860 mm) above the lowest level of fire department vehicle access or more than 30 feet (9144 mm) below the level of exit discharge serving such floor levels shall be a smokeproof enclosure or pressurized stairway in accordance with Section 909.20.*

Discussion and Commentary: In those buildings where vertical egress travel is extensive, an additional level of protection is mandated, primarily to address the hazard of smoke and toxic gases that are produced in a fire. Through ventilation or pressurization, the potential for smoke and gases to enter the enclosure is dramatically reduced.

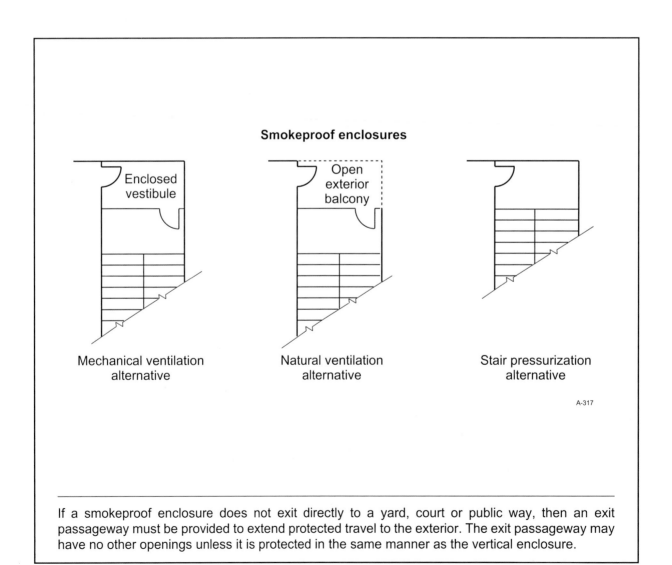

Smokeproof enclosures

Enclosed vestibule

Mechanical ventilation alternative

Open exterior balcony

Natural ventilation alternative

Stair pressurization alternative

A-317

If a smokeproof enclosure does not exit directly to a yard, court or public way, then an exit passageway must be provided to extend protected travel to the exterior. The exit passageway may have no other openings unless it is protected in the same manner as the vertical enclosure.

Code Text: *An exit passageway shall not be used for any purpose other than a means of egress. Exit passageway enclosures shall have walls, floor and ceiling of not less than 1-hour fire-resistance rating, and not less than that required for any connecting exit enclosure. Exit passageways shall be constructed as fire barriers in accordance with Section 706.*

Discussion and Commentary: An exit passageway is defined as an exit component that is separated from all other interior spaces of a building or structure by fire-resistance-rated construction and opening protectives, and provides for a protected path of egress travel in a horizontal direction to the exit discharge or the public way. It is an egress component of a higher level than a fire-resistance-rated corridor, based primarily on limited openings and penetrations, and increased fire ratings.

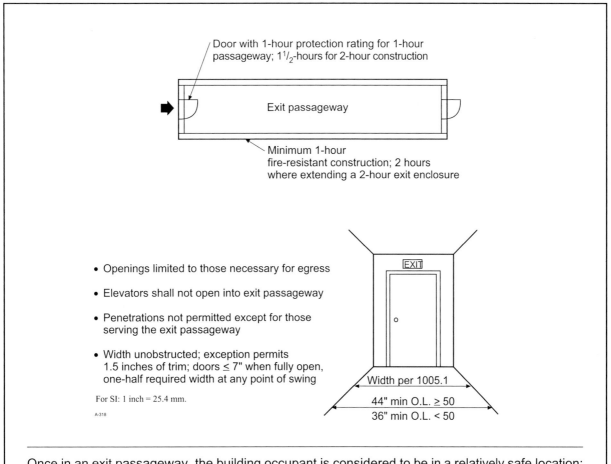

Door with 1-hour protection rating for 1-hour passageway; 1¹/₂-hours for 2-hour construction

Exit passageway

Minimum 1-hour
fire-resistant construction; 2 hours
where extending a 2-hour exit enclosure

- Openings limited to those necessary for egress

- Elevators shall not open into exit passageway

- Penetrations not permitted except for those serving the exit passageway

- Width unobstructed; exception permits 1.5 inches of trim; doors ≤ 7" when fully open, one-half required width at any point of swing

For SI: 1 inch = 25.4 mm.

A-318

EXIT

Width per 1005.1

44" min O.L. ≥ 50
36" min O.L. < 50

Once in an exit passageway, the building occupant is considered to be in a relatively safe location; thus, travel distances within the exit passageway are unregulated. Simply put, an exit passageway is a horizontal exit enclosure, with conditions and limitations similar to those of a vertical exit enclosure.

Code Text: *Except as permitted in Section 402.4.6 (covered mall buildings), openings in exit enclosures (and exit passageways) other than unprotected exterior openings shall be limited to those necessary for exit access to the enclosure from normally occupied spaces and for egress from the enclosure (or exit passageway). Penetrations into and openings through an exit enclosure are prohibited except for required exit doors, equipment and duct work necessary for independent pressurization, sprinkler piping, standpipes, electrical raceway serving the fire department communication and electrical conduit serving the exit enclosure (or exit passageway) and terminating at a steel box not exceeding 16 square inches (0.010 m^2).*

Discussion and Commentary: Given the importance of an exit enclosure in the means of egress system, no unnecessary openings or penetrations are permitted to breach the fire-resistant separation. The provisions are essentially the same for both vertical exit enclosures and exit passageways.

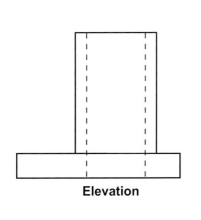

Elevation

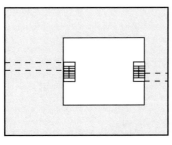

Plan

Enclosure construction:

- Four or more stories—2-hour fire resistance
- Less than four stories—1-hour fire resistance

Openings and penetrations:

- Permitted exterior openings (704)
- Egress from normally occupied spaces
- Egress from enclosure
- Standpipe and sprinklers
- Ductwork for independent pressurization
- Limited electrical conduit

Doors: (715)

- Self-closing or automatic closing
- 1-hour rating in 1-hour construction
- 1$^1/_2$-hour rating in 2-hour construction
- Temperature rise limit of 450°F above ambient

A-319

Several methods are set forth in the code to provide for ventilation of a vertical exit enclosure. In general, penetrations for ductwork must enter directly from the building's exterior or from an interior space separated from the remainder of the building by a shaft enclosure.

Code Text: *A horizontal exit shall not serve as the only exit from a portion of a building, and where two or more exits are required, not more than one-half of the total number of exits or total width of exit width shall be horizontal exits.* See exceptions for Group I-2 and I-3 occupancies. *The refuge area of a horizontal exit shall be a space occupied by the same tenant or public areas and each such area of refuge shall be adequate to house the original occupant load of the refuge space plus the occupant load anticipated from the adjoining compartment.*

Discussion and Commentary: A horizontal exit is defined as a path of egress travel from one building to an area in another building on approximately the same level, or a path of egress travel through or around a wall or partition to an area on approximately the same level in the same building, which affords safety from fire and smoke from the area of incidence and areas communicating therewith. Constructed as a fire wall or a minimum 2-hour fire barrier, a horizontal exit is an exit component of the means of egress system.

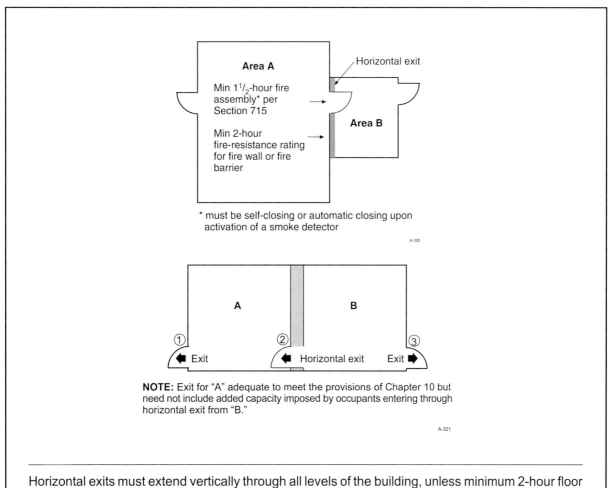

Horizontal exits must extend vertically through all levels of the building, unless minimum 2-hour floor assemblies with no unprotected openings are provided. The horizontal exit walls are to extend continuously from exterior wall to exterior wall in order to completely divide the floor.

Code Text: *Exterior exit ramps and stairways serving as an element of a required means of egress shall be open on at least one side. An open side shall have a minimum of 35 square feet (3.3 m²) of aggregate open area adjacent to each floor level and the level of each intermediate landing. The required open area shall be located not less than 42 inches (1067 mm) above the adjacent floor or landing level.*

Discussion and Commentary: For a stairway or ramp to be considered exterior, it must be open enough to the outside so that smoke and toxic gases will not tend to corrupt the exit route. An exterior exit ramp or exterior exit stairway is considered an exit component and is permitted as an egress element in all occupancies except Group I-2. Where permitted as an element of a required means of egress, an exterior exit stairway is limited to 6 stories and to 75 feet in height.

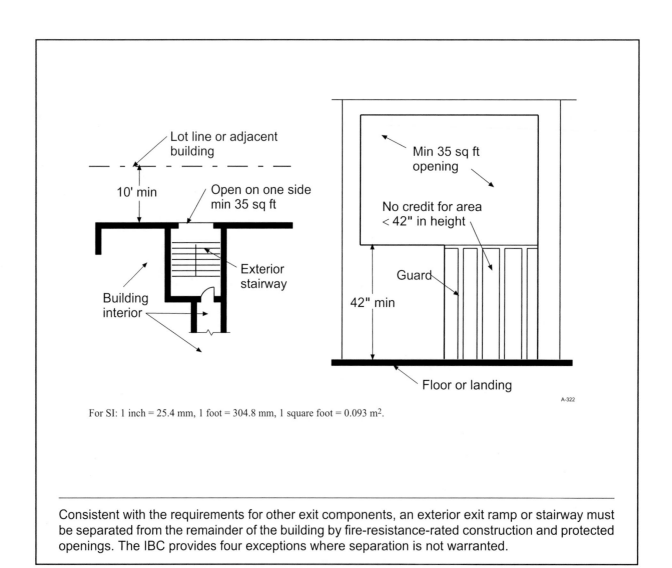

For SI: 1 inch = 25.4 mm, 1 foot = 304.8 mm, 1 square foot = 0.093 m².

Consistent with the requirements for other exit components, an exterior exit ramp or stairway must be separated from the remainder of the building by fire-resistance-rated construction and protected openings. The IBC provides four exceptions where separation is not warranted.

Code Text: The exit discharge is *that portion of a means of egress system between the termination of an exit and a public way. Exits shall discharge directly to the exterior of the building. See exceptions for discharge level spaces and vestibules. The exit discharge shall be at grade or shall provide direct access to grade. The exit discharge shall not reenter a building.* See three exceptions.

Discussion and Commentary: Exit discharge travel typically takes place outside of the building, where hazards to the occupants are greatly reduced. Although concern over the accumulation of smoke and toxic gases is eliminated, there is still a risk to the occupants, which is eliminated only at a point of considerable distance from the structure, generally the public way.

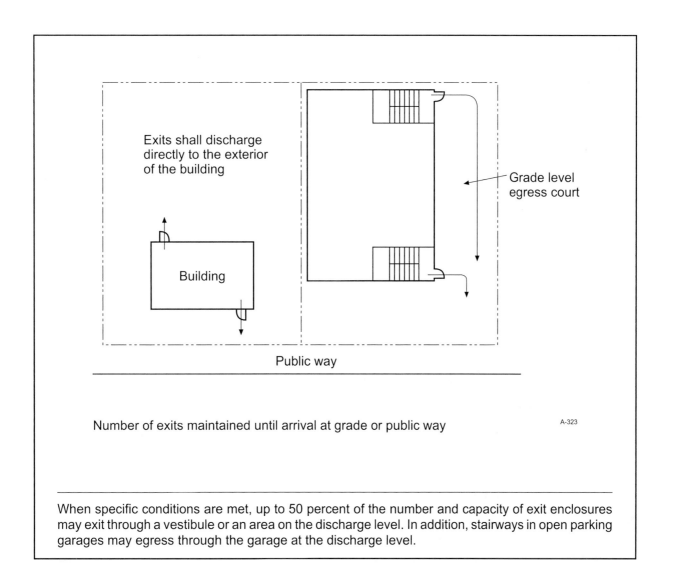

Exits shall discharge directly to the exterior of the building

Building

Grade level egress court

Public way

Number of exits maintained until arrival at grade or public way

A-323

When specific conditions are met, up to 50 percent of the number and capacity of exit enclosures may exit through a vestibule or an area on the discharge level. In addition, stairways in open parking garages may egress through the garage at the discharge level.

Code Text: *Where an egress court serving a building or portion thereof is less than 10 feet (3048 mm) in width, the egress court walls shall be not less than 1-hour fire-resistance-rated construction for a distance of 10 feet (3048 mm) above the floor of the court. Openings within such walls shall be protected by opening protectives having a fire protection rating of not less than $^3/_4$ hour.* See exceptions for small occupant loads and dwellings.

Discussion and Commentary: An egress court is defined as a court or yard which provides access to a public way for one or more exits. Because an egress court is an element of the exit discharge, occupants must be afforded sufficient protection from a fire within the structure to be reasonably sure that once outside they will reach the safety of a public way.

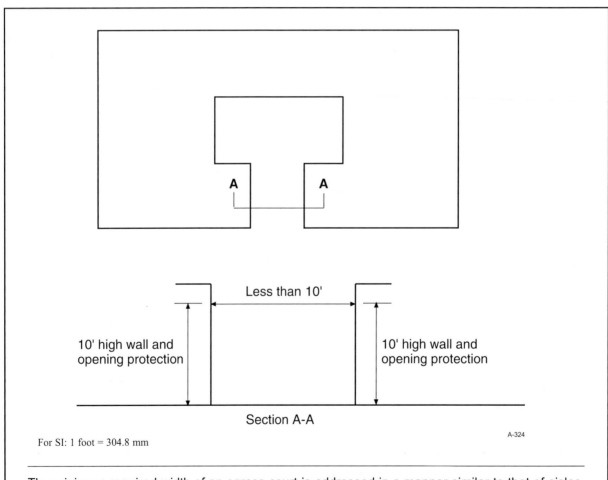

Section A-A

For SI: 1 foot = 304.8 mm

A-324

The minimum required width of an egress court is addressed in a manner similar to that of aisles, corridors and stairways. The width must accommodate the calculated capacity, based on occupant load served; however, in no case may it be less than a specified width of 44" (36" in Group R-3).

Topic: Access to a Public Way

Category: Means of Egress

Reference: IBC 1024.6

Subject: Exit Discharge

Code Text: *The exit discharge shall provide a direct and unobstructed access to a public way.* See exception for designation of a safe dispersal area where access to a public way cannot be provided.

Discussion and Commentary: The means of egress is not complete until the occupants of the building have reached a safe place. Such a place is typically a public way, in that it is relatively unobstructed, and more importantly, continuous. It is always possible to continue egress travel along a public way until the necessary level of safety is achieved. The path to reach the public way, as for all other components of the means of egress, must be free of obstructions and other concerns that would cause the occupants' egress travel to be delayed, restricted or unavailable. It is important that the travel path outside of the building be continuously maintained in order to keep the means of egress in a complying condition.

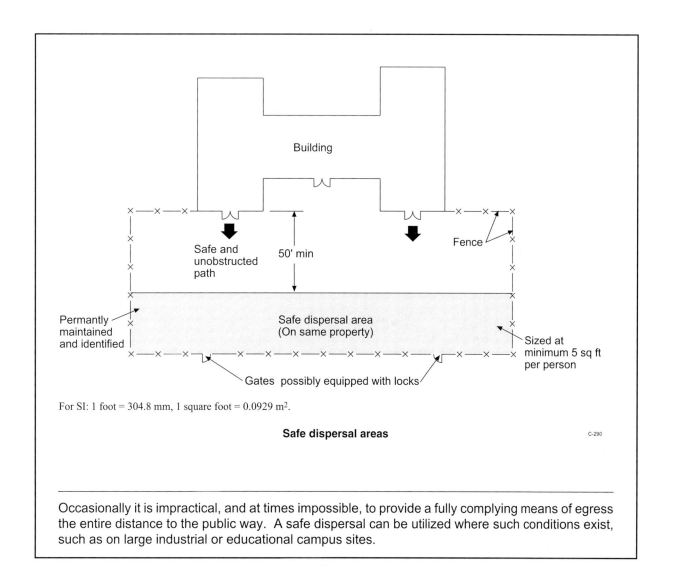

For SI: 1 foot = 304.8 mm, 1 square foot = 0.0929 m².

Safe dispersal areas

C-290

Occasionally it is impractical, and at times impossible, to provide a fully complying means of egress the entire distance to the public way. A safe dispersal can be utilized where such conditions exist, such as on large industrial or educational campus sites.

Topic: Assembly Main Exit

Reference: IBC 1025.2

Category: Means of Egress

Subject: Assembly Seating

Code Text: *Group A occupancies that have an occupant load of greater than 300 shall be provided with a main exit. The main exit shall be of sufficient width to accommodate not less than one-half of the occupant load, but such width shall not be less than the total required width of all means of egress leading to the exit.*

Discussion and Commentary: In most assembly-type uses, the occupants tend to enter the room or space at a single location. It is expected that under emergency conditions, most of the occupants will attempt to exit at the same point. Therefore, the main entrance/exit must be wide enough to handle a sizeable percentage of the occupants. If there are multiple main entrance/exits, the exit width can be distributed among the exits around the perimeter of the building.

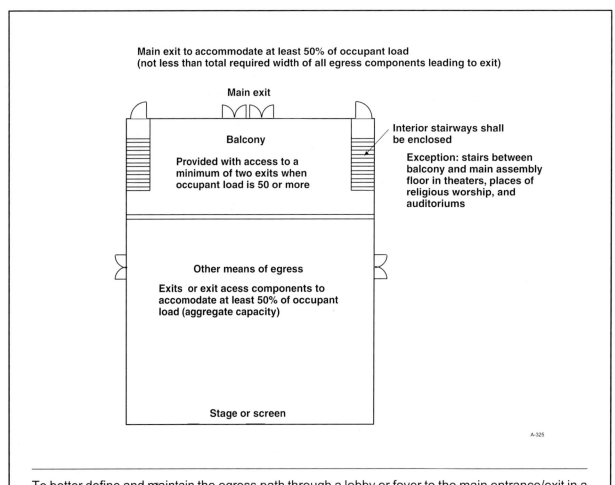

To better define and maintain the egress path through a lobby or foyer to the main entrance/exit in a Group A-1 occupancy, the code mandates that the waiting area be separated from the required egress width. Where seats are not available, the use of railings or partitions is specified.

Topic: Assembly Aisle Walking Surfaces   **Category:** Means of Egress
Reference: IBC 1025.11   **Subject:** Assembly Seating

Code Text: *Aisles with a slope not exceeding one unit vertical in eight units horizontal (12.5-percent slope) shall consist of a ramp having a slip-resistant walking surface. Aisles with a slope exceeding one unit vertical in eight units horizontal (12.5-percent slope) shall consist of a series of risers and treads that extends across the full width of aisles and complies with Sections 1025.11.1 through 1025.11.3. Tread depths shall be a minimum of 11 inches (279 mm) and shall have dimensional uniformity.* See exception allowing a 0.188 inch tolerance between adjacent treads. *Where the gradient of aisle stairs is to be the same as the gradient of adjoining seating areas, the riser height shall not be less than 4 inches (102 mm) nor more than 8 inches (203 mm) and shall be uniform within each flight.* See exceptions allowing riser height nonuniformity and maximum riser heights of 9 inches where needed to maintain sightlines.

Discussion and Commentary: Where aisles have a relatively flat slope, 1:8 or flatter, steps in the aisle are prohibited because occupants in low-slope aisles have a tendency to not notice steps as readily as they would in steeper aisles. Continuous surfaces provide a safer path of travel.

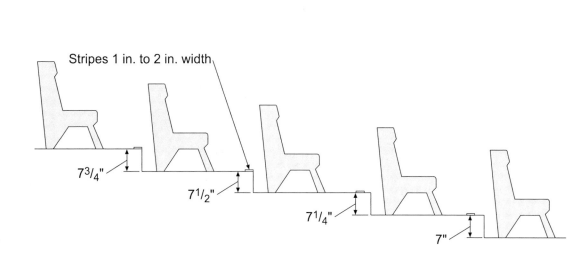

Stripes 1 in. to 2 in. width

7³/₄"

7¹/₂"

7¹/₄"

7"

Risers 4 in. min, 8 in. max
Variation not to exceed 0.188 in. unless striped

For SI: 1 inch = 25.4 mm.

One provision that helps the user of an aisle notice a step requires that a contrasting strip or other approved marking be placed on the leading edge of each tread. Designed to identify the edge of the tread, this making strip may be omitted where it can be shown that the location of each tread is readily apparent when viewed in descent.

Code Text: *Ramped aisles having a slope exceeding one unit vertical in 15 units horizontal (6.7-percent slope) and aisle stairs shall be provided with handrails located either at the side or within the aisle width.* See exceptions for 1) ramped aisles with a slope no greater than 1:8 with seating on both sides, and 2) guards that comply with the graspability requirements of handrails. *Where there is seating on both sides of the aisle, the handrails shall be discontinuous with gaps or breaks at intervals not exceeding five rows to facilitate access to seating and to permit crossing from one side of the aisle to the other.*

Discussion and Commentary: Where seating is located on both sides of an aisle, the required handrails may be placed on either side of, or down the center of, the aisle served and are permitted to project into the required width no more than $4^{1}/_{2}$ inches.

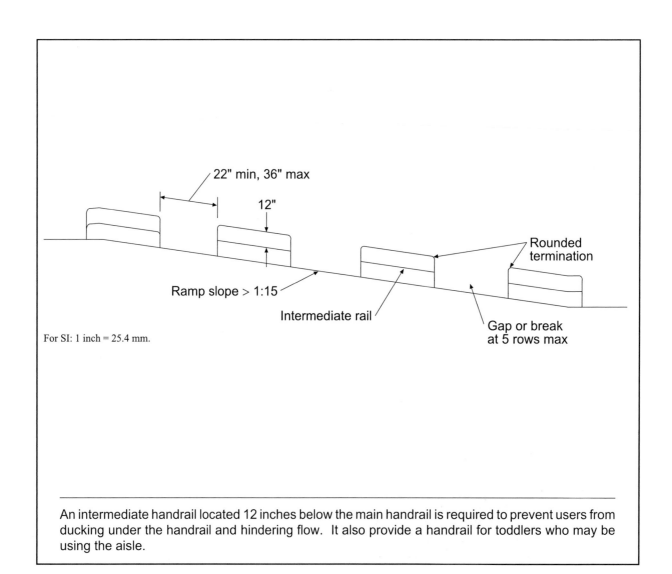

For SI: 1 inch = 25.4 mm.

An intermediate handrail located 12 inches below the main handrail is required to prevent users from ducking under the handrail and hindering flow. It also provide a handrail for toddlers who may be using the aisle.

Topic: Required Openings   **Category:** Means of Egress

Reference: IBC 1026   **Subject:** Emergency Escape and Rescue

Code Text: *In addition to the means of egress required by Chapter 10, provisions shall be made for emergency escape and rescue in Group R and Group I-1 occupancies. Basements and sleeping rooms below the fourth story above grade plane shall have at least one exterior emergency escape and rescue opening in accordance with Section 1025.* See seven exceptions, including one for sprinklered buildings and one for buildings with a complying corridor. *Such openings shall open directly into a public way or a yard or court that opens to a public way.*

Discussion and Commentary: In those occupancies where persons are sometimes sleeping, a fire will often spread quickly and block the normal egress routes. By requiring a sizeable opening directly from the sleeping room to the exterior, rescue can be more easily accomplished, or alternatively, the occupants may escape without having to travel through the building. In other than Group R-3 occupancies, such openings are not required if the building is fully sprinklered.

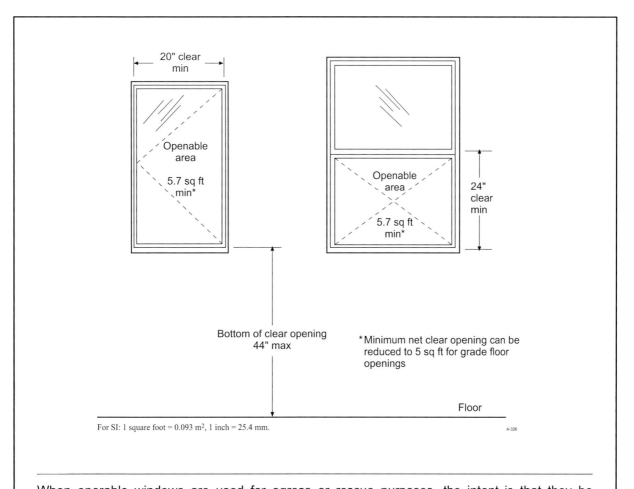

For SI: 1 square foot = 0.093 m², 1 inch = 25.4 mm.

A-326

When operable windows are used for egress or rescue purposes, the intent is that they be double-hung, horizontal sliding or casement styles operated by a simple operation. Special types other than those listed must be evaluated for compliance with the operational constraint limitations.

Code Text: *An emergency escape and rescue opening with a finished sill height below the adjacent ground level shall be provided with a window well in accordance with Section 1026.5.1 and 1026.5.2. The minimum horizontal area of the window well shall be 9 square feet (0.84 m²), with a minimum dimension of 36 inches (914 mm). The area of the window well shall allow the emergency escape and rescue opening to be fully opened. Window wells with a vertical depth of more than 44 inches (1118 mm) shall be equipped with an approved permanently affixed ladder or steps.*

Discussion and Commentary: It is important that persons who travel through an emergency escape and rescue opening located below grade be provided with adequate space in the window well to allow for escape or rescue from the area. Therefore, a complying window well is mandated that provides an adequate cross-sectional area for escape and rescue operations.

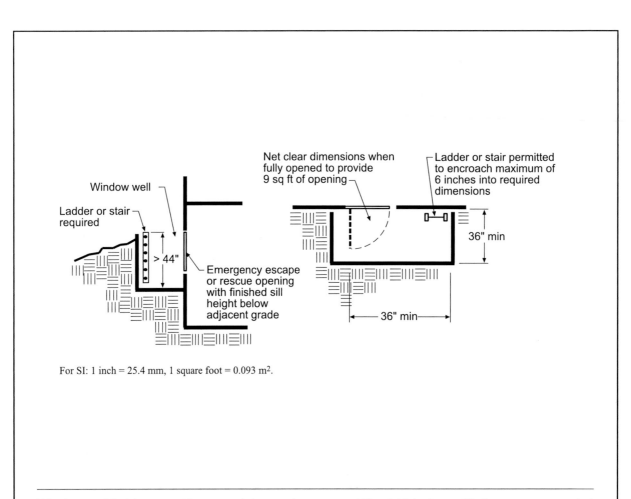

For SI: 1 inch = 25.4 mm, 1 square foot = 0.093 m².

Window well ladders must have a minimum clear rung width of 12 inches with the rungs spaced at maximum 18-inch intervals vertically. The ladder or steps cannot encroach into the required dimensions of the window well more than 6 inches.

Quiz

Study Session 12
IBC Sections 1018 – 1026

1. A building having an occupant load of 720 persons shall be provided with a minimum of _____ approved independent exits.

   a. two   b. three

   c. four   d. five

   Reference_____

2. A single exit is permitted from what type of parking garage?

   a. open   b. enclosed

   c. commercial trucks and buses   d. vehicles are mechanically parked

   Reference_____

3. A single exit from the second floor of a nonspirinklered Group B office building is permitted if the maximum occupant load is _____ persons and the maximum travel distance is _____ feet.

   a. 10, 75   b. 30, 75

   c. 29, 200   d. 49, 200

   Reference_____

4. Vertical exit enclosures a minimum of _____ stories in height shall be 2-hour fire-resistance rated.

    a. three   b. four

    c. five   d. six

Reference_____

5. In other than Group H and I occupancies, a stairway serving a maximum occupant load of _____ persons need not be enclosed where located not more than one story above the level of exit discharge.

    a. 9   b. 29

    c. 30   d. 49

Reference_____

6. Where a smokeproof enclosure exits into an exit passageway, the exit passageway shall be separated from the remainder of the building by minimum _____-hour fire-resistance-rated construction.

    a. $^3/_4$ b. 1

    c. 2   d. 3

Reference_____

7. Stairway floor number signs shall be provided in interior vertical exit enclosures connecting a minimum of _____ stories.

    a. 2   b. 3

    c. 4   d. 6

Reference_____

8. Smokeproof enclosures are not required for exits serving stories a maximum of _____ feet above the lowest level of fire department vehicle access.

    a. 30   b. 55

    c. 75   d. 160

Reference_____

9. Exit passageways shall be constructed as minimum _____.

    a. 1-hour fire partitions   b. 1-hour fire barriers

    c. 2-hour fire barriers   d. 2-hour fire walls

Reference_____

10. In other than Group I occupancies, a refuge area serving a horizontal exit shall be sized for capacity based on _____ square feet for each occupant to be accommodated.

    a. 3   b. 6

    c. 15   d. 30

Reference_____

11. In other than a Group I-2 occupancy, exterior exit stairways are permitted as means of egress elements for buildings a maximum of _____ stories above grade plane.

    a. three   b. four

    c. five   d. six

Reference_____

12. An exterior exit stairway serving as an element of a required means of egress must be open on at least one side, with such open area a minimum of _____ square feet and located a minimum of _____ inches above the floor or landing level.

    a. 9, 36   b. 16, 42

    c. 20, 36   d. 35, 42

Reference_____

13. Egress balconies and exterior exit stairways shall be located a minimum of _____ feet from adjacent lot lines.

    a. 3   b. 5

    c. 10   d. 20

Reference_____

14. What is the minimum required width of an egress court serving 40 occupants from an office building?

    a. 36 inches   b. 44 inches

    c. 48 inches   d. 60 inches

Reference_____

15. Exterior walls enclosing an egress court shall be of minimum 1-hour fire-resistance-rated construction to a minimum height of _____ where the egress court is less than 10 feet in width.

    a. 6 feet, 8 inches   b. 7 feet, 0 inches

    c 8 feet, 0 inches   d. 10 feet, 0 inches

Reference_____

16. A main exit shall be provided for all Group A occupancies having a minimum occupant load of _____ persons.

    a. 51   b. 301

    c. 501   d. 1,001

Reference_____

17. At least two means of egress shall be provided from assembly seating areas in balconies having a minimum seating capacity of _____ persons.

    a. 10   b. 30

    c. 50   d. 100

Reference_____

18. In assembly areas not provided with smoke protection, the minimum width for stairs having 8-inch risers is based on a minimum of _____ inch of width for each occupant served.

    a. 0.2   b. 0.22

    c. 0.3   d. 0.35

Reference_____

19. In a 15,000-seat arena provided with smoke-protected assembly seating, the minimum width for a 1:12 ramped aisle is based on a minimum of _____ inch of width for each occupant served.

   a. 0.300   b. 0.200

   c. 0.120   d. 0.070

   Reference_____

20. The smoke level in a means of egress serving a smoke-protected assembly seating area is intended to be located a minimum of _____ feet above the floor of the means of egress.

   a. 6   b. 8

   c. 12   d. 15

   Reference_____

21. What is the maximum travel distance for seating in open-air assembly structures of Type I or II construction?

   a. 200 feet   b. 300 feet

   c. 400 feet   d. unlimited

   Reference_____

22. What is the minimum required clear width of aisle stairs having seating on only one side?

   a. 32 inches   b. 36 inches

   c. 42 inches   d. 44 inches

   Reference_____

23. In aisle stairs, the maximum tolerance permitted between adjacent treads is _____ inch.

   a. 0.125   b. 0.188

   c. 0.25   d. 0.375

   Reference_____

24. Where the foot of the aisle is more than 30 inches above the floor or grade below, a railing or fascia system at the foot of aisles in an assembly occupancy shall be a minimum of _____ in height.

    a. 26 inches   b. 32 inches

    c. 36 inches   d. 42 inches

Reference_____

25. What is the minimum net clear opening required for emergency escape and rescue grade floor openings?

    a. 20 inches by 22 inches   b. 20 inches by 24 inches

    c. 5.0 square feet   d. 5.7 square feet

Reference_____

26. Where horizontal exits are used in the means of egress in a Group I-2 hospital, the horizontal exits can be used for a maximum of _____ of the required exits.

    a. one   b. two

    c. one-half   d. two-thirds

Reference_____

27. Along with other limitations, the exit discharge may occur through a vestibule, provided the vestibule has a maximum depth from the exterior of the building of _____ feet and a maximum length of _____ feet.

    a. 8, 15   b. 10, 20

    c. 10, 30   d. 15, 30

Reference_____

28. Egress need not extend to a public way where a safe dispersal area is provided a minimum of _____ feet from the building.

    a. 30   b. 50

    c. 75   d. 100

Reference_____

29. An emergency escape and rescue opening shall be located so that the bottom of the clear opening is a maximum of _____ inches above the floor surface.

    a. 36   b. 42

    c. 44   d. 48

Reference_____

30. Where a window well is provided to serve an emergency escape and rescue opening, it shall have a minimum horizontal area of _____ square feet.

    a. 5.0   b. 5.7

    c. 9.0   d. 10.0

Reference_____

31. The separation between refuge areas connected by a horizontal exit will be provided by a fire wall or fire barrier having a minimum fire-resistance rating of _____ hour(s).

    a. 1   b. 2

    c. 3   d. 4

Reference _____

32. Where seating rows in an assembly seating area have 14 or fewer seats, the clear aisle accessway width will be a minimum of _____ inches measured horizontally from the back of the row ahead to the nearest projection of the row behind.

    a. 12   b. 16

    c. 19   d. 22

Reference _____

33. Where there is assembly seating on both sides of an aisle, the handrails shall be discontinuous with gaps or breaks a minimum of _____ inches and a maximum of _____ inches in width, measured horizontally.

    a. 16, 32   b. 20, 34

    c. 22, 36   d. 27, 42

Reference _____

34. An emergency escape and rescue opening is not required from a basement without habitable spaces, provided the floor area of the basement is a maximum of _____ square feet.

    a. 120   b. 200

    c. 240   d. 400

Reference _____

35. An emergency escape and rescue opening shall be a minimum of _____ inches in net clear opening height and a minimum of _____ inches in net clear opening width.

    a. 24, 20   b. 24, 22

    c. 20, 26   d. 20, 28

Reference _____

13

2006 IBC Chapter 11
Accessibility

OBJECTIVE: To become familiar with the scoping provisions relating to the design and construction of accessible buildings, facilities and elements.

REFERENCE: Chapter 11, 2006 *International Building Code*

KEY POINTS:
- What is the scope of the provisions regulating accessibility and usability?
- How does ICC A117.1 relate to the IBC?
- Are temporary buildings regulated for accessibility?
- How are employee work areas viewed for accessibility?
- Which types of uses are not required to be accessible?
- Where within employee work areas is an accessible route not required?
- What is an accessible route? Which site elements must be connected by an accessible route to an accessible building entrance?
- Which areas within a building must be connected by an accessible route?
- Under which conditions is a vertical accessible route not required?
- How should the accessible route be provided in relationship to the general circulation path?
- How many entrances to a building must be accessible? To a tenant space within a building?
- What percentage of parking spaces must be designed as accessible?
- When is a van-accessible space required?
- Where must accessible parking spaces be located?
- When is an accessible loading zone required?
- What is a Type A dwelling unit? Type B unit? In which types of buildings are such units located?
- Where dwelling units and sleeping units are required to be accessible, what accessible features are required in each Accessible unit, Type A unit and Type B unit?
- In a theater, auditorium or similar assembly area, how is the minimum number of required wheelchair spaces determined?
- How shall wheelchair spaces be distributed throughout a multilevel assembly facility?

- What is the function of an assistive listening system? Under which conditions are assistive listening systems required?
- Which types of dining areas must be accessible?
- What specific areas of a judicial facility must be accessible?
- What percentage of toilet rooms are required to be accessible? Bathing facilities?
- What are unisex toilet rooms and bathing rooms? When are such rooms required?
- What are the features required in a unisex toilet room? Bathing room?
- Where drinking fountains are provided, how many must be accessible? Storage lockers? Fitting rooms? Check-out aisles?
- Under what conditions is a platform (wheelchair) lift permitted to be a part of a required accessible route?
- Where are detectable warnings required?
- What type of operating mechanisms or controls are regulated for usability?
- When must a stairway be considered an accessible element?
- What is the International Symbol of Accessibility? Where are such signs required?
- Where is directional signage mandated?

| **Topic:** Scope | **Category:** Accessibility |
| **Reference:** IBC 1101 | **Subject:** General Provisions |

Code Text: *The provisions of Chapter 11 shall control the design and construction of facilities for accessibility to physically disabled persons. Buildings and facilities shall be designed and constructed to be accessible in accordance with* the IBC *and ICC A117.1.*

Discussion and Commentary: Chapter 11 of the *International Building Code* sets forth the scoping provisions that identify where and to what degree access must be provided. Once it has been determined that accessible elements are required, the ICC design standard *Accessible and Usable Buildings and Facilities* sets forth the specific technical criteria. As with any other provision of the code, alternative designs, products or technologies that provide equivalent or superior compliance may be accepted by the building official.

Scope
The provisions of Chapter 11 shall control the design and construction of facilities for accessibility to physically disabled persons.

Design
Buildings and facilities shall be designed and constructed in accordance with the IBC and the ICC A117.1-2003 standard.

A-327

Although space requirements can vary greatly depending on the nature of the disability and the physical functions of the individual, it is generally accepted that spaces designed to accommodate persons using wheelchairs will be functional for most people.

Code Text: *Sites, buildings, structures, facilities, elements and spaces, temporary or permanent, shall be accessible to persons with physical disabilities.* See 15 general exceptions.

Discussion and Commentary: In general, all portions of all buildings are to be provided with elements that will make them fully accessible to individuals with disabilities. There are, however, a number of general exceptions that reduce or eliminate accessibility requirements. Specific areas that are not required to be accessible include: individual employee work areas; detached dwellings and their accessory structures; construction sites; raised security or safety areas, such as observation galleries or fire towers; nonoccupiable spaces and equipment spaces, including elevator pits and transformer vaults; and single-occupant structures accessed at other than grade, such as toll booths.

Specific requirements. Where not required per Sections 1104 through 1110.

Existing buildings. Existing buildings shall comply with Section 3409.

Employee work areas. Need only comply with fire alarm, accessible means of egress and common use circulation path provisions. Must be able to approach, enter and exit the work area.

Detached dwellings. Detached one- and two-family dwellings and accessory structures, and their associated sites and facilities.

Utility buildings. Group U are exempt except:
1. In agricultural buildings, access is required to paved work areas and areas open to the general public.
2. Private garages or carports that contain required accessible parking.

Construction sites. Structures, sites and equipment directly associated with the actual processes of construction.

Raised areas. Raised areas used primarily for purpose of security, life safety or fire safety.

Limited access spaces. Nonoccupiable spaces accessed only by ladders, catwalks, crawl spaces, freight elevators, very narrow passageways or tunnels.

Equipment spaces. Spaces frequented only by personnel for maintenance, repair or monitoring of equipment.

Single occupant structures. Single occupant structures accessed only by passageways below grade or elevated above grade.

Residential Group R-1. Buildings of Group R-1 containing not more than five sleeping units for rent or hire which are also occupied as the residence of the proprietor.

Day care facilities. Where part of a dwelling unit.

Detention and correctional facilities. Common use areas not serving accessible cells.

Fuel-dispensing systems. Fuel-dispensing devices.

Walk-in coolers and freezers. Coolers and freezers intended for employee use only.

Other than those residential occupancies exempt from the accessibility provisions, most buildings will require some level of accessibility. Only those specific areas identified by the code are exempt, whereas the remainder of the structure is regulated for complying accessibility and usability.

Topic: Connected Spaces   **Category:** Accessibility
Reference: IBC 1104.3   **Subject:** Accessible Route

Code Text: *When a building, or portion of a building, is required to be accessible, an accessible route shall be provided to each portion of the building, to accessible building entrances connecting accessible pedestrian walkways and the public way.* See exceptions for fixed-seating assembly areas.

Discussion and Commentary: An accessible route is defined as a continuous, unobstructed path that complies with Chapter 11. It includes corridors, aisles, ramps, elevators, platform (wheelchair) lifts and clear floor space at fixtures. Chapter 4 of ICC A117.1 addresses the design and construction specifications for the elements of an accessible route.

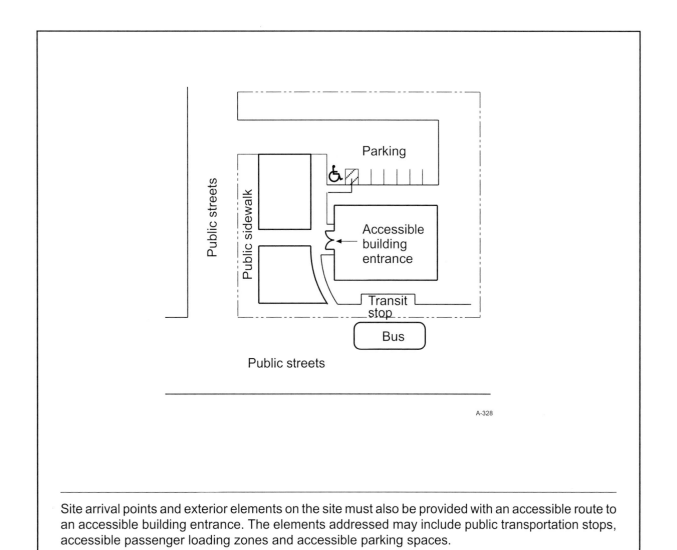

A-328

Site arrival points and exterior elements on the site must also be provided with an accessible route to an accessible building entrance. The elements addressed may include public transportation stops, accessible passenger loading zones and accessible parking spaces.

Code Text: *At least one accessible route shall connect each accessible level, including mezzanines, in multistory buildings and facilities.* See exceptions for specific occupancies with small floor areas, areas without accessible elements, air traffic control towers and two-story buildings with a small occupant load on one story.

Discussion and Commentary: Access must be provided both horizontally and vertically throughout a building. Under most conditions, multilevel facilities will contain an accessible elevator to extend an accessible route to the other levels. Ramps also can be used where the elevation change is not excessive. In some occupancies, an accessible route is not required for levels above or below an accessible level, provided that the aggregate size of the inaccessible levels does not exceed 3,000 square feet. It has been determined that it is not feasible to require elevator service for such small spaces.

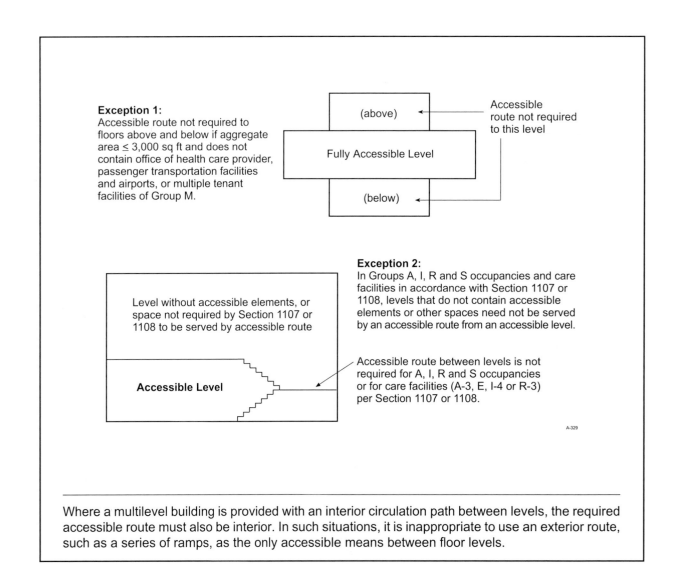

Exception 1:
Accessible route not required to floors above and below if aggregate area ≤ 3,000 sq ft and does not contain office of health care provider, passenger transportation facilities and airports, or multiple tenant facilities of Group M.

(above)

Fully Accessible Level

(below)

Accessible route not required to this level

Level without accessible elements, or space not required by Section 1107 or 1108 to be served by accessible route

Accessible Level

Exception 2:
In Groups A, I, R and S occupancies and care facilities in accordance with Section 1107 or 1108, levels that do not contain accessible elements or other spaces need not be served by an accessible route from an accessible level.

Accessible route between levels is not required for A, I, R and S occupancies or for care facilities (A-3, E, I-4 or R-3) per Section 1107 or 1108.

A-329

Where a multilevel building is provided with an interior circulation path between levels, the required accessible route must also be interior. In such situations, it is inappropriate to use an exterior route, such as a series of ramps, as the only accessible means between floor levels.

Code Text: *In addition to accessible entrances required by Sections 1105.1.1 through 1105.1.6 (parking garage entrances, entrances from tunnels or elevated walkways, restricted entrances, entrances for inmates or detainees, service entrances, and entrances to tenant spaces, dwelling units and sleeping units), at least 60 percent of all public entrances shall be accessible.* See exceptions for entrances to areas not required to be accessible and loading/service entrances that are not the only tenant space entrance.

Discussion and Commentary: To provide accessibility to all buildings and tenant spaces, a minimum of one accessible entrance is required. Where additional entrances are provided, often for convenience purposes, at least 60 percent of the total number of entrances must be accessible. Elements to be considered at entrances include the slope of exterior surfaces, door hardware and clear floor space for maneuvering clearances.

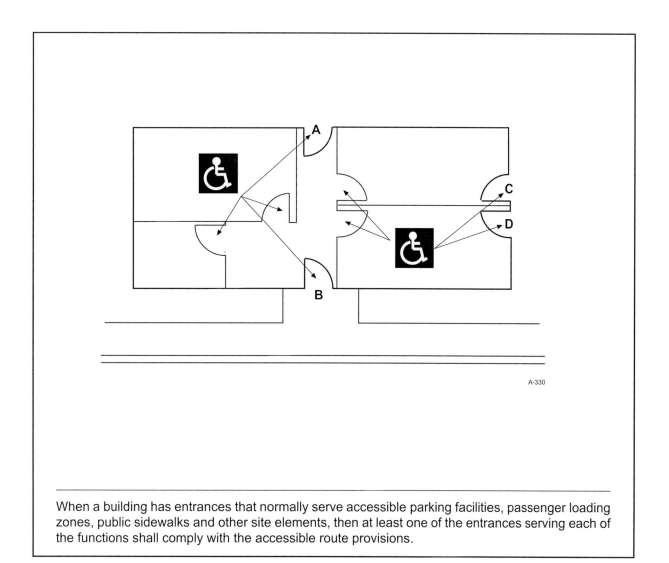

When a building has entrances that normally serve accessible parking facilities, passenger loading zones, public sidewalks and other site elements, then at least one of the entrances serving each of the functions shall comply with the accessible route provisions.

Code Text: *Where parking is provided, accessible parking spaces shall be provided in compliance with Table 1106.1 except as required by Sections 1106.2 (Groups R-2 and R-3), 1106.3, (hospital outpatient facilities) and 1106.4 (rehabilitation facilities). For every six or fraction of six accessible parking spaces, at least one shall be a van-accessible parking space. Accessible parking spaces shall be located on the shortest accessible route of travel from adjacent parking to an accessible building entrance.*

Discussion and Commentary: The number of required accessible parking spaces is based on the total number of spaces in the lot or garage. In hospital outpatient facilities where it is anticipated that more accessible spaces will be needed, at least one in ten parking spaces must be accessible. All accessible parking spaces shall be located as close as possible to an accessible building entrance to reduce the distance of travel for those individuals with mobility limitations.

Where parking is provided, accessible parking spaces shall be provided in compliance with Table 1106.1.

Exceptions:

1. Where Group R-2 and R-3 occupancies are required to have accessible dwelling units, 2% of the parking spaces shall be accessible.

   Where parking is provided within or beneath a building, accessible parking spaces shall also be provided within or beneath the building.

2. Where patient and visitor parking spaces serve hospital outpatient facilities, 10% of the spaces shall be accessible.

3. At rehabilitation facilities and outpatient physical therapy facilities, 20% of patient and visitor parking spaces shall be accessible.

TABLE 1106.1
ACCESSIBLE PARKING SPACES

TOTAL PARKING SPACES PROVIDED	REQUIRED MINIMUM NUMBER OF ACCESSIBLE SPACES
1 to 25	1
26 to 50	2
51 to 75	3
76 to 100	4
101 to 150	5
151 to 200	6
201 to 300	7
301 to 400	8
401 to 500	9
501 to 1,000	2% of total
1,001 and over	20, plus one for each 100, or fraction thereof, over 1,000

Every parking facility with at least one accessible parking space must provide for accessible van parking. Based on a percentage of the total number of accessible spaces, the van space or spaces must have a vertical clearance of at least 98 inches and a minimum 8-foot access aisle.

Topic: Group I Occupancies **Category:** Accessibility
Reference: IBC 1107.5 **Subject:** Dwelling Units and Sleeping Units

Code Text: *In Group I-1 occupancies, at least 4 percent, but not less than one, of the dwelling units and sleeping units shall be Accessible units. In nursing homes of Group I-2 occupancies, at least 50 percent but not less than one of each type of the dwelling sleeping units shall be Accessible units. In general-purpose hospitals, psychiatric facilities, detoxification facilities and residential care/assisted living facilities of Group I-2 occupancies, at least 10 percent, but not less than one, of the dwelling units and sleeping units shall be Accessible units. In hospitals and rehabilitation facilities of Group I-2 occupancies which specialize in treating conditions that affect mobility, or units within either which specialize in treating conditions that affect mobility, 100 percent of the dwelling and sleeping units shall be Accessible units. In Group I-3 occupancies, at least 2 percent, but not less than one, of the dwelling units and sleeping units shall be Accessible units.*

Discussion and Commentary: All Group I occupancies are regulated for accessibility. In Groups I-1, I-2 and I-3, it is not necessary that every dwelling or sleeping unit be an Accessible unit. The percentage of required Accessible units varies based on the anticipated need for such units because of the specifics of the institutional use.

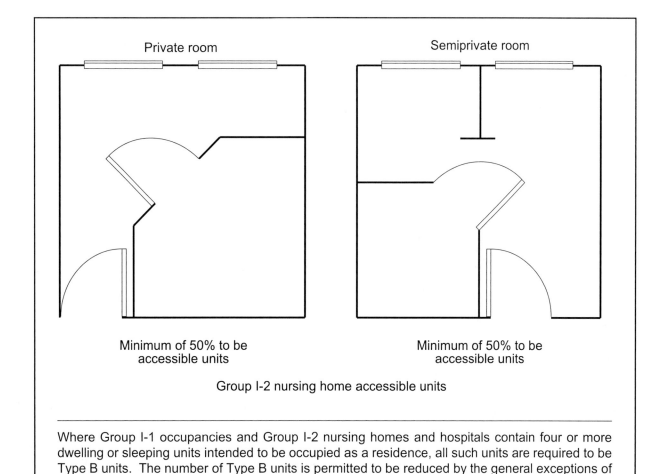

Private room

Semiprivate room

Minimum of 50% to be
accessible units

Minimum of 50% to be
accessible units

Group I-2 nursing home accessible units

Where Group I-1 occupancies and Group I-2 nursing homes and hospitals contain four or more dwelling or sleeping units intended to be occupied as a residence, all such units are required to be Type B units. The number of Type B units is permitted to be reduced by the general exceptions of Section 1107.7.

Code Text: *In Group R-1 occupancies, Accessible dwelling units and sleeping units shall be provided in accordance with Table 1107.6.1.1. In Group R-2 occupancies* (limited to apartment houses, monasteries and convents) *containing more than 20 dwelling units or sleeping units, at least 2 percent but not less than one of the units shall be a Type A unit. Where there are four or more dwelling units or sleeping units intended to be occupied as a residence in a single structure, every dwelling and sleeping unit intended to be occupied as a residence shall be a Type B unit.* See general exceptions in Section 1107.7.

Discussion and Commentary: Accessible, Type A and Type B dwelling units and sleeping units are defined in IBC Section 1102 and described in ICC A117.1. Accessible units are generally regarded as fully accessible. Type A units provide a considerable degree of accessibility, whereas Type B units are only required to have specific accessible elements. A dwelling unit designed and constructed as a Type B unit is intended to comply with the technical requirements for Fair Housing required by federal law.

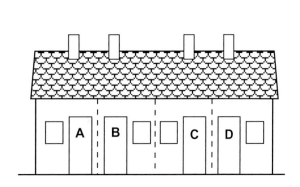

In R-2 and R-3 occupancies where there are ≥ 4 dwelling units in a single structure, every unit shall be a Type B dwelling unit. (Type A units may be substituted for Type B.)

In R-2 occupancies containing > 20 dwelling units, at least 2 percent but not less than 1 shall be a Type A dwelling unit.

Exceptions: Five exceptions are provided that are dependent on limited or no elevator service.

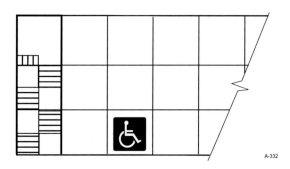

Where Group R-2 and R-3 occupancies contain public or common-use areas, such areas must be accessible if they serve accessible dwelling units. Any recreational facilities serving these occupancies must also must be accessible to a limited degree.

Code Text: *In theaters, bleachers, grandstands, stadiums, arenas and other fixed seating assembly areas, accessible wheelchair spaces complying with ICC A117.1 shall be provided in accordance with Sections 1108.2.2.1 through 1108.2.2.4. Wheelchair spaces shall be provided in accordance with Table 1108.2.2.1. In multilevel assembly seating areas, wheelchair spaces shall be provided on the main floor level and on one of each two additional floor or mezzanine levels.* See two exceptions where all wheelchair spaces may be located on the main level.

Discussion and The unique features of assembly occupancies dictate special accessibility features. In
Commentary: addition to the requirements for wheelchair spaces and assistive listening devices, the code requires accessible seating throughout all dining areas. Specific provisions address fixed seating at booths and tables, as well as at counters.

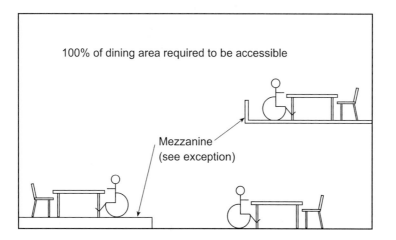

In dining areas, the total floor area allotted for seating and tables shall be accessible.

Exception: In buildings without elevators, an accessible route to a mezzanine seating area is not required, provided that the mezzanine contains less than 25 percent of the total area and the same services are provided in the accessible area.

100% of dining area required to be accessible

Mezzanine
(see exception)

A-331

Under limited conditions, a dining area may have a mezzanine level that is not served by an accessible route. The mezzanine must be limited in size to 25 percent of the total floor area, and the services provided on the mezzanine must be available on the accessible level.

Code Text: *Each assembly area where audible communications are integral to the use of the space shall have an assistive listening system.* See exception for spaces where no audio amplification system is installed. *Receivers shall be provided for assistive listening systems in accordance with Table 1108.2.7.1.* See exceptions for 1) buildings with multiple assembly areas, and 2) assembly areas where all seats are served by an induction loop system.

Discussion and Commentary: These provisions are intended to accommodate people with a hearing impairment. In these assembly areas, audible communication is often integral to the use and full enjoyment of the space. This requirement offers the possibility for individuals with hearing impairments to attend functions in these facilities without having to give advance notice and without disrupting the event in order to have a portable assistive listening system set up and made ready for use.

International symbol of access for
hearing loss

A-333b

TABLE 1108.2.6.1
RECEIVERS FOR ASSISTIVE LISTENING SYSTEMS

CAPACITY OF SEATING IN ASSEMBLY AREAS	MINIMUM REQUIRED NUMBER OF RECEIVERS	MINIMUM NUMBER OF RECEIVERS TO BE HEARING-AID COMPATIBLE
50 or less	2	2
51 to 200	2, plus 1 per 25 seats over 50 seats*	2
201 to 500	2, plus 1 per 25 seats over 50 seats.*	1 per 4 receivers*
501 to 1,000	20, plus 1 per 33 seats over 500 seats*	1 per 4 receivers*
1,001 to 2,000	35, plus 1 per 50 seats over 1,000 seats*	1 per 4 receivers*
Over 2,000	55, plus 1 per 100 seats over 2,000 seats*	1 per 4 receivers*

NOTE: * = or fraction thereof

There are three primary types of listening systems available: induction loop, AM/FM and infrared. Each type of system has certain advantages and disadvantages that should be taken into consideration when choosing the system that is most appropriate for the intended application.

Code Text: *Accessible building features and facilities shall be provided in accordance with Sections 1109.2 through 1109.14. See exception for Type A and Type B dwelling and sleeping units.*

Discussion and Commentary: Where elements such as sinks, drinking fountains, storage lockers, fitting rooms and check-out aisles are provided, a portion, but not less than one of each type of element, must be accessible. In general, all toilet rooms must be accessible. Within each toilet room, at least one water closet and lavatory must be accessible. When other elements are provided, such as mirrors and towel fixtures, at least one must be accessible. Operating mechanisms intended for occupant operation, such as light switches and convenience outlets, shall be useable by persons with physical disabilities.

In other than Type A and Type B dwelling units, accessible building features and facilities shall be provided as required in Section 1109. This includes:

- Toilet and bathing facilities
- Sinks
- Kitchens, kitchenettes and wet bars
- Drinking fountains
- Elevators
- Lifts
- Storage
- Detectable warnings
- Assembly area seating
- Seating at tables, counters and work surfaces
- Service facilities
- Controls, operating mechanisms and hardware
- Recreational facilities

Certain elements addressed in ICC/ANSI A117.1, including telephones and automatic teller machines, have not been included in the scoping provisions of Chapter 11. However, scoping requirements for such features are set forth in Appendix E of the IBC.

Code Text: *Toilet rooms and bathing facilities shall be accessible. Where a floor level is not required to be connected by an accessible route, the only toilet rooms or bathing facilities provided within the facility shall not be located on the inaccessible floor. At least one of each type of fixture, element, control or dispenser in each accessible toilet room and bathing facility shall be accessible.* See exceptions for toilet rooms and bathing rooms 1) accessed through a private office and intended for use by a single occupant; 2) that serve a dwelling unit or sleeping unit not required to be accessible; 3) clustered in a single location; and 4) part of critical care or intensive care patient sleeping rooms. An additional exception provides that where only one urinal is provided in a toilet room or bathing facility, it need not be accessible.

Discussion and Commentary: As a general rule, all toilet rooms and bathing facilities must provide for accessibility. There are limited conditions under which some facilities need not be made accessible.

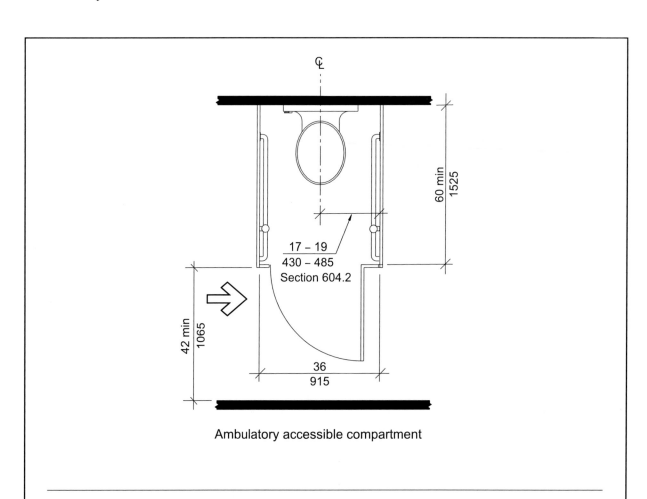

Ambulatory accessible compartment

In those toilet rooms and bathing facilities where water closet compartments are provided, a minimum of one wheelchair-accessible compartment is required. If the total number of water closet compartments and urinals provided is six or more, an ambulatory-accessible water closet compartment is mandated in addition to the wheelchair-accessible compartment.

Topic: Unisex Toilet and Bathing Rooms

Category: Accessibility

Reference: IBC 1109.2.1

Subject: Features and Facilities

Code Text: *In assembly and mercantile occupancies, an accessible unisex toilet room shall be provided where an aggregate of six or more male and female water closets is required. In recreational facilities where separate-sex bathing rooms are provided, an accessible unisex bathing room shall be provided.* See exception for single-fixture bathing rooms. *Fixtures located within unisex toilet and bathing rooms shall be included in determining the number of fixtures provided in an occupancy.*

Discussion and Commentary: The primary issue relative to unisex toilet/bathing facilities is that some people with disabilities require assistance to utilize them. If the attendant is of the opposite sex, a facility that can accommodate both persons is required. The provisions are applicable only to those types of transient uses where it is expected such facilities are frequently required.

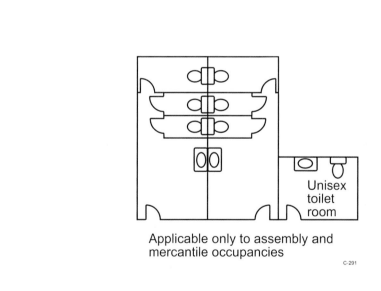

Applicable only to assembly and
mercantile occupancies

C-291

These types of facilities have also been identified as "family bathrooms." For example, a person requiring assistance can also be a small child. A parent shopping or attending an event with a child of the opposite sex can utilize the facility without disrupting the separate-sex toilet rooms.

Code Text: *Required accessible elements shall be identified by the International Symbol of Accessibility at the following locations: 1) accessible parking spaces required by Section 1106.1 except where the total number of parking spaces provided is four or less, 2) accessible passenger loading zones, 3) accessible areas of refuge required by Section 1007.6, 4) accessible rooms where multiple single-user toilet or bathing rooms are clustered at a single location, 5) accessible entrances where not all entrances are accessible, 6) accessible check-out aisles where not all aisles are accessible, 7) unisex toilet and bathing rooms, and 8) accessible dressing, fitting, and locker rooms where not all such rooms are accessible.*

Discussion and Commentary: Those site or building elements that need to be identified as accessible for convenience, clarification or life safety purposes are specified in the code. Special signage is also required for assistive listening capabilities and at every exit stairway door.

International Symbol of Accessibility

A-333a

Where building entrances are not accessible, directional signage must be installed indicating the travel route to the nearest accessible entrance. Signs must also be provided at inaccessible public toilets directing occupants to the nearest accessible toilet facilities.

Quiz

Study Session 13
IBC Chapter 11

1. For the purpose of accessibility, which of the following areas in a bank is considered an employee work area?

   a.   vault

   b.   toilet room

   c.   corridor

   d.   break room

   Reference_____

2. A continuous, unobstructed path complying with the provisions of IBC Chapter 11 is considered a(n) _____.

   a.   accessible route

   b.   accessible means of egress

   c.   circulation path

   d.   public entrance

   Reference_____

3. Which of the following Group U occupancies is exempt from the accessibility provisions?

   a.   paved work areas at accessible buildings

   b.   private garages containing required accessible parking

   c.   areas in agricultural buildings open to the public

   d.   stables and/or livestock shelters

   Reference_____

4. When occupied as the residence for the proprietor, Group R-1 occupancies containing a maximum of _____ sleeping units for rent or hire are not required to be accessible.

    a. 5   b. 6

    c. 10   d. 15

Reference_____

5. In other than an Accessible, Type A or Type B dwelling unit, where only one accessible route is provided, the route is permitted to pass through a _____.

    a. kitchen   b. laboratory

    c. storage room   d. toilet room

Reference_____

6. Where four public entrances are provided to a building, a minimum of _____ entrance(s) shall be accessible.

    a. one   b. two

    c. three   d. four

Reference_____

7. Where a 1,625-space parking garage serves a covered mall building, a minimum of _____ accessible parking spaces must be provided.

    a. 20   b. 27

    c. 33   d. 42

Reference_____

8. Where a 282-space parking lot serves a cluster of Group R-2 apartment buildings, a minimum of _____ accessible parking spaces must be provided.

    a. 3   b. 6

    c. 7   d. 14

Reference_____

9. A minimum of _____ van-accessible parking space(s) shall be provided for a parking facility containing 16 accessible parking spaces.

   a. one   b. two

   c. three   d. six

   Reference_____

10. A theater with a total seating capacity of 840 persons shall be provided with a minimum of _____ accessible wheelchair spaces.

    a. 4   b. 8

    c. 9   d. 10

    Reference_____

11. Where cubicles are provided in the visiting area of a judicial facility, a minimum of _____ of the cubicles shall be accessible.

    a. 5 percent   b. 10 percent
    c. 25 percent   d. 50 percent

    Reference_____

12. An ambulatory-accessible water closet compartment is required in a toilet room where the total number of water closet compartments and urinals provided in the room is a minimum of _____ fixtures.

    a. 4   b. 6

    c. 10   d. 12

    Reference_____

13. In a Group R-2 fraternity house having a minimum of _____ sleeping units, every unit intended to be occupied as a residence shall be a Type B unit.

    a. 4   b. 10

    c. 16   d. 20

    Reference_____

14. In a 250-seat motion picture theater, a minimum of _____ receivers shall be provided for the assistive listening system.

    a. 2   b. 6

    c. 10   d. 14

Reference_____

15. In a Group I-2 nursing home containing 120 patient sleeping units, what is the minimum number of such units that must be Accessible units?

    a. 5   b. 12

    c. 60   d. 120

Reference_____

16. In a 185-room Group R-1 hotel, a minimum of _____ sleeping units shall be Accessible units.

    a. two   b. four

    c. seven   d. eight

Reference_____

17. A minimum of _____ Accessible units in a 410-room hotel shall be provided with roll-in showers.

    a. 4   b. 8

    c. 12   d. 13

Reference_____

18. A Group R-2 apartment building having 16 dwelling units shall be provided with a minimum of _____ Type A dwelling units.

    a. zero (no Type A units are required)

    b. one

    c. two

    d. four

Reference_____

19. A self-storage facility containing 160 storage spaces shall be provided with a minimum of _____ accessible individual self-storage spaces.

    a. one   b. two

    c. five   d. eight

Reference_____

20. Based on the number of required water closets, an accessible unisex toilet room may be required in which of the following occupancies?

    a. assembly and educational   b. assembly and mercantile

    c. business and mercantile   d. factory and storage

Reference_____

21. The maximum distance of the accessible route from any separate-sex toilet room to a unisex toilet room shall be _____.

    a. 200 feet   b. 300 feet

    c. 500 feet   d. unlimited

Reference_____

22. In a large manufacturing plant, a minimum of how many of the 24 drinking fountains shall be accessible?

    a. 3   b. 4

    c. 12   d. 24

Reference_____

23. A minimum of _____ percent of seating at fixed tables in an accessible space shall be accessible.

    a. 5   b. 10

    c. 20   d. 50

Reference_____

24. A 4,000-square-foot retail sales room with six check-out aisles requires a minimum of _____ accessible check-out aisle(s).

    a. one

    b. two

    c. three

    d. four

Reference_____

25. An International Symbol of Accessibility sign is not required at an accessible parking space in parking facilities having a maximum of _____ total parking spaces.

    a. one, the sign is always required

    b. two

    c. four

    d. five

Reference_____

26. Permanently defined common use circulation paths that are located within an employee work area are not required to be regulated as accessible routes, provided the work area is less than _____ square feet in floor area.

    a. 300

    b. 500

    c. 1,500

    d. 3,000

Reference_____

27. A press box serving bleachers is not required to be served by an accessible route where it is a maximum of _____ square feet in floor area and has its points of entry at only one level.

    a. 500

    b. 1,000

    c. 1,500

    d. 3,000

Reference_____

28. What is the minimum number of accessible parking spaces required to be provided in a 40-space parking lot serving an outpatient physical therapy facility?

    a. one

    b. two

    c. four

    d. eight

Reference_____

29. Where a dining surface is provided at a counter for the consumption of food or drink, a minimum of _____ seating and standing space(s) shall be accessible.

    a. zero   b. one

    c. two   d. four

Reference_____

30. At which of the following locations is directional signage not required to indicate the route to the nearest like accessible element?

    a. inaccessible building entrances

    b. elevators not serving an accessible route

    c. inaccessible dressing rooms

    d. inaccessible public toilet facilities

Reference_____

31. Where passenger loading zones are provided, at least one accessible passenger loading zone is required in every continuous _____ feet maximum of loading zone space.

    a. 100   b. 150

    c. 200   d. 300

Reference _____

32. All required wheelchair spaces are permitted to be located on the main level in multilevel assembly spaces used for religious worship where the mezzanine level contains a maximum of _____ percent of the total seating capacity.

    a. 10   b. 20

    c. 25   d. $33^{1}/_{3}$

Reference _____

33. Where multiple single-user toilet rooms are clustered at a single location, a minimum of _____ percent, but not less than one room, shall be accessible.

    a. 5   b. 10

    c. 25   d. 50

Reference _____

34. Where a total of eight accessible drinking fountains are provided, _____ drinking fountain(s) shall comply with the requirements for standing persons.

    a. one   b. two

    c. four   d. five

Reference _____

35. In a 40-lane bowling facility, an accessible route shall be provided to a minimum of _____ bowling lane(s).

    a. one   b. two

    c. four   d. ten

Reference _____

2006 IBC Chapter 4
Detailed Occupancy Requirements

OBJECTIVE: To obtain an understanding of special building types, features and uses, including covered mall buildings, high-rise buildings, atriums, underground buildings, motor-vehicle-related occupancies, stages and platforms, concealed combustible storage areas, hazardous materials, dwelling unit and sleeping unit seperations, and Groups I-2, I-3 and H.

REFERENCE: Chapter 4, 2006 *International Building Code*

KEY POINTS:
- How are covered mall buildings, anchor buildings and food courts defined?
- In a covered mall building, which special conditions relate to the automatic sprinkler system, standpipe system and smoke-control system?
- How is the occupant load determined for a covered mall building?
- Which specific provisions relate to covered mall buildings in determining the number and arrangement of means of egress?
- What qualifies a structure as a high-rise building?
- In a high-rise building, how are the type-of-construction provisions modified?
- What are the specific provisions relating to smoke detection, fire alarms and communication systems in high-rise buildings?
- What are the various required elements in a fire command center?
- In a high-rise structure, which types of standby power, light and emergency systems are required?
- What defines an atrium? What are the limits of an atrium's use? What fire protection features must be provided in a building containing an atrium?
- How must adjacent spaces be separated from an atrium?
- What is the maximum travel distance when a required exit path enters the atrium space?
- What are the conditions that create an underground building?
- In an underground building, when is compartmentalization required?
- What is the maximum floor area permitted for a private garage?
- How must exterior openings be sized and distributed for a garage to qualify as open?
- How do the provisions differ for an open parking garage in either a single-use or multiple-occupancy building?

KEY POINTS:
(Cont'd)

- What are the benefits of an open parking garage as opposed to an enclosed garage?
- Which special criteria must be applied to motor vehicle service stations? Repair garages?
- How does a platform differ from a stage?
- What are the minimum construction requirements for stages and platforms?
- When must a stage area be provided with a means for emergency ventilation?
- How shall the proscenium opening between a stage and an auditorium be protected?
- What type of fire separation is required for combustible storage in concealed spaces?
- When are smoke compartments required in Group I-2 occupancies?
- Which special considerations are given to detention facilities because of their unique characteristics?
- How are Group I-3 occupancies classified according to their occupancy condition?
- What is the function of a control area? How many control areas are permitted in a building?
- How must control areas be separated from other portions of the building?
- At what distance must Group H occupancies be set back from property lines?
- Where the spraying of flammable finishes occurs, what special conditions must be met?
- How must dwelling units or sleeping units be separated from other such units in Group I-1, R-1, R-2 and R-3 occupancies?

Code Text: *The area of any covered mall building, including anchor buildings of Types I, II, III and IV construction shall not be limited provided the covered mall building and attached anchor buildings and parking garages are surrounded on all sides by a permanent open space of not less than 60 feet (18 288 mm) and the anchor buildings do not exceed three stories in height. The covered mall building and buildings connected shall be provided throughout with an automatic sprinkler system in accordance with Section 903.3.1.1.*

Discussion and Commentary: A covered mall building is defined as *a single building enclosing a number of tenants and occupants such as retail stores, drinking and dining establishments, entertainment and amusement facilities, passenger transportation terminals, offices, and other similar uses wherein two or more tenants have a main entrance into one or more malls.* Because of its unique character, a covered mall building is more highly regulated for fire protection and egress.

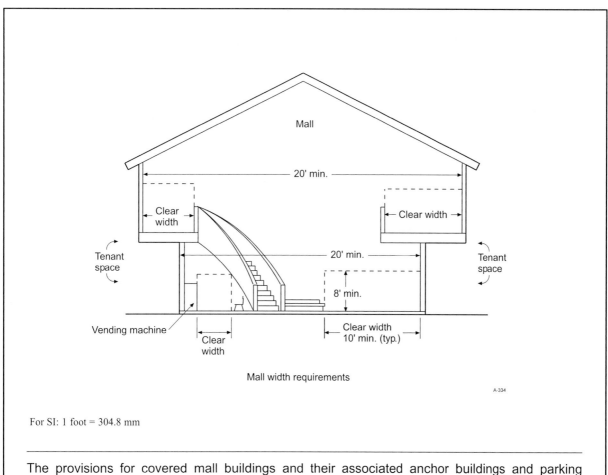

Mall width requirements

A-334

For SI: 1 foot = 304.8 mm

The provisions for covered mall buildings and their associated anchor buildings and parking structures allow for an alternative method of design for structures that have these specific features. Where compliant with the provisions of Section 402, similar requirements found elsewhere in the code can be superceded.

Code Text: *The provisions of Section 403 shall apply to buildings with an occupied floor located more than 75 feet (22 860 mm) above the lowest level of fire department vehicle access.* See five exceptions where high-rise provisions are not applicable. *Buildings and structures shall be equipped throughout with an automatic sprinkler system in accordance with Section 903.3.1.1 and a secondary water supply where required by Section 903.3.5.2.*

Discussion and Commentary: A high-rise building is characterized by several features: 1) it is impractical to completely evacuate the building in a timely manner,) prompt rescue and firefighting operations are difficult, 3) the occupant load is relatively high and 4) a potential exists for stack effect. The special provisions of Section 403 are designed to address these concerns.

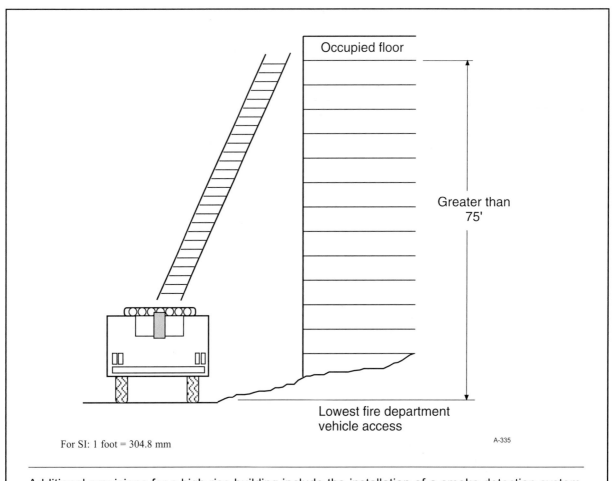

Occupied floor

Greater than 75'

Lowest fire department vehicle access

For SI: 1 foot = 304.8 mm

A-335

Additional provisions for a high-rise building include the installation of a smoke detection system, emergency voice/alarm and fire department communications systems, a fire command center for use by fire department personnel, and standby power, light and emergency systems.

Code Text: *In other than Group H occupancies, and where permitted by Exception 5 in Section 707.2, the provisions of Section 404 shall apply to buildings or structures containing vertical openings defined herein as "Atriums." An approved automatic sprinkler system shall be installed throughout the entire building. A smoke-control system shall be installed in accordance with Section 909. Atrium spaces shall be separated from adjacent spaces by a 1-hour fire barrier wall.* See exceptions to the sprinkler system, smoke-control system and fire barrier separation requirements.

Discussion and Commentary: An atrium is defined as an opening connecting two or more stories other than enclosed stairways, elevators, hoistways, escalators, plumbing, electrical, air-conditioning or other equipment, which is closed at the top and not defined as a mall. The concept of developing atriums is to maintain equivalence in safety to that of an open court, as well as to provide protection of a shaft enclosure.to maintain equivalence in safety to that of an open court, as well as to provide protection of a shaft enclosure. The provisions for atriums are only applicable where Exception 5 of Section 707.2 is utilized to eliminate a required shaft enclosure. Where another exception to Section 707.2 is used, such as Exception 7 for an opening connecting only two stories, the atrium provisions of Section 404 are not to be applied.

Sprinkler system throughout—prevents spread of fire.

Smoke-control system—keeps building and atrium clear of smoke so that safe exiting may be accomplished through the atrium.

Atrium concept

A-336

Atriums are permitted based on alternative methods of protecting the building from vertical spread of fire, smoke and toxic gases. Additional protection is provided through 1) limited travel distance, 2) standby power, 3) smoke detection and 4) interior finish regulation.

Code Text: *The provisions of Section 405 apply to building spaces having a floor level used for human occupancy more than 30 feet (9144 mm) below the lowest level of exit discharge. See five exceptions for uses not regulated as underground buildings. The underground portion of the building shall be of Type I construction. The highest level of exit discharge serving the underground portions of the building and all levels below shall be equipped with an automatic sprinkler system installed in accordance with Section 903.3.1.1.*

Discussion and Commentary: An underground building is highly regulated for many of the same reasons as is a high-rise building. In the case of a structure substantially below ground level, fire department access and fire-fighting operations are often even more difficult. Therefore, the code mandates the installation of multiple fire protection systems, including a smoke-control system.

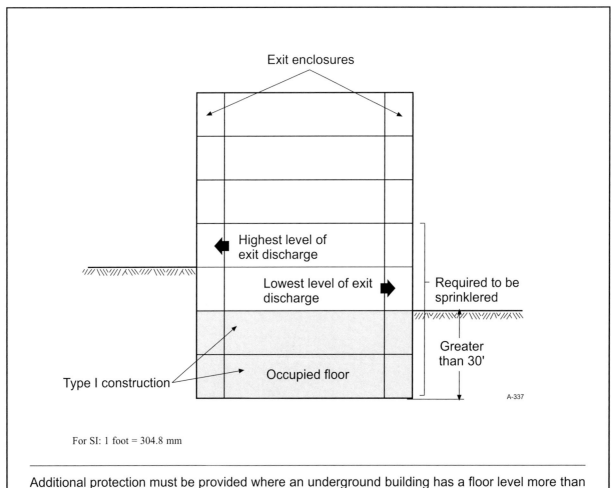

For SI: 1 foot = 304.8 mm

Additional protection must be provided where an underground building has a floor level more than 60 feet below the lowest level of exit discharge. The required creation of multiple compartments assists in both occupant egress and fire department operations.

Code Text: *Parking garages shall be classified as either open, as defined in Section 406.3, or enclosed and shall meet the appropriate criteria in Section 406.4. Motor vehicle service stations shall be constructed in accordance with the* International Fire Code *and Section 406.5. Repair garages shall be constructed in accordance with the* International Fire Code *and Section 406.6.*

Discussion and Commentary: Where motor vehicles are located within a structure, varying degrees of hazard are involved. In a small, private garage or carport, the hazard is relatively low. A moderate level of hazard exists in open parking garages, increasing where the parking garage is enclosed. In structures where vehicles are being fueled or repaired, a relatively high hazard exists. Specific provisions of the IBC address each type of motor-vehicle-related use.

Exterior walls must have uniformly distributed openings on two or more sides.

Interior wall and column lines shall be at least 20 percent open (area) with uniformly distributed openings.

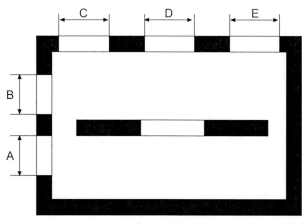

General case

1. Area: A + B + C + D + E ≥ 20% total perimeter area of each tier
2. Length: A + B + C + D + E ≥ 40% total perimeter area of each tier

Open parking garages

A-338

Private garages classified as Group U occupancies are usually limited to 1,000 square feet in floor area. However, such structures are permitted to be 3,000 square feet maximum where no repair work is done and no fuel dispensed, and where the exterior wall and opening protection is specifically regulated.

Code Text: *Smoke barriers shall be provided to subdivide every story used by patients for sleeping or treatment and to divide other stories with an occupant load of 50 or more persons, into at least two smoke compartments. Such stories shall be divided into smoke compartments with an area of not more than 22,500 square feet (2092 m²) and the travel distance from any point in a smoke compartment to a smoke barrier door shall not exceed 200 feet (60 960 mm).*

Discussion and Commentary: Hospitals, nursing homes and similar uses must have unique life-safety characteristics due to the immobility or limited mobility of most of the occupants. By providing multiple refuge areas on each story of the building, occupants can be moved horizontally into an adjoining smoke compartment that provides protection from adjacent areas.

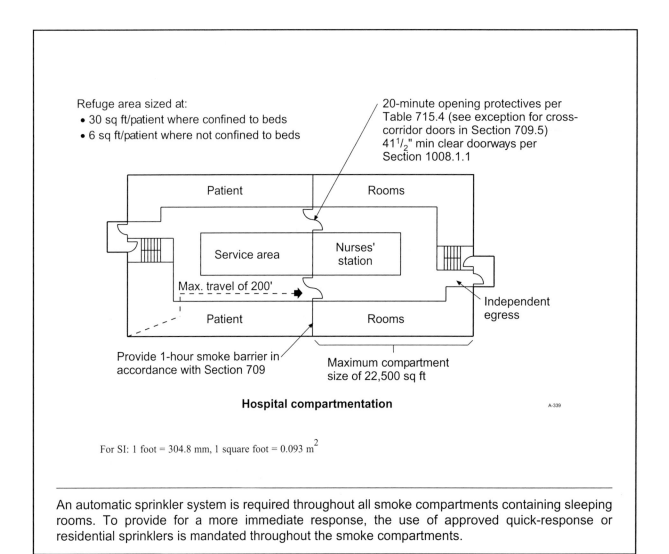

Refuge area sized at:
- 30 sq ft/patient where confined to beds
- 6 sq ft/patient where not confined to beds

20-minute opening protectives per Table 715.4 (see exception for cross-corridor doors in Section 709.5) 41½" min clear doorways per Section 1008.1.1

Patient   Rooms

Service area   Nurses' station

Max. travel of 200'

Independent egress

Patient   Rooms

Provide 1-hour smoke barrier in accordance with Section 709

Maximum compartment size of 22,500 sq ft

Hospital compartmentation   A-339

For SI: 1 foot = 304.8 mm, 1 square foot = 0.093 m²

An automatic sprinkler system is required throughout all smoke compartments containing sleeping rooms. To provide for a more immediate response, the use of approved quick-response or residential sprinklers is mandated throughout the smoke compartments.

Code Text: *Where security operations necessitate the locking of required means of egress, provisions shall be made for the release of occupants at all times. Egress doors are permitted to be locked in accordance with the applicable use conditions. Exits are permitted to discharge into a fenced or walled courtyard. Enclosed yards or courts shall be of a size to accommodate all occupants, a minimum of 50 feet (15 240 mm) from the building with a net area of 15 square feet (1.4 m²) per person.*

Discussion and Commentary: The need for restraint or security in specific types of uses such as jails, prisons and detention centers makes it necessary to install locking devices that are usually unacceptable for a means of egress. The code recognizes such a need and provides alternative design methods to balance the desire for both safety and security.

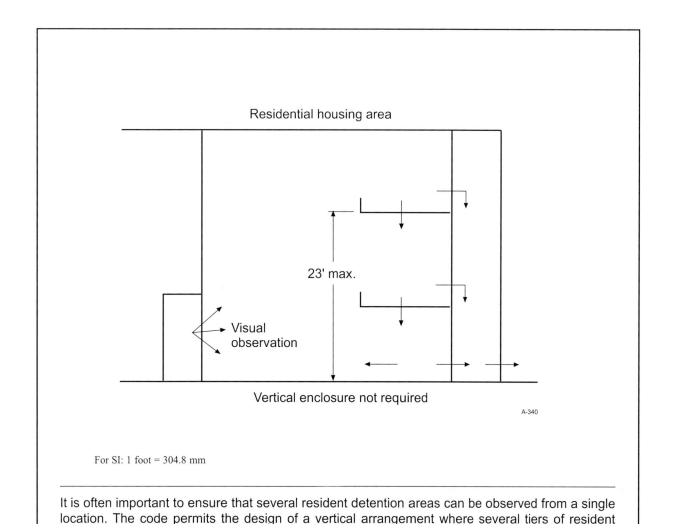

Residential housing area

23' max.

Visual observation

Vertical enclosure not required

A-340

For SI: 1 foot = 304.8 mm

It is often important to ensure that several resident detention areas can be observed from a single location. The code permits the design of a vertical arrangement where several tiers of resident housing areas can be open to each other and to the supervisory area.

Code Text: *A stage is a space within a building utilized for entertainment or presentations, which includes overhead hanging curtains, drops, scenery or stage effects other than lighting or sound. Stages shall be constructed of materials as required for floors for the type of construction of the building in which such stages are located. See three exceptions. Where the stage height is greater than 50 feet (15 240 mm), all portions of the stage shall be completely separated from the seating area by a proscenium wall with not less than a 2-hour fire-resistance rating extending continuously from the foundation to the roof.*

Discussion and Commentary: Given the increased potential for fire hazards in an assembly occupancy with a stage, such a use is regulated for certain elements. The stage must be separated from accessory spaces by fire barriers; ventilation of the stage must be accomplished through smoke control or roof vents; and the proscenium opening must be protected with a curtain of approved materials or an approved water curtain.

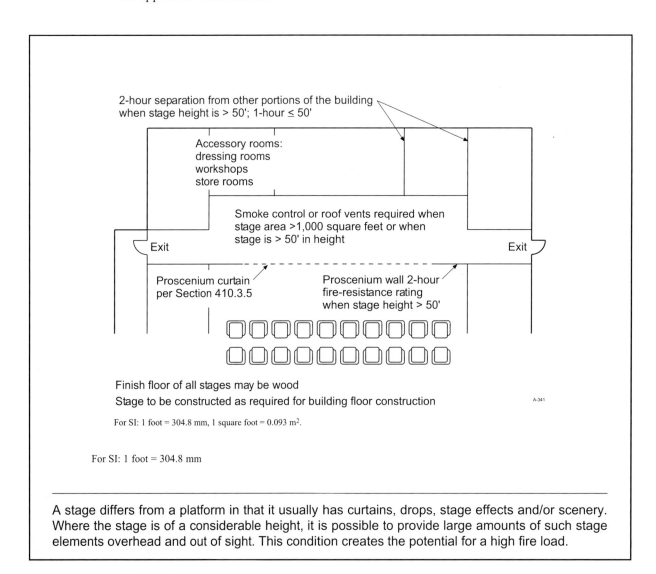

2-hour separation from other portions of the building when stage height is > 50'; 1-hour ≤ 50'

Accessory rooms:
dressing rooms
workshops
store rooms

Smoke control or roof vents required when stage area >1,000 square feet or when stage is > 50' in height

Exit   Exit

Proscenium curtain per Section 410.3.5

Proscenium wall 2-hour fire-resistance rating when stage height > 50'

Finish floor of all stages may be wood
Stage to be constructed as required for building floor construction

A-341

For SI: 1 foot = 304.8 mm, 1 square foot = 0.093 m².

For SI: 1 foot = 304.8 mm

A stage differs from a platform in that it usually has curtains, drops, stage effects and/or scenery. Where the stage is of a considerable height, it is possible to provide large amounts of such stage elements overhead and out of sight. This condition creates the potential for a high fire load.

Code Text: *A platform is a raised area within a building used for worship, the presentation of music, plays or other entertainment; the head table for special guests; the raised area for lecturers and speakers; boxing and wresting rings; theater-in-the-round stages; and similar purposes wherein there are no overhead hanging curtains, drops, scenery or stage effects other than lighting and sound. Permanent platforms shall be constructed of materials as required for the type of construction of the building in which the permanent platform is located.* See allowances for use of fire-retardant-treated wood for Types I, II or IV construction.

Discussion and Commentary: Few requirements are placed on platforms. However, a minimum 1-hour fire-resistant platform floor construction is required when the area below the platform is used for storage or a similar purpose, because of concern about combustibles being stored within a concealed space below a raised area.

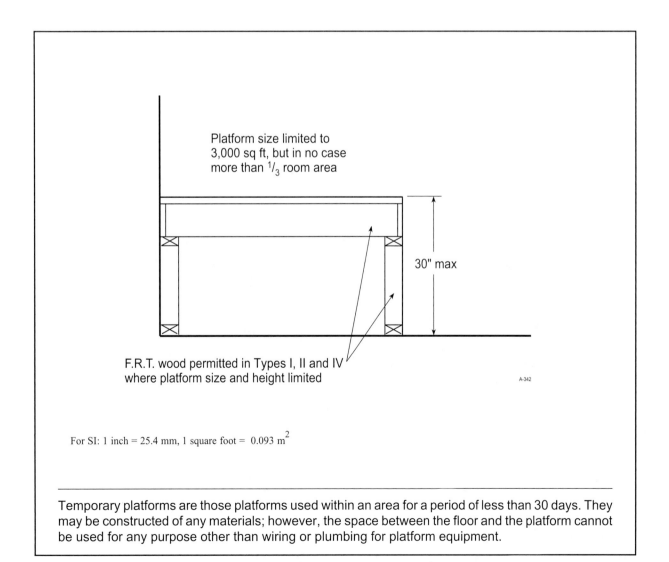

Platform size limited to 3,000 sq ft, but in no case more than $1/_3$ room area

30" max

F.R.T. wood permitted in Types I, II and IV where platform size and height limited

A-342

For SI: 1 inch = 25.4 mm, 1 square foot = 0.093 m^2

Temporary platforms are those platforms used within an area for a period of less than 30 days. They may be constructed of any materials; however, the space between the floor and the platform cannot be used for any purpose other than wiring or plumbing for platform equipment.

Code Text: *A special amusement building is any temporary or permanent building or portion thereof that is occupied for amusement, entertainment or educational purposes and that contains a device or system that conveys passengers or provides a walkway along, around or over a course in any direction so arranged that the means of egress path is not readily apparent due to visual or audio distractions or is intentionally confounded or is not readily available because of the nature of the attraction or mode of conveyance through the building or structure.*

Discussion and Commentary: In most cases, an amusement building will be classified as a Group A occupancy. The hazards associated with such a unique use are addressed through provisions for the detection of fire, the illumination of the exit path, the presence of an alarm and emergency voice/alarm communications system and the sprinklering of the structure.

Rapid detection and notification of a fire condition, as well as the discernment of the exit path, are critical in an amusement building. Actuation of either the sprinkler system or the fire detection system shall automatically activate the approved egress directional markings.

Code Text: *Exterior walls located less than 30 feet (9144 mm) from property lines, lot lines or a public way shall have a fire-resistance rating not less than 2 hours. Heating equipment shall be placed in another room separated by 2-hour fire-resistance-rated construction. Entrance shall be from the outside or by means of a vestibule providing a two-doorway separation.* See two exceptions.

Discussion and Commentary: Although most commercial aircraft hangars will not be limited in height (Section 504.1) or in area (Section 507) based on the presence of an automatic sprinkler system, they must be regulated in regards to exterior-wall fire-resistance ratings, basement limitations, floor surfaces, heating equipment separation and finishing restrictions. All of these provisions serve to abate the hazards associated with large aircraft and their integral fuel tanks to acceptable fire safety levels.

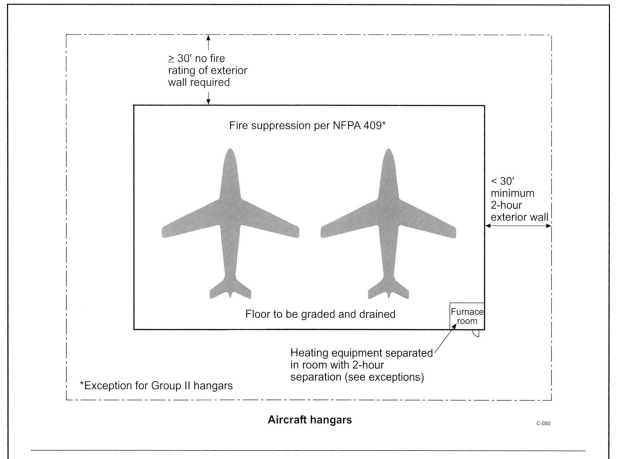

≥ 30' no fire rating of exterior wall required

Fire suppression per NFPA 409*

< 30' minimum 2-hour exterior wall

Floor to be graded and drained

Furnace room

Heating equipment separated in room with 2-hour separation (see exceptions)

*Exception for Group II hangars

Aircraft hangars

C-050

Although the provisions of Section 903.2.9 do not require automatic sprinkler protection for Group S-2 occupancies (aircraft hangars), a fire suppression system per NFPA 409 is mandated. The foam suppression requirements are exempted for Group II hangars where no major maintenance is performed.

Code Text: *Attic, under-floor and concealed spaces used for storage of combustible materials shall be protected on the storage side as required for 1-hour fire-resistant construction. Openings shall be protected by assemblies that are self-closing and are of noncombustible construction or solid wood core not less than $1^3/_4$ inch (45 mm) in thickness.* See exceptions for 1) areas protected by an automatic sprinkler system, and 2) Group R-3 and U occupancies.

Discussion and Commentary: Those areas in a building that tend to be unoccupied present a potential fire hazard where combustible goods are being stored. The presence of a considerable fire load, coupled with the probable delay in recognition of the fire, makes it necessary to provide some degree of protection. A sprinkler system or a fire-resistant separation can provide ample protection.

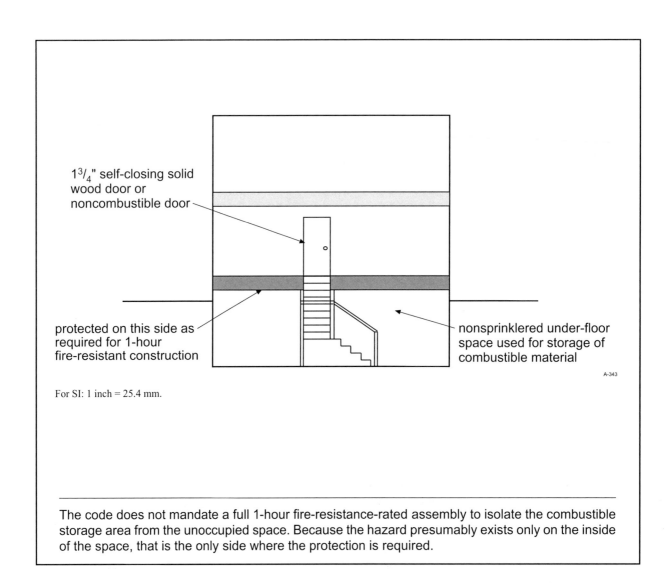

$1^3/_4$" self-closing solid wood door or noncombustible door

protected on this side as required for 1-hour fire-resistant construction

nonsprinklered under-floor space used for storage of combustible material

A-343

For SI: 1 inch = 25.4 mm.

The code does not mandate a full 1-hour fire-resistance-rated assembly to isolate the combustible storage area from the unoccupied space. Because the hazard presumably exists only on the inside of the space, that is the only side where the protection is required.

Code Text: *Control areas are spaces within a building where quantities of hazardous materials not exceeding the maximum allowable quantities per control area are stored, dispensed, used or handled. Control areas shall be separated from each other by fire barriers constructed in accordance with Section 706 or horizontal assemblies constructed in accordance with Section 711, or both.*

Discussion and Commentary: The use of control areas provides an alternative method for the use and storage of hazardous materials without classifying the building or structure as a high-hazard (Group H) occupancy. This concept is based on regulating the allowable quantities of hazardous materials per control area rather than per building area by giving credit for further compartmentation through the use of fire-resistance-rated fire barrier walls and horizontal assemblies.

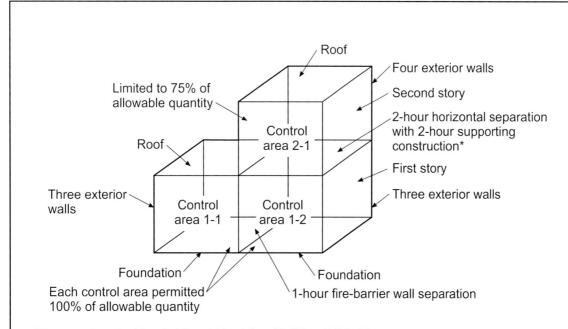

*Exception allows for 1-hour in fully sprinklered Type IIA, IIIA and VA buildings no more than three stories in height.

Multistory control areas

A-344

The maximum quantities of hazardous materials within a given control area cannot exceed the quantities for a given material listed in either Table 307.1 (1) for physical hazards and Table 307.1(2) for health hazards, as modified by Table 414.2.2 for location within the building.

Code Text: *The percentage of maximum allowable quantities of hazardous materials per control area permitted at each floor level within a building shall be in accordance with Table 414.2.2. The maximum number of control areas within a building shall be in accordance with Table 414.2.2. The required fire-resistance rating for fire barriers shall be in accordance with Table 414.2.2. The floor construction of the control area and the construction supporting the floor of the control area shall have a minimum 2-hour fire-resistance rating.* See exception permitting a 1-hour floor separation in fully-sprinklered Type IIA, IIIA and VA buildings not exceeding three stories in height.

Discussion and Commentary: By distributing hazardous materials in multiple fire-resistant compartments throughout a structure, the amount of material exposed to an immediate fire event is limited. The code allows such limited quantities in buildings of other than Group H occupancy; hence, the use of control areas is an effective method for reducing the occupancy classification by reducing the hazard.

[F] TABLE 414.2.2
DESIGN AND NUMBER OF CONTROL AREAS

FLOOR LEVEL		PERCENTAGE OF THE MAXIMUM ALLOWABLE QUANTITY PER CONTROL AREAa	NUMBER OF CONTROL AREAS PER FLOOR	FIRE-RESISTANCE RATING FOR FIRE BARRIERS IN HOURSb
Above grade plane	Higher than 9	5	1	2
	7-9	5	2	2
	6	12.5	2	2
	5	12.5	2	2
	4	12.5	2	2
	3	50	2	1
	2	75	3	1
	1	100	4	1
Below grade plane	1	75	3	1
	2	50	2	1
	Lower than 2	Not Allowed	Not Allowed	Not Allowed

a. Percentages shall be of the maximum allowable quantity per control area shown in Tables 307.1(1) and 307.1(2), with all increases allowed in the notes to those tables.

b. Fire barriers shall include walls and floors as necessary to provide separation from other portions of the building.

The purpose of a control area is to allow the building to be classified according to its general occupancy instead of being classified as a Group H occupancy. A building may comprise a single control area where the amount of hazardous materials in the entire structure is compliant.

Code Text: *Regardless of any other provisions, buildings containing Group H occupancies shall be set back to the minimum fire seperation distance as set forth in Items 1 through 4 below.* See specific setback requirements for Groups H-1, H-2 and H-3 occupancies. *Distances shall be measured from the walls enclosing the occupancy to lot lines, including those on a public way.*

Discussion and Commentary: Because of the potentially volatile nature of hazardous materials, specific setback requirements are necessary for Group H occupancies. These provisions take precedence over Table 602 regarding the minimum fire separation distance based on building construction type and exposure. The listed conditions are dependent on the type of materials that are indicative of the specified Group H occupancies, the size of the hazardous material storage area and whether a detached building is required.

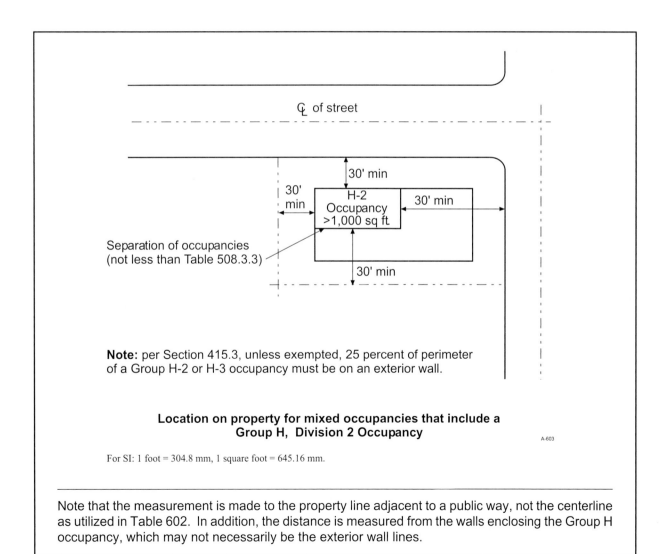

Note: per Section 415.3, unless exempted, 25 percent of perimeter of a Group H-2 or H-3 occupancy must be on an exterior wall.

**Location on property for mixed occupancies that include a
Group H, Division 2 Occupancy**

A-603

For SI: 1 foot = 304.8 mm, 1 square foot = 645.16 mm.

Note that the measurement is made to the property line adjacent to a public way, not the centerline as utilized in Table 602. In addition, the distance is measured from the walls enclosing the Group H occupancy, which may not necessarily be the exterior wall lines.

Code Text: *Group H-1 occupancies shall be in buildings used for no other purpose, shall not exceed one story in height and be without basements, crawl spaces or other under-floor spaces. Group H-2 and H-3 occupancies containing quantities of hazardous materials in excess of those set forth in Table 415.3.2 shall be in buildings used for no other purpose, shall not exceed one story in height and be without basements, crawl spaces or other under-floor spaces.*

Discussion and Commentary: Because of the explosion hazard potential associated with Group H-1 materials, Group H-1 occupancies are required to be in separate detached structures. The limitation of one story is based on the need for a building with a detonation hazard to be exited as soon as possible. Higher-level Group H-2 and Group H-3 occupancies where the quantities of hazardous materials pose an extremely high risk must also be located in buildings with no other uses.

TABLE 415.3.2
REQUIRED DETACHED STORAGE

DETACHED STORAGE IS REQUIRED WHEN THE QUANTITY OF MATERIAL EXCEEDS THAT LISTED HEREIN			
Material	**Class**	**Solids and Liquids (tons)[a]**	**Gases (cubic feet)[a][b]**
Explosives	Division 1.1	Maximum Allowable Quantity	Not Applicable
	Division 1.2	Maximum Allowable Quantity	
	Division 1.3	Maximum Allowable Quantity	
	Division 1.4	Maximum Allowable Quantity	
	Division 1.4[c]	1	
	Division 1.5	Maximum Allowable Quantity	
	Division 1.6	Maximum Allowable Quantity	
Oxidizers	Class 4	Maximum Allowable Quantity	Maximum Allowable Quantity
Unstable (reactives) detonable	Class 3 or 4	Maximum Allowable Quantity	Maximum Allowable Quantity
Oxidizer, liquids and solids	Class 3	1,200	Not Applicable
	Class 2	2,000	Not Applicable
Organic peroxides	Detonable	Maximum Allowable Quantity	Not Applicable
	Class I	Maximum Allowable Quantity	Not Applicable
	Class II	25	Not Applicable
	Class III	50	Not Applicable
Unstable (reactives) nondetonable	Class 3	1	2,000
	Class 2	25	10,000
Water reactives	Class 3	1	Not Applicable
	Class 2	25	Not Applicable
Pyrophoric gases	Not Applicable	Not Applicable	2,000

For SI: 1 ton = 906 kg, 1 cubic foot = 0.02832 m^3, 1 pound = 0.454 kg.

a. For materials that are detonable, the distance to other buildings or lot lines shall be as specified in Table 415.3.1 based on trinitrotoluene (TNT) equivalence of the material. For materials classified as explosives, see Chapter 33 the International Fire Code. For all other materials, the distance shall be as indicated in Section 415.3.1.

b. "Maximum Allowable Quantity" means the maximum allowable quantity per control area set forth in Table 307.7(1).

c. Limited to Division 1.4 materials and articles, including articles packaged for shipment, that are not regulated as an explosive under Bureau of Alcohol, Tobacco and Firearms (BATF) regulations or unpackaged articles used in process operations that do not propagate a detonation or deflagration between articles, providing the net explosive weight of individual articles does not exceed 1 pound.

The need for detached storage is a function of the type, physical state and quantity of material. Because such a single-use structure must be located an adequate distance from surrounding lot lines and other buildings, exterior walls and exterior openings need not be protected for exposure.

Code Text: *The provisions of Section 416 shall apply to the construction, installation and use of buildings and structures, or parts thereof, for the spraying of flammable paints, varnishes and lacquers or other flammable materials or mixtures or compounds used for painting, varnishing, staining or similar purposes. Such construction and equipments shall comply with the* International Fire Code. *Spray rooms shall be enclosed with fire barriers with not less than a 1-hour fire-resistance rating. Floors shall be waterproofed and drained in an approved manner..*

Discussion and Commentary: The primary hazards associated with paint spraying and spray booths originate from the presence of flammable liquids or powders and their vapors or mists. The requirements address such issues as the ventilation, sprinkler protection and interior surfaces of spray rooms and spraying spaces.

Spray rooms	Spray booths	Limited spray space
• Designed and constructed per IBC Section 416 • Separated from remainder of building by 1-hour fire barriers • Automatic fire-extinguishing system required	• Designed and constructed per IFC Section 1504.3.1 • Constructed of approved noncombustible materials • Limited in size and location • Automatic fire-extinguishing system required	• Aggregate surface area to be sprayed limited to 9 sq ft • Spraying operations not to be continuous in nature • Ventilation and wiring regulated • Portable fire extinguishers required

Chapter 15 of the *International Fire Code* provides comprehensive requirements for the application of flammable finishes, including detailed provisions for spray booths. Spray finishing operations in Group A, E, I or R occupancies shall be located in a spray room. In other occupancies, such operations may occur in a spray room, spray booth or spraying space approved for such use.

Code Text: *Occupancies in Groups I-1, R-1, R-2 and R-3 shall comply with the provisions of Section 419 and other applicable provisions of the IBC. Walls separating dwelling units in the same building and walls separating sleeping units in the same building shall comply with Section 708 (fire partitions). Floor/ceiling assemblies separating dwelling units in the same buildings and floor/ceiling assemblies separating sleeping units in the same building shall be constructed in accordance with Section 711 (horizontal assemblies).*

Discussion and Commentary: In residential-type occupancies, it is important that some degree of fire-resistive separation be provided to isolate each individual living unit from all others in the building. It is intended that should a fire initiate within one of the dwelling units or sleeping units, the occupants and contents of the other units would be adequately protected.

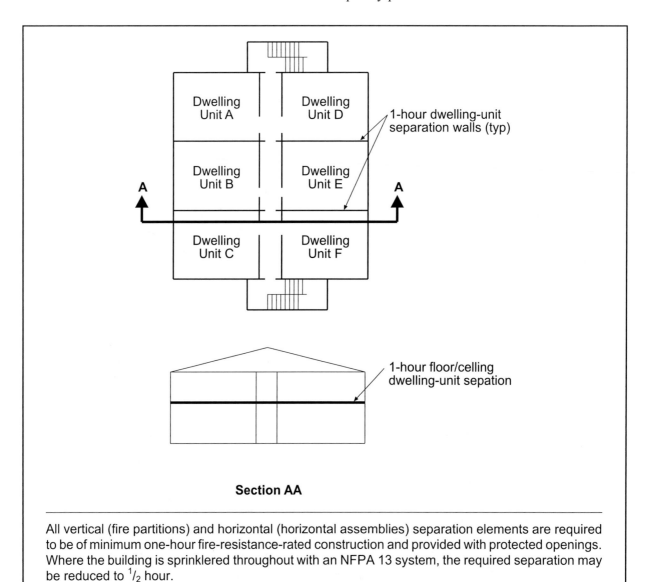

Section AA

All vertical (fire partitions) and horizontal (horizontal assemblies) separation elements are required to be of minimum one-hour fire-resistance-rated construction and provided with protected openings. Where the building is sprinklered throughout with an NFPA 13 system, the required separation may be reduced to $^1/_2$ hour.

Quiz

Study Session 14
IBC Chapter 4

1. The maximum distance from any point within a mall to an exit shall be _____ feet.

    a. 75   b. 200

    c. 250   d. 300

    Reference_____

2. In a covered mall building, groupings of kiosks shall be separated a minimum of _____ feet from other structures within the mall.

    a. 6   b. 10

    c. 20   d. 30

    Reference_____

3. A high-rise building is defined as those buildings having occupied floors located more than _____ feet above the lowest level of fire department vehicle access.

    a. 55   b. 75

    c. 120   d. 160

    Reference_____

4. In a high-rise building, if standby power is a generator within the building, it shall be located in a separate room enclosed by minimum _____ fire-resistance-rated assemblies or horizontal assemblies, or both.

    a. 1-hour fire barrier   b. 1 hour fire-partition

    c. 2-hour fire wall   d. 2- hour fire barrier

Reference_____

5. What is the minimum classification for the interior finish of atrium walls and ceilings?

    a. A   b. B

    c. C   d. DOC FF-1

Reference_____

6. At other than the lowest level of the atrium, the maximum permitted travel distance within an atrium space is _____ feet.

    a. 75   b. 100

    c. 150   d. 200

Reference_____

7. An underground building need not be divided into at least two compartments when a floor level is located a maximum of _____ feet below the lowest level of exit discharge.

    a. 30   b. 60

    c. 75   d. 120

Reference_____

8. In parking facilities, vehicle barriers need not be provided where the difference in adjacent floor elevation is a maximum of _____ inches.

    a. 12   b. 30

    c. 48   d. 60

Reference_____

9. Canopies under which fuels are dispensed shall have a minimum clearance of _____ above the surface of the drive-through area.

    a. 12 feet, 0 inches   b. 13 feet, 6 inches

    c. 14 feet, 6 inches   d. 16 feet, 0 inches

    Reference_____

10. Unless a permitted increase is applied, what is the maximum number of tiers permitted in a ramp-accessed single-use open parking garage of Type IIA construction?

    a. 8   b. 10

    c. 12   d. 15

    Reference_____

11. Smoke compartments in a Group I-2 occupancy shall have a maximum floor area of _____ square feet.

    a. 5,000   b. 10,000

    c. 12,000   d. 22,500

    Reference_____

12. On floors housing patients confined to beds, a minimum of _____ net square feet per patient shall be provided in refuge areas of Group I-2 occupancies.

    a. 3   b. 6

    c. 15   d. 30

    Reference_____

13. Doors to resident sleeping units in Group I-3 occupancies shall have a minimum clear width of _____ inches.

    a. 22   b. 26

    c. 28   d. 32

    Reference_____

14. In a Group I-3 occupancy, what is the maximum permitted number of residents in any single smoke compartment?

    a. 30   b. 50

    c. 100   d. 200

Reference_____

15. The minimum permitted size of a motion picture projection room containing 8 projecting machines shall be _____ square feet.

    a. 320   b. 360

    c. 600   d. 640

Reference_____

16. An approved fire curtain or water curtain need not be provided at the proscenium opening of stages having a maximum height of _____ feet.

    a. 35   b. 50

    c. 55   d. 75

Reference_____

17. Stages exceeding _____ square feet in floor area shall be provided with emergency ventilation.

    a. 200   b. 400

    c. 500   d. 1,000

Reference_____

18. What is the minimum interior finish classification for walls and ceilings in a special amusement building?

    a. A   b. B

    c. C   d. DOC FF-1

Reference_____

19. What is the occupancy classification of an aircraft paint hangar?

    a. Group F-1   b. Group H-2

    c. Group S-1   d. Group S-2

    Reference_____

20. How many control areas are permitted on the sixth floor above grade plane of a research and development building?

    a. zero   b. one

    c. two   d. four

    Reference_____

21. A Group H-2 liquid use, dispensing and mixing room need not be located on the outer perimeter of a building when limited to a maximum floor area of _____ square feet.

    a. 100   b. 200

    c. 500   d. 1,000

    Reference_____

22. A Group H-3 occupancy required to be in a detached building shall be located a minimum of _____ feet from all lot lines.

    a. 10   b. 20

    c. 30   d. 50

    Reference_____

23. Dead ends in service corridors of Group H-5 occupancies shall be a maximum of _____ feet in length.

    a. 4   b. 15

    c. 20   d. 30

    Reference_____

24. In a drying room, all overhead heating piping shall be located a minimum of _____ inches from combustible contents in the dryer.

    a. 1   b. 2

    c. 4   d. 6

Reference_____

25. Nitrocellulose storage located in a building where organic coatings are manufactured shall be enclosed in a room of minimum _____ fire barriers.

    a. 1-hour   b. 2-hour

    c. 3-hour   d. 4-hour

Reference_____

26. Where the stage height exceeds 50 feet, workshops and storerooms serving the stage shall be separated from the stage by minimum _____.

    a. 1-hour fire partitions   b. 1-hour fire barriers

    c. 2-hour fire barriers   d. 2-hour fire walls

Reference_____

27. In a nonsprinklered office building, an attic space used for the storage of combustible material shall be protected on the storage side with _____.

    a. $^1/_2$-inch gypsum board

    b. $^1/_2$-inch Type X gypsum board

    c. $^5/_8$-inch gypsum board

    d. any material approved for one-hour construction

Reference_____

28. A minimum _____ fire-resistance rating is required for the floor construction separating multiple control areas within a Type IIIB building.

    a. $^1/_2$-hour   b. 1-hour

    c. 2-hour   d. 3-hour

Reference_____

29. Where weather protection shelters an outdoor hazardous material storage area, the sheltered area can only be considered outdoor storage for purposes of the code where the overhead structure is of noncombustible construction and the maximum allowable area of the structure is _____ square feet, plus any permitted increases.

    a. 1,000   b. 1,500

    c. 2,000   d. 3,000

Reference_____

30. What is the minimum level of construction required to enclose spray rooms used for the application of flammable paints?

    a. noncombustible, nonrated   b. 1-hour fire partitions

    c. 1-hour fire barriers   d. 2-hour fire barriers

Reference_____

31. Where a building contains only a private garage or carport classified as a Group U occupancy, the exterior wall does not require a fire-resistance rating where the fire separation distance is a minimum of _____ feet.

    a. 3   b. 5

    c. 10   d. 20

Reference _____

32. A temporary platform shall be installed for a maximum of _____ days.

    a. 10   b. 30

    c. 90   d. 180

Reference _____

33. Unless located a minimum of _____ feet from lot lines or a public way, exterior walls of an aircraft hangar shall have a minimum 2-hour fire-resistance rating.

    a. 20   b. 30

    c. 40   d. 60

Reference _____

34. Walls separating sleeping units in the same buildings shall comply with the provisions for _____.

    a.   fire partitions   b.   fire barriers

    c.   smoke partitions   d.   smoke barriers

Reference _____

35. The gas detection system required in a hydrogen cutoff room shall be designed to activate when the level of flammable gas exceeds _____ percent of the lower flammability limit for the gas or mixtures present at their anticipated temperature and pressure.

    a.   5   b.   10

    c.   15   d.   25

Reference _____

2006 IBC Chapters 14, 15 and 18
Exterior Wall Coverings, Roofs and Foundations

OBJECTIVE: To obtain an understanding of the requirements for exterior wall coverings, including weather-resistant coverings and veneer; roofing assemblies, roof coverings and rooftop structures; and footings and foundations.

REFERENCE: Chapters 14, 15 and 18, 2006 *International Building Code*

KEY POINTS:
- What are the components of a weather-resistant wall envelope?
- What material is considered acceptable as a water-resistive barrier?
- Where is flashing to be installed?
- What is the minimum required thickness of wood veneers on exterior walls of Type I, II, III and IV buildings?
- What is anchored masonry veneer? Adhered masonry veneer?
- How must metal veneers be attached? Glass veneer? Stone veneer?
- What limitations are placed on the sill height of operable windows?
- Under what conditions is vinyl siding permitted?
- How is fiber cement siding to be applied?
- How are combustible exterior wall coverings regulated for ignition resistance? Fireblocking?
- What level of fire resistance is required of combustible balconies and similar projections?
- When is fire-retardant-treated wood permitted for the construction of balconies, porches, decks and exterior stairways in Types I and II construction?
- How shall bay windows and oriel windows be constructed?
- What is a roof assembly? Roof covering?
- What is the purpose of a roof assembly?
- How are roof assemblies classified?
- What is the effectiveness of a Class A roof assembly? Class B? Class C?
- What is a nonclassified roof? A special purpose roof? Where are such roofs permitted?
- How must roof covering materials be identified?
- How shall asphalt shingles be installed? Clay and concrete tile? Wood shakes and shingles? Metal roof panels and roof shingles?

KEY POINTS:
(Cont'd)

- What is the maximum permitted height of a penthouse? Tower or spire?
- How is reroofing addressed?
- What methods of frost protection are approved for foundation walls, piers and other permanent foundations of buildings?
- What is the minimum depth of a footing below the undisturbed ground surface?
- How shall backfill in the excavated area adjacent to the foundation be placed?
- What special considerations are applicable to footings adjacent to ascending slopes? Descending slopes?
- If not specifically designed, how must footings supporting walls of light-frame construction be constructed?
- How must concrete footings be protected from frost?
- Where is dampproofing required? Waterproofing?
- How are pier and pile foundations regulated?

Topic: Weather Protection

Category: Exterior Walls

Reference: IBC 1403.2

Subject: Performance Requirements

Code Text: *Exterior walls shall provide the building with a weather-resistant exterior wall envelope. The exterior envelope shall include flashing, described in Section 1405.3. The exterior wall envelope shall be designed and constructed in such a manner as to prevent the accumulation of water within the wall assembly by providing a water-resistant barrier behind the exterior veneer, as described in Section 1404.2, and a means for draining water that enters the assembly to the exterior.* See two exceptions.

Discussion and Commentary: The code considers it necessary to apply at least one layer of No. 15 asphalt felt, complying with ASTM D 226 for Type 1 felt, in order to provide the weather-resistant barrier between the sheathing and exterior wall veneer. The use of flashing in conjunction with the felt will provide a continuous barrier against water penetration.

TABLE 1405.2
MINIMUM THICKNESS OF WEATHER COVERINGS

COVERING TYPE	MINIMUM THICKNESS (inches)
Adhered masonry veneer	0.25
Aluminum siding	0.019
Anchored masonry veneer	2.625
Asbestos-cement boards	0.125
Asbestos shingles	0.156
Cold-rolled copper[d]	0.0216 nominal
Copper shingles[d]	0.0162 nominal
Exterior plywood (with sheathing)	0.313
Exterior plywood (without sheathing)	See Section 2304.6
Fiber cement lap siding	0.25[c]
Fiber cement panel siding	0.25[c]
Fiberboard siding	0.5
Glass-fiber reinforced concrete panels	0.375
Hardboard siding[c]	0.25
High-yield copper[d]	0.0162 nominal
Lead-coated copper[d]	0.0216 nominal
Lead-coated high-yield copper	0.0162 nominal
Marble slabs	1
Particleboard (with sheathing)	See Section 2304.6
Particleboard (without sheathing)	See Section 2304.6
Precast stone facing	0.625
Steel (approved corrosion resistant)	0.0149
Stone (cast artificial)	1.5

(Continued)

For all exterior walls other than those constructed of concrete or masonry, the IBC requires the installation of a weather-resistant exterior wall envelope. Any other approved method to resist condensation and moisture leakage is also acceptable.

Code Text: *Flashing shall be installed in such a manner so as to prevent moisture from entering the wall or to redirect it to the exterior. Flashing shall be installed at the perimeters of exterior door and window assemblies, penetrations and terminations of exterior wall assemblies, exterior wall intersections with roofs, chimneys, porches, decks, balconies and similar projections and at built-in gutters and similar locations where moisture could enter the wall. Flashing with projecting flanges shall be installed on both sides and the ends of copings, under sills and continuously above projecting trim.*

Discussion and Commentary: In general, the code requires that all intersections of exterior surfaces and/or components be flashed to prevent water intrusion. Roof and wall intersections and parapets are especially troublesome, as are exterior wall openings exposed to weather and, in particular, wind-driven rain.

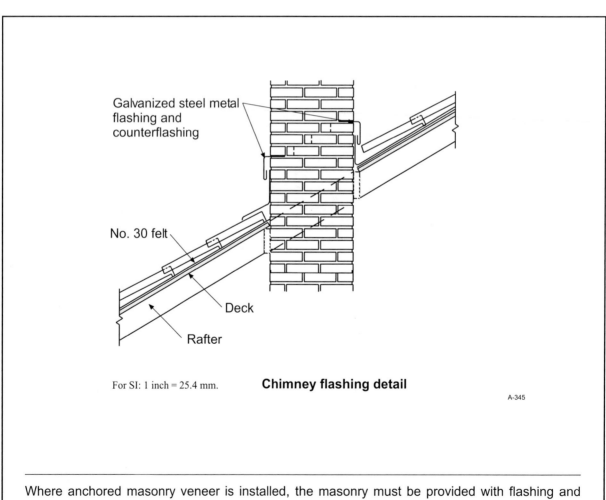

Galvanized steel metal flashing and counterflashing

No. 30 felt

Deck

Rafter

For SI: 1 inch = 25.4 mm.

Chimney flashing detail

A-345

Where anchored masonry veneer is installed, the masonry must be provided with flashing and weepholes in the first course above finished ground level above the foundation wall or slab, as well as at other points of support.

Code Text: *Veneer is a facing attached to a wall for the purpose of providing ornamentation, protection, or insulation, but not counted as adding strength to the wall. Adhered masonry veneer is veneer secured and supported through the adhesion of an approved bonding material applied to an approved backing. Anchored masonry veneer is veneer secured with approved mechanical fasteners to an approved backing. Wood veneers on exterior walls of buildings of Type I, II, III and IV construction shall be not less than 1 inch (25 mm) nominal thickness, 0.438-inch (11.1 mm) exterior hardboard siding or 0.375-inch (9.5 mm) exterior-type wood structural panels or particleboard. Metal veneers shall not be less than 0.0149-inch (0.378 mm) nominal thickness sheet steel mounted on wood or metal furring strips or approved sheathing on the wood construction.*

Discussion and Commentary: Years ago, veneer was considered an ornamental facing for a masonry wall. Today, the IBC regulates a variety of veneer materials: wood, anchored masonry, stone, slab-type, terra cotta, adhered masonry, metal, glass and, in Chapter 26, plastic. The code regulates material size, type and attachment, as well as other concerns that would cause the veneer to fail.

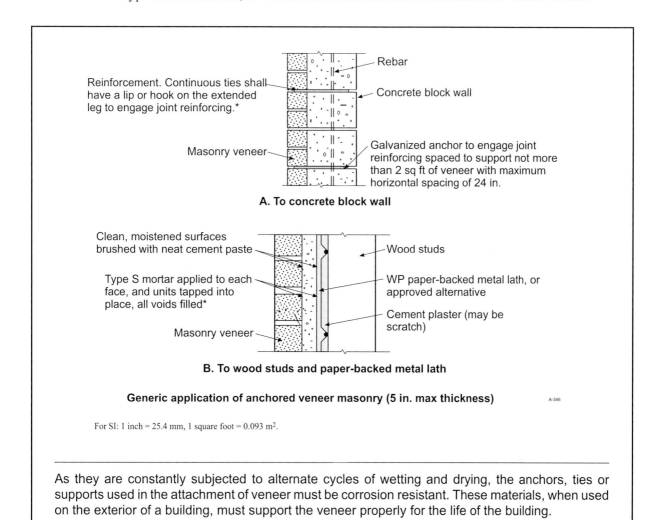

A. To concrete block wall

B. To wood studs and paper-backed metal lath

Generic application of anchored veneer masonry (5 in. max thickness)   A-346

For SI: 1 inch = 25.4 mm, 1 square foot = 0.093 m².

As they are constantly subjected to alternate cycles of wetting and drying, the anchors, ties or supports used in the attachment of veneer must be corrosion resistant. These materials, when used on the exterior of a building, must support the veneer properly for the life of the building.

Topic: Window Sills

Reference: IBC 1405.12.2

Category: Exterior Walls

Subject: Exterior Windows and Doors

Code Text: *In Occupancy Groups R-2 and R-3, one- and two-family and multiple-family dwellings, where the opening of the sill portion of an operable window is located more than 72 inches (1829 mm) above the finished grade or other surface below, the lowest part of the clear opening of the window shall be a minimum of 24 inches (610 mm) above the finished floor surface of the room in which the window is located. Glazing between the floor and a height of 24 inches (610 mm) shall be fixed or have openings such that a 4-inch (102 mm) diameter sphere cannot pass through.* See exception for openings protected with window guards.

Discussion and Commentary: The minimum sill height requirement for operable windows located a considerable height above the surface below is intended to reduce the number of falls by children from such windows. The minimum height is established at a point above the center of gravity of small children.

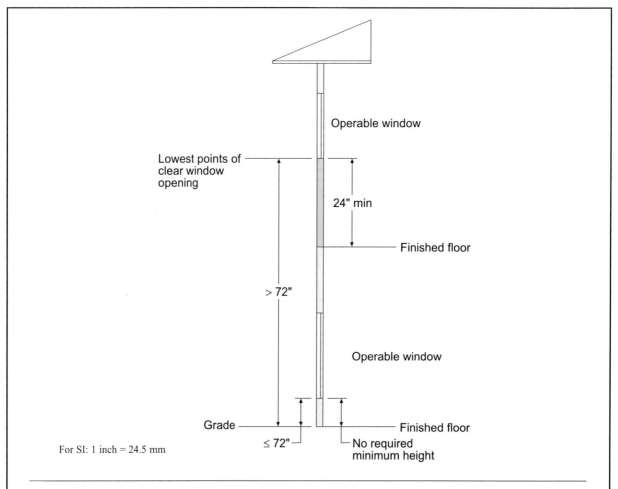

Operable window

Lowest points of clear window opening

24" min

Finished floor

> 72"

Operable window

Grade

Finished floor

≤ 72"

No required minimum height

For SI: 1 inch = 24.5 mm

The measurements on both the interior and exterior sides of the building are taken from the lowest part of the clear window opening, providing for consistent application. Where the lower window panel is inoperable, the measurement is to be taken to the lowest point of the lowest operable panel.

Code Text: *Balconies and similar projections of combustible construction other than fire-retardant-treated wood shall be fire-resistance rated in accordance with Table 601 for floor construction or shall be of Type IV construction in accordance with Section 602.4. The aggregate length shall not exceed 50 percent of the building's perimeter on each floor.* See four exceptions.

Discussion and Commentary: Although projections may extend beyond the floor area of a building, they still pose some degree of hazard due to their materials of construction. Combustible projections are generally limited to buildings of Types III, IV and V construction; however, fire-retardant-treated wood is acceptable for balconies, porches, decks and exterior stairways not used as required exits in Type I or II buildings that are three stories or less in height.

Balconies and similar projections, when combustible, to have fire-resistive rating as for floor, or be of Type IV construction, except:

- Fire-retardant-treated wood in Types I and II, limited to 3 stories in height, and not used for exiting

- Permitted to be Type V in buildings of Type III, IV or V with no fire rating where sprinkler protection extended to projections

- Guard devices, such as pickets and rails, limited to 42 inches in height

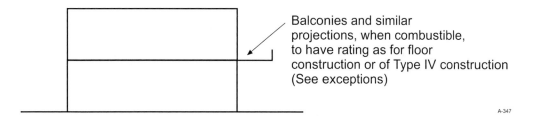

Balconies and similar projections, when combustible, to have rating as for floor construction or of Type IV construction (See exceptions)

A-347

If not constructed of fire-retardant-treated wood, combustible projections must be of heavy timber or must provide the same degree of fire resistance as the building's floor construction. However, in sprinklered buildings of combustible construction, no fire-resistance rating is required.

Code Text: *A roof assembly is a system designed to provide weather protection and resistance to design loads. The system consists of a roof covering and roof deck or a single component serving as both the roof covering and the roof deck. A roof assembly includes the roof deck, vapor retarder, substrate or thermal barrier, insulation, vapor retarder and roof covering. Roof covering is the covering applied to the roof deck for weather resistance, fire classification or appearance.*

Discussion and Commentary: There are many components of a roof assembly. Viewed as a unit, a roof assembly is regulated for its resistance to wind, weathering, impact and fire. In addition, roof coverings must be designed, installed and maintained to protect the building from the weather.

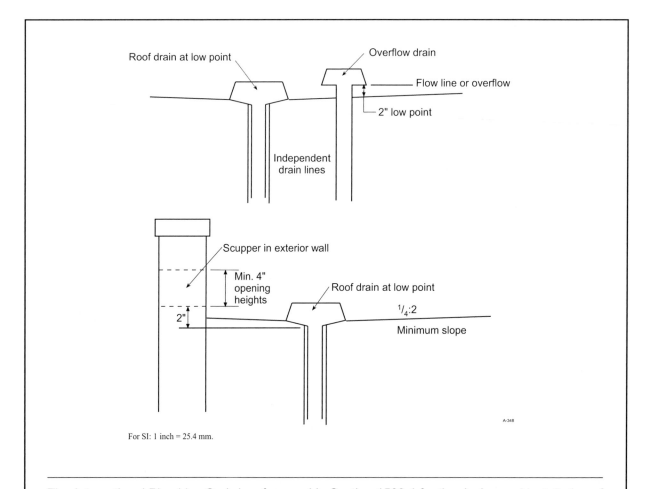

For SI: 1 inch = 25.4 mm.

The *International Plumbing Code* is referenced in Section 1503.4 for the design and installation of roof drainage systems. Where water on the roof is not intended to flow over the roof edge, provisions for roof drainage will include primary roof drains supplemented by overflow drains.

Code Text: *Roof decks shall be covered with approved roof coverings secured to the building or structure in accordance with the provisions of Chapter 15. Roof coverings shall be designed, installed and maintained in accordance with the IBC and the approved manufacturer's installation instructions such that the roof covering shall serve to protect the building or structure. Roof coverings shall be applied in accordance with the applicable provisions of Section 1507 and the manufacturer's instructions.*

Discussion and Commentary: The IBC contains installation requirements for a number of types of roof covering materials and systems. Selectively included in the provisions are deck requirements, limitations on roof slope, underlayment, materials, fasteners and attachment, flashings and application methods.

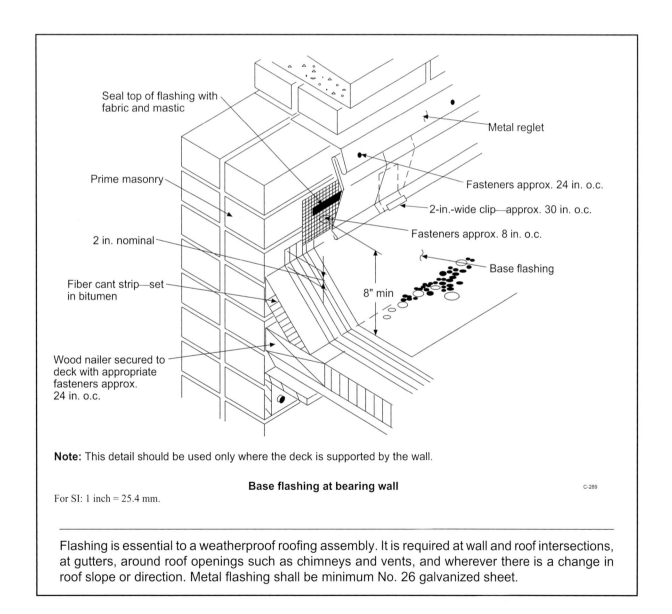

Seal top of flashing with fabric and mastic

Metal reglet

Prime masonry

Fasteners approx. 24 in. o.c.

2-in.-wide clip—approx. 30 in. o.c.

Fasteners approx. 8 in. o.c.

2 in. nominal

Base flashing

Fiber cant strip—set in bitumen

8" min

Wood nailer secured to deck with appropriate fasteners approx. 24 in. o.c.

Note: This detail should be used only where the deck is supported by the wall.

Base flashing at bearing wall

C-289

For SI: 1 inch = 25.4 mm.

Flashing is essential to a weatherproof roofing assembly. It is required at wall and roof intersections, at gutters, around roof openings such as chimneys and vents, and wherever there is a change in roof slope or direction. Metal flashing shall be minimum No. 26 galvanized sheet.

Code Text: *Gravel or stone shall not be used on the roof of a building located in a hurricane-prone region as defined in Section 1609.2, or on any other building with a mean roof height exceeding that permitted by Table 1504.8 based on the exposure category and basic wind speed at the building site.*

Discussion and Commentary: Field assessments of damage to buildings caused by high-wind events have shown that gravel or stone blown from the roofs of buildings has increased the damage to other buildings due to the breakage of glass. Once the glass is broken, higher internal pressures are created within the building, often resulting in substantial structural damage. In addition, breakage of windows will generally result in considerable wind and water damage to the building's interior and contents.

TABLE 1504.8
MAXIMUM ALLOWABLE MEAN ROOF HEIGHT PERMITTED FOR
BUILDINGS WITH GRAVEL OR STONE ON THE ROOF IN AREAS
OUTSIDE A HURRICANE-PRONE REGION

BASIC WIND SPEED FROM FIGURE 1609 (mph)[b]	MAXIMUM MEAN ROOF HEIGHT (ft)[a,c]		
	Exposure category		
	B	C	D
85	170	60	30
90	110	35	15
95	75	20	NP
100	55	15	NP
105	40	NP	NP
110	30	NP	NP
115	20	NP	NP
120	15	NP	NP
Greater than 120	NP	NP	NP

For SI: 1 foot = 304.8 mm; 1 mile per hour = 0.447 m/s.
a. Mean roof height in accordance with Section 1609.2.
b. For intermediate values of basic wind speed, the height associated with the next higher value of wind speed shall be used, or direct interpolation is permitted.
c. NP = gravel and stone not permitted for any roof height.

Gravel and stone is prohibited on roofs of buildings in hurricane-prone regions, defined as areas along the U.S. Atlantic Ocean and Gulf of Mexico coasts where the basic wind speed exceeds 90 miles per hour, as well as the islands of Hawaii, Puerto Rico, Guam, Virgin Islands and American Samoa.

Topic: General Requirements   **Category:** Roof Assemblies and Rooftop Structures
Reference: IBC 1505.1   **Subject:** Fire Classification

Code Text: *Roof assemblies shall be divided into the classes defined below. Class A, B and C roof assemblies and roof coverings required to be listed by Section 1505 shall be tested in accordance with ASTM E 108 or UL 790. The minimum roof coverings installed on buildings shall comply with Table 1505.1 based on the type of construction of the building. See exception for skylights and sloped glazing.*

Discussion and Commentary: The various required roof covering classifications are related directly to the type of construction of the building. Based on Table 1505.1, a minimum level of fire protection is assigned to address external fire exposures. The exposures are generally created by fires in adjoining structures, wild fires and fire from the subject building that may extend up the exterior wall and onto the top surface of the roof.

TABLE 1505.1[a,b]
MINIMUM ROOF COVERING CLASSIFICATION
FOR TYPES OF CONSTRUCTION

IA	IB	IIA	IIB	IIIA	IIIB	IV	VA	VB
B	B	B	C[c]	B	C[c]	B	B	C[c]

For SI: 1 foot = 304.8 mm, 1 square foot = 0.0929 m^2.

a. Unless otherwise required in accordance with the International Wildland-Urban Interface Code or due to the location of the building within a fire district in accordance with Appendix D.

b. Nonclassified roof coverings shall be permitted on buildings of Group R-3 and Group U occupancies, where there is a minimum fire-separation distance of 6 feet measured from the leading edge of the roof.

c. Buildings that are not more than two stories in height and having not more than 6,000 square feet of projected roof area and where there is a minimum 10-foot fire-separation distance from the leading edge of the roof to a lot line on all sides of the building, except for street fronts or public ways, shall be permitted to have roofs of No. 1 cedar or redwood shakes and No. 1 shingles.

In addition to the Class A, B and C listed roof assemblies, the IBC permits the use of nonclassified roofing and special purpose roofs under limited conditions. These types of roof coverings are limited, respectively, to Group R-3 occupancies and small nonfire-rated buildings.

Code Text: *Class A roof assemblies are those that are effective against severe fire-test exposure. Class B roof assemblies are those that are effective against moderate fire-test exposure. Class C roof assemblies are those that are effective against light fire-test exposure. Nonclassified roofing is approved material that is not listed as a Class A, B or C roof covering.*

Discussion and Commentary: Traditional Class A roof coverings include masonry, concrete, slate, tile and cement-asbestos. Additionally, any assembly that is tested as Class A in accordance with ASTM E 108 by an approved testing agency and is listed and identified by that agency is included. Traditional Class B roof coverings include metal sheets and shingles, as well as those tested, listed and identified as Class B. There are no specific materials that qualify as Class C; therefore all Class C roof covering are listed as such.

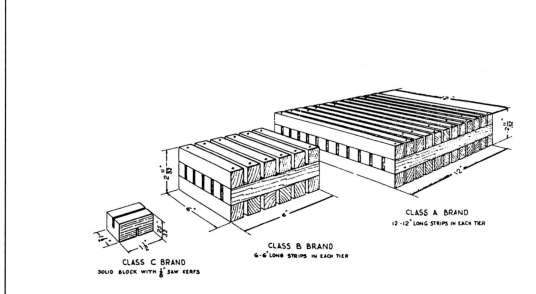

CLASS A BRAND
12-12" LONG STRIPS IN EACH TIER

CLASS B BRAND
6-6" LONG STRIPS IN EACH TIER

CLASS C BRAND
SOLID BLOCK WITH ⅛" SAW KERFS

Brands for Classes A, B, and C Tests

Several test methods are included as a part of the fire-test-response standard ASTM E 108, including the intermittent flame exposure test, spread of flame test, burning brand test, flying brand test and rain test. It also is critical that the roof coverings do not slip from position.

Topic: Identification	**Category:** Roof Assemblies and Rooftop Structures
Reference: IBC 1506.4	**Subject:** Roof Covering Materials

Code Text: *Roof covering materials shall be delivered in packages bearing the manufacturer's identifying marks and approved testing agency labels required in accordance with Section 1505. Bulk shipments of materials shall be accompanied with the same information issued in the form of a certificate or on a bill of lading by the manufacturer.*

Discussion and Commentary: Roof covering materials must comply with the appropriate quality standards set forth in Section 1507 for each different type of material. The materials must be compatible with the building or structure to which they are applied. In addition, identification of the roof covering materials is mandatory to verify that they comply with the quality standard.

Asphalt Shingles	Section 1507.2
Clay and Concrete Tile	Section 1507.3
Metal Roof Panels	Section 1507.4
Metal Roof Shingles	Section 1507.5
Mineral-surfaced Roll Roofing	Section 1507.6
Slate Shingles	Section 1507.7
Wood Shingles	Section 1507.8
Wood Shakes	Section 1507.9
Built-up Roofs	Section 1507.10
Modified Bitumen Roofing	Section 1507.11
Thermoset Single-ply Roofing	Section 1507.12
Thermoplastic Single-ply Roofing	Section 1507.13
Sprayed Polyurethane Foam Roofing	Section 1507.14
Liquid-applied Coatings	Section 1507.15

Where there are not applicable standards for a specific roof covering material, or where the materials are of questionable suitability, the building official must ask for testing by an approved agency to determine the material's character, quality and limitations of application.

Code Text: *Asphalt shingles shall be fastened to solidly sheathed decks. Asphalt shingles shall only be used on roof slopes of two units vertical in 12 units horizontal (17-percent slope) or greater. Asphalt shingles shall be secured to the roof with not less than four fasteners per strip shingle or two fasteners per individual shingle. For roof slopes from two units vertical in 12 units horizontal (17-percent slope), up to four units vertical in 12 units horizontal (33-percent slope), underlayment shall be two layers . . . . Distortions in the underlayment of the shingles shall not interfere with the ability of the shingles to seal. Provide drip edge at eaves and gables of shingle roofs.*

Discussion and Commentary: There are two fundamental types of asphalt shingles: strip shingles (the most common type) such as three-tab shingles, and individual interlocking shingles such as t-lock shingles. In addition to three-tab, other strip shingles include random or multi-tab, no-cut-out and laminated architectural.

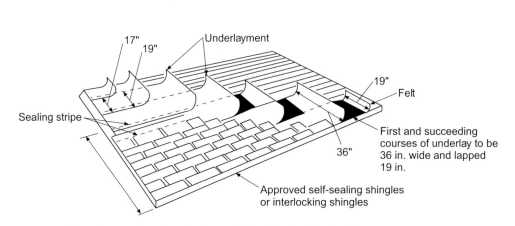

Note: In areas where there has been a history of ice forming along the eaves causing a backup of water, felt plies of underlayment should be cemented up from eaves far enough to overlie a point 24 in. inside the wall line of the building.

source NRCA

For SI: 1 inch = 25.4 mm, °C = [(°F)-32/1.8].

A-674

Application of asphalt shingle on slopes between 2:12 and 4:12

Asphalt shingles are typically classified in two types, either cellulose felt reinforced (i.e., organic shingles) and fiberglass mat reinforced (i.e., fiberglass shingles). Consistent with the requirements of Section 1203.2, the roofing industry recommends the attic space below asphalt shingle roofs be properly ventilated.

Topic: Wood Shakes	**Category:** Roof Assemblies and Rooftop Structures
Reference: IBC 1507.9	**Subject:** Roof Coverings

Code Text: *Wood shakes shall only be used on solid or spaced sheathing. Wood shakes shall only be used on slopes of four units vertical in 12 units horizontal (33-percent slope) or greater. Interlayment shall comply with ASTM D 226, Type I. Fasteners for wood shakes shall be corrosion resistant with a minimum penetration of 0.75 inch (19.1 mm) into the sheathing. Wood shakes shall be laid with a side lap not less than 1.5 inches (38 mm) between joints in adjacent courses. Spacing between shakes in the same course shall be 0.375 to 0.625 (9.5 to 15.9 mm) inches for shakes and taper sawn shakes of naturally durable wood and shall be 0.25 to 0.375 inch (6.4 to 9.5 mm) for preservative taper sawn shakes. Weather exposure for wood shakes shall not exceed those set in Table 1507.9.7.*

Discussion and Commentary: Wood shakes, which are defined as roofing products split from logs and then shaped as required by the individual manufacturers, differ from wood shingles in that shingles are defined as sawed wood products featuring a uniform butt thickness per individual length.

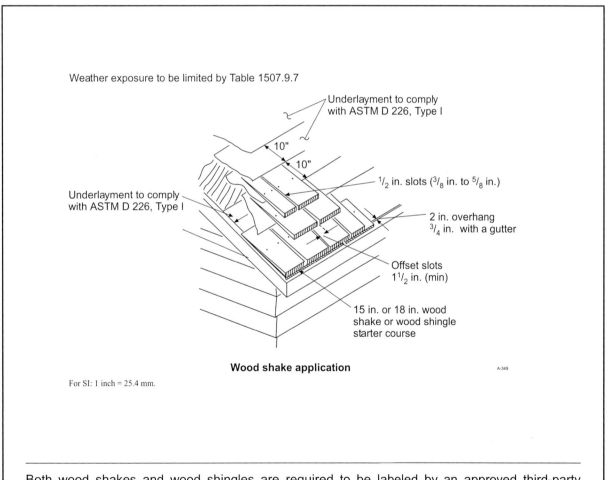

Wood shake application

A-349

For SI: 1 inch = 25.4 mm.

Both wood shakes and wood shingles are required to be labeled by an approved third-party inspection agency. The applicable set of grading rules, required of the quality control program, is typically prescribed by the Cedar Shake and Shingle Bureau.

Code Text: *A penthouse or other projection above the roof in structures of other than Type I construction shall not exceed 28 feet (8534 mm) above the roof where used as an enclosure for tanks or for elevators that run to the roof and in all other cases shall not extend more than 18 (5486 mm) feet above the roof. The aggregate area of penthouses and other rooftop structures shall not exceed one-third the area of the supporting roof. A penthouse, bulkhead or any other similar projection above the roof shall not be used for purposes other than shelter of mechanical equipment or shelter of vertical shaft openings in the roof.*

Discussion and Commentary: The general premise is that a penthouse be treated no differently than any other portion of the building. However, the reductions in the general requirements for a story recognize the lack of occupant load or fire loading, as well as the reduced exposure of penthouses when the exterior wall is recessed from the exterior wall of the building.

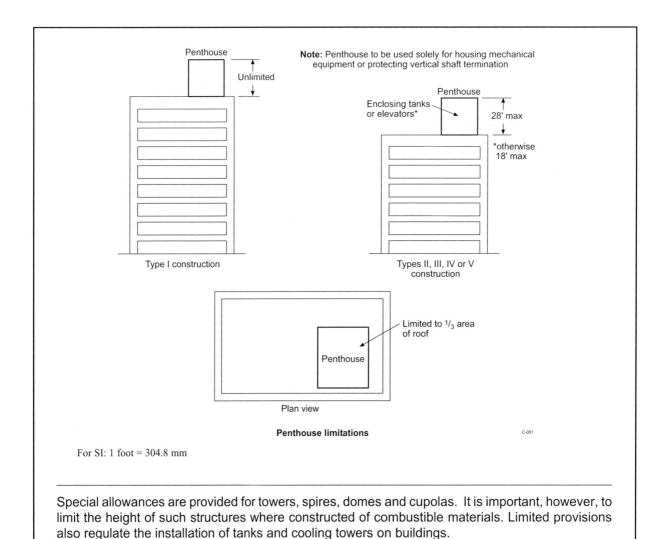

Penthouse

Note: Penthouse to be used solely for housing mechanical equipment or protecting vertical shaft termination

Unlimited

Penthouse

Enclosing tanks or elevators*

28' max

*otherwise 18' max

Type I construction

Types II, III, IV or V construction

Limited to ⅓ area of roof

Penthouse

Plan view

Penthouse limitations

C-051

For SI: 1 foot = 304.8 mm

Special allowances are provided for towers, spires, domes and cupolas. It is important, however, to limit the height of such structures where constructed of combustible materials. Limited provisions also regulate the installation of tanks and cooling towers on buildings.

Code Text: *Footings and foundations shall be built on undisturbed soil, compacted fill material, or CLSM (controlled low-strength material). The top surface of footings shall be level. The bottom surface of footings are permitted to have a slope not exceeding 1 unit vertical in 10 units horizontal (10-percent slope). Footings shall be stepped where it is necessary to change the elevation of the top surface of the footing or where the surface of the ground slopes more than 1 unit vertical in 10 units horizontal (10-percent slope).*

Discussion and Commentary: If compacted fill material is used to support a footing, the material must be in compliance with the provisions of an approved report. The code identifies seven issues that must be addressed in the report, including specifications for both the site preparation and the material to be used as fill.

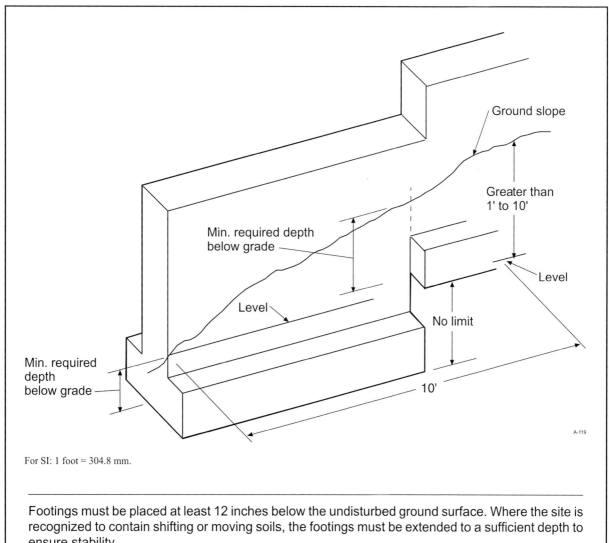

Ground slope

Greater than 1' to 10'

Min. required depth below grade

Level

No limit

Min. required depth below grade

Level

10'

A-119

For SI: 1 foot = 304.8 mm.

Footings must be placed at least 12 inches below the undisturbed ground surface. Where the site is recognized to contain shifting or moving soils, the footings must be extended to a sufficient depth to ensure stability.

Code Text: *The minimum depth of footings below the undisturbed ground surface shall be 12 inches (305 mm). Except where otherwise protected from frost, foundation walls, piers and other permanent supports of buildings and structures shall be protected from frost by one or more of the following methods: 1) extending below the frost line of the locality, 2) construction in accordance with ASCE-32* (Design and Construction of Frost Protected Shallow Foundations)*, or 3) erecting on solid rock.* See exception for small free-standing structures. *Footings shall not bear on frozen soil unless such frozen condition is of a permanent character.*

Discussion and Commentary: In winter, frost action can raise the ground level (frost heave), whereas in springtime, the same area will soften and settle back. If foundations are constructed on soils that can freeze, then the heave or vertical movement of the ground, which is rarely uniform, can cause serious damage to buildings and other structures.

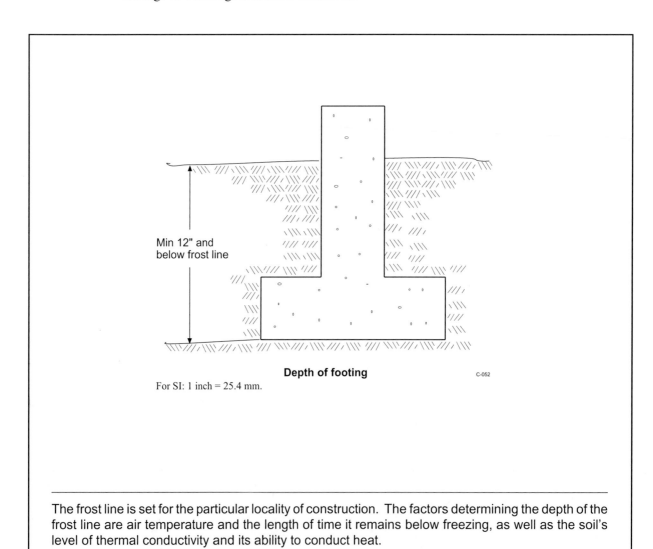

Min 12" and below frost line

Depth of footing

C-052

For SI: 1 inch = 25.4 mm.

The frost line is set for the particular locality of construction. The factors determining the depth of the frost line are air temperature and the length of time it remains below freezing, as well as the soil's level of thermal conductivity and its ability to conduct heat.

Topic: Foundation Elevation

Category: Soils and Foundations

Reference: IBC 1805.3.4

Subject: Foundations

Code Text: *On graded sites (where buildings are placed on or adjacent to slopes), the top of any exterior foundation shall extend above the elevation of the street gutter at point of discharge or the inlet of an approved drainage device a minimum of 12 inches (305 mm) plus 2 percent. Alternate elevations are permitted subject to the approval of the building official, provided it can be demonstrated that required drainage to the point of discharge and away from the structure is provided at all locations on the site.*

Discussion and Commentary: Where natural drainage away from a building is not available, the site must be graded so that water will not drain toward, or accumulate at, the exterior foundation wall. A prescriptive elevation is set forth that will ensure positive drainage to a street gutter or other drainage point; however, any other method that moves water away from the building can be accepted by the building official.

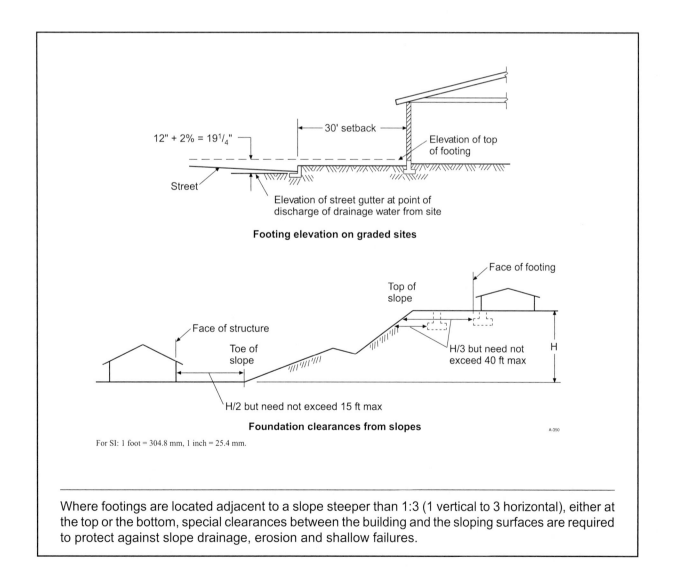

12" + 2% = 19¹/₄"

30' setback

Elevation of top of footing

Street

Elevation of street gutter at point of discharge of drainage water from site

Footing elevation on graded sites

Face of footing

Top of slope

Face of structure

Toe of slope

H/3 but need not exceed 40 ft max

H

H/2 but need not exceed 15 ft max

Foundation clearances from slopes

A-350

For SI: 1 foot = 304.8 mm, 1 inch = 25.4 mm.

Where footings are located adjacent to a slope steeper than 1:3 (1 vertical to 3 horizontal), either at the top or the bottom, special clearances between the building and the sloping surfaces are required to protect against slope drainage, erosion and shallow failures.

Code Text: *Where a specific design is not provided, concrete footings supporting walls of light-frame construction are permitted to be designed in accordance with Table 1805.4.2.*

Discussion and Commentary: In lieu of an engineered design, Table 1805.4.2 provides a prescriptive method for determining footing size criteria that can be used in conjunction with conventional light-framed construction. The minimum thickness of the foundation wall, as well as the minimum width and thickness of the footing, are specified based on the number of floors supported. The minimum depth below undisturbed ground surface is also addressed. Unless protected from frost or erected on solid rock, the footings must also extend below the frost line. The table is based on anticipated loads on the footings and foundations due to wall, floor and roof systems.

TABLE 1805.4.2
FOOTINGS SUPPORTING WALLS OF LIGHT-FRAME CONSTRUCTION[a, b, c, d, e]

NUMBER OF FLOORS SUPPORTED BY THE FOOTING[f]	WIDTH OF FOOTING (inches)	THICKNESS OF FOOTING (inches)
1	12	6
2	15	6
3	18	8[g]

For SI: 1 inch = 25.4 mm, 1 foot = 304.8 mm.

a. Depth of footings shall be in accordance with Section 1805.2.

b. The ground under the floor is permitted to be excavated to the elevation of the top of the footing.

c. Interior-stud-bearing walls are permitted to be supported by isolated footings. The footing width and length shall be twice the width shown in this table, and footings shall be spaced not more than 6 feet on center.

d. See Section 1908 for additional requirements for footings of structures assigned to Seismic Design Category C, D, E or F.

e. For thickness of foundation walls, see Section 1805.5.

f. Footings are permitted to support a roof in addition to the stipulated number of floors. Footings supporting roof only shall be as required for supporting one floor.

g. Plain concrete footings for Group R-3 occupancies are permitted to be 6 inches thick.

Although Table 1805.4.2 is normally used for continuous footings, it can also be used for isolated footings that support interior-stud bearing walls. The footings shall be spaced a maximum of 6 feet on center, with their widths and lengths being twice that shown in the table.

Study Session 15
IBC Chapters 14, 15 and 18

1. An exterior wall is defined as a building enclosing wall that has a minimum slope of _____ degrees with the horizontal plane.

    a. 45   b. 60

    c. 75   d. 90

Reference_____

2. Veneer secured and supported through the adhesion of an approved bonding material applied to an approved backing is considered _____ masonry veneer.

    a. adhered   b. anchored

    c. attached   d. spandrel

Reference_____

3. A minimum of _____ shall be attached to exterior sheathing in order to provide a continuous water-resistant barrier behind the exterior wall veneer.

    a. one layer of No. 15 asphalt felt

    b. one layer of No. 30 asphalt felt

    c. two layers of No. 15 asphalt felt applied shingle fashion

    d. two layers of No. 30 asphalt felt applied shingle fashion

Reference_____

4. Precast stone facing shall be a minimum of _____ inch in thickness in order to be acceptable as an approved weather covering.

    a. 0.50   b. 0.625

    c. 0.75   d. 1.00

Reference_____

5. Interior adhered masonry veneer shall have a maximum weight of _____ psf.

    a. 20   b. 25

    c. 30   d. 40

Reference_____

6. What is the minimum permitted thickness for metal veneer mounted on approved sheathing on wood construction?

    a. 0.0149 inch   b. 0.0224 inch

    c. 0.0478 inch   d. 0.0630 inch

Reference_____

7. Combustible exterior wall coverings are permitted in Type II buildings having a maximum height of three stories or _____ feet above grade plane.

    a. 25   b. 30

    c. 35   d. 40

Reference_____

8. Roofing interlayment shall have a minimum width of _____ inches.

    a. 12   b. 18

    c. 34   d. 36

Reference_____

9. What is the minimum roof covering classification for a roof assembly on a building of Type IA construction?

    a. Class A   b. Class B

    c. Class C   d. nonclassified

Reference_____

10. Roof assemblies consisting of metal sheets or shingles are considered _____ roof assemblies.

    a. Class A   b. Class B

    c. Class C   d. special purpose

Reference_____

11. Double underlayment application is required beneath asphalt shingles on roofs having a slope of 2:12 to _____.

    a. $2^1/_2$:12   b. 3:12

    c. 4:12   d. 5:12

Reference_____

12. In areas subject to high winds, underlayment beneath asphalt shingles shall be fastened along the overlap at a maximum spacing of _____ inches on center.

    a. 2   b. 18

    c. 24   d. 36

Reference_____

13. Fasteners for concrete or clay roof tiles shall penetrate the deck a minimum of _____ inch or through the thickness of the deck, whichever is less.

    a. $^1/_2$ b. $^5/_8$

    c. $^3/_4$ d. 1

Reference_____

14. What is the minimum permitted roof slope for the installation of metal roof shingles?

    a. 2:12   b. 3:12

    c. 4:12   d. 5:12

Reference_____

15. Wood shakes shall be applied to a roof with a minimum side lap of _____ inch(es) between joints in adjacent courses.

    a. $^3/_8$ b. $^1/_2$

    c. $^3/_4$ d. $1^1/_2$

Reference_____

16. Where 18-inch-long No. 1 wood shingles are installed on a roof of 12:12 pitch, the maximum weather exposure shall be _____ inches.

    a. $3^3/_4$ b. $4^1/_4$

    c. $4^1/_2$ d. $5^1/_2$

    Reference_____

17. Sprayed polyurethane foam roofs shall have a minimum design slope of _____ for drainage purposes.

    a. $^1/_8$:12   b. $^1/_4$:12

    c. 1:12   d. 2:12

    Reference_____

18. In other than Type I construction, a penthouse shall extend a maximum of _____ feet above the roof when used for the protection of rooftop mechanical equipment.

    a. 8   b. 12

    c. 18   d. 28

    Reference_____

19. The bottom surface of footings shall have a maximum slope of _____.

    a. $^1/_2$:12   b. 1:12

    c. 1:10   d. 1:20

    Reference_____

20. The bottom of a footing shall be located a minimum of _____ inches below the undisturbed ground surface.

    a. 6   b. 12

    c. 15   d. 18

    Reference_____

21. Unless data is submitted that substantiates the use of a higher value, the maximum allowable foundation pressure for silty gravel supporting soil shall be _____ pounds per square foot.

    a. 12,000   b. 4,000

    c. 3,000   d. 2,000

Reference_____

22. At 28 days, concrete in footings shall have a minimum specified compressive strength of _____ pounds per square inch.

    a. 1,500   b. 2,000

    c. 2,500   d. 3,000

Reference_____

23. Concrete footings shall be protected from freezing during depositing and for a minimum time period of _____ thereafter.

    a. 12 hours   b. 24 hours

    c. 3 days   d. 5 days

Reference_____

24. Unless a specific design is provided, concrete footings supporting two floors of light-frame construction shall be a minimum of _____ inches in width and _____ inches in thickness.

    a. 12, 6   b. 15, 6

    c. 15, 7   d. 18, 8

Reference_____

25. At the girder supports at the top of hollow masonry foundation walls, a minimum of _____ inches of solid masonry shall be provided.

    a. 3   b. 4

    c. 6   d. 8

Reference_____

26. For which of the following types of construction is vinyl siding permitted to be installed on the exterior walls of buildings?

    a. Type IIA   b. Type IIB

    c. Type IIIB   d. Type VA

    Reference_____

27. The construction of a special purpose wood shake roof mandates a minimum underlayment of _____ placed under the roof sheathing.

    a. 0.5-inch wood structural panel

    b. 0.5-inch gypsum wallboard

    c. 0.625-inch Type X gypsum sheathing

    d. 0.5-inch water-resistant gypsum backing board

    Reference_____

28. Slate shingles shall be installed only on roof decks having a slope of _____ or greater.

    a. 2:12   b. 2 $^1/_2$:12

    c. 3:12   d. 4:12

    Reference_____

29. A cupola used as an architectural embellishment, where built of combustible construction, is limited to a maximum height of _____ feet above grade plane.

    a. 45   b. 50

    c. 60   d. 85

    Reference_____

30. Unless warranted by climatic or soil conditions, a minimum ground slope of _____ is required away from a building's foundation wall for a minimum distance of _____ feet.

    a. 1:48, 10   b. 1:20, 10

    c. 1:15, 5   d. 1:12, 5

    Reference_____

31. In a Group R-2 multiple-family dwelling, where the sill portion of an operable window is located more than 72 inches above the finished grade or surface below, the lowest part of the clear window openings shall be a minimum of _____ inches above the finished floor surface of the room in which the window is located.

    a. 18   b. 24

    c. 30   d. 44

Reference _____

32. Fiber cement horizontal lap siding shall be installed with a minimum lap of _____ inch(es).

    a. $^3/_4$ b. 1

    c. $1^1/_4$ d. $1^1/_2$

Reference _____

33. Fire-retardant-treated wood is permitted for the construction of balconies, porches, decks and exterior stairways of Type I and II buildings, provided the building is a maximum of _____ in height.

    a. 3 stories   b. 4 stories

    c. 55 feet   d. 75 feet

Reference _____

34. Where a building is located on a site with a basic wind speed of 90 mph and an exposure category of B, gravel or stone is permitted on the roof, provided the building has a maximum height of _____ feet.

    a. 0 (gravel and stone roof covering materials prohibited in Exposure Category B)

    b. 15

    c. 35

    d. 110

Reference _____

35. Grout used in micropiles shall have a minimum 28-day specified compressive strength of _____ psi.

    a. 2,500   b. 3,000

    c. 3,500   d. 4,000

Reference _____

2006 IBC Chapters 16, 17, 19, 21, 22 and 23
Special Inspections, Concrete, Masonry and Wood

OBJECTIVE: To identify the provisions relating to general structural forces and engineered design; applicable structural tests; special inspections and structural observation; and specific materials of construction, including concrete, masonry, steel and wood.

REFERENCE: Chapters 16, 17, 19, 21, 22 and 23, 2006 *International Building Code*

KEY POINTS:
- In structural design, what is considered a live load? A dead load?
- How is the minimum design live load for a floor system determined? Concentrated loads? Partition loads?
- How shall design roof loads be determined? When should snow loads be considered?
- What is the design criteria for wind loads? Seismic loads? Flood loads?
- What is special inspection? What are the duties and responsibilities of a special inspector?
- Which types of work shall be inspected by a special inspector?
- What is structural observation? When is structural observation required?
- What information is required in the statement of special inspections?
- How does Chapter 19 (Concrete) relate to the provisions of ACI 318?
- How shall concrete be evaluated and accepted?
- Where are the general requirements for concrete mixing, conveying, depositing and curing located?
- In masonry construction, how shall mortar and grout be regulated?
- How shall masonry be prepared, constructed and protected in cold weather? In hot weather?
- How can the lateral stability of masonry walls be accomplished?
- How must a masonry chimney be designed, anchored, supported and reinforced?
- What are the minimum required thicknesses of masonry fireplace walls and firebox walls?
- What is the minimum clearance required between combustible materials and fireplace or chimney walls? Between combustible materials and the fireplace opening?
- How must the hearth be constructed for a masonry fireplace?
- Hearth extensions must be of what minimum size?

KEY POINTS: • At what minimum height must a masonry chimney terminate?
(Cont'd) • How can the minimum capacity of structural wood-framing members be established?
• For decay and termite protection, what manner of under-floor clearance is required between exposed ground and wood girders, joists or structural floors?
• How must structural floor and roof sheathing be designed?
• What is considered "conventional light-frame construction"?
• How shall girders be supported?
• What are the limitations on the notching and boring of holes in floor joists, ceiling joists and roof rafters?
• What are braced wall panels? Where are such panels required?
• How shall rafters be framed at the ridge?
• What is a purlin? How shall a purlin system be constructed?

Code Text: *The live loads used in the design of buildings and other structures shall be the maximum loads expected by the intended use or occupancy but shall in no case be less than the minimum uniformly distributed unit loads required by Table 1607.1. Floors and other similar surfaces shall be designed to support the uniformly distributed live loads prescribed in Section 1607.3 or the concentrated load, in pounds, given in Table 1607.1, whichever produces the greater load effects.*

Discussion and Commentary: The anticipated live loads are based on the daily use of the building, as well as any temporary loading conditions such as remodeling activities, large group gatherings and short-term storage. They also reflect that, within the general use category, changes will likely occur in furniture layout, traffic patterns, etc. Concentrated loads take into account more specific types of loading consistent with the use of the building.

TABLE 1607.1
MINIMUM UNIFORMLY DISTRIBUTED LIVE LOADS AND MINIMUM CONCENTRATED LIVE LOADS[g]

OCCUPANCY OR USE	UNIFORM (psf)	CONCENTRATED (lbs.)	OCCUPANCY OR USE	UNIFORM (psf)	CONCENTRATED (lbs.)
1. Apartments (see residential)	—	—	23. Libraries		
			Corridors above first floor	80	1,000
2. Access floor systems			Reading rooms	60	1,000
Office use	50	2,000	Stack rooms	150[b]	1,000
Computer use	100	2,000	24. Manufacturing		
3. Armories and drill rooms	150	—	Heavy	250	3,000
			Light	125	2,000
4. Assembly areas and theaters			25. Marquees	75	—
Fixed seats (fastened to floor)	60				
Follow spot, projections and control rooms	50	—	26. Office buildings		
Lobbies	100		Corridors above first floor	80	2,000
Movable seats	100		File and computer rooms shall be designed for heavier loads based on anticipated occupancy	—	—
Stages and platforms	125		Lobbies and first-floor corridors	100	2,000
5. Balconies	100		Offices	50	2,000
On one- and two-family residences only, and not exceeding 100 sq ft	60	—	27. Penal institutions		
			Cell blocks	40	
6. Bowling alleys	75	—	Corridors	100	
7. Catwalks	40	300	28. Residential		
8. Dance halls and ballrooms	100	—	One- and two-family dwellings		
			Uninhabitable attics without storage[i]	10	
9. Decks	Same as occupancy served[h]	—	Uninhabitable attics with limited storage[i, j, k]	20	
			Habitable attics and sleeping areas	30	
			All other areas except balconies and decks	40	—

(continued)

The code does not require the concurrent application of uniform live load and concentrated live load. The load to be utilized in the structural design of the building would be of the type that produces the greater stress in the structural elements.

Code Text: *Basement, foundation and retaining walls shall be designed to resist lateral soil loads. Soil loads specified in Table 1610.1 shall be used as the minimum design lateral soil loads unless specified otherwise in a soil investigation report approved by the building official. Design lateral pressure from surcharge loads shall be added to the lateral earth pressure load. Design lateral pressure shall be increased if soils with expansion potential are present at the site.*

Discussion and Commentary: Design lateral loads are established in Table 1610.1 for various soil types. The indicated loads address both at-rest pressure and active pressure conditions. It is noted that expansive soils are not to be used as backfill, as such materials can exert very high pressures against walls. Special soil testing is required to determine the magnitude of these pressures.

TABLE 1610.1
SOIL LATERAL LOAD

DESCRIPTION OF BACKFILL MATERIAL[c]	UNIFIED SOIL CLASSIFICATION	DESIGN LATERAL SOIL LOAD[a] (pound per square foot per foot of depth)	
		Active pressure	At-rest pressure
Well-graded, clean gravels; gravel-sand mixes	GW	30	60
Poorly graded clean gravels; gravel-sand mixes	GP	30	60
Silty gravels, poorly graded gravel-sand mixes	GM	40	60
Clayey gravels, poorly graded gravel-and-clay mixes	GC	45	60
Well-graded, clean sands; gravelly sand mixes	SW	30	60
Poorly graded clean sands; sand-gravel mixes	SP	30	60
Silty sands, poorly graded sand-silt mixes	SM	45	60
Sand-silt clay mix with plastic fines	SM-SC	45	100
Clayey sands, poorly graded sand-clay mixes	SC	60	100
Inorganic silts and clayey silts	ML	45	100
Mixture of inorganic silt and clay	ML-CL	60	100
Inorganic clays of low to medium plasticity	CL	60	100
Organic silts and silt clays, low plasticity	OL	Note b	Note b
Inorganic clayey silts, elastic silts	MH	Note b	Note b
Inorganic clays of high plasticity	CH	Note b	Note b
Organic clays and silty clays	OH	Note b	Note b

For SI: 1 pound per square foot per foot of depth = 0.157 kPa/m, 1 foot = 304.8 mm.

a. Design lateral soil loads are given for moist conditions for the specified soils at their optimum densities. Actual field conditions shall govern. Submerged or saturated soil pressures shall include the weight of the buoyant soil plus the hydrostatic loads.

b. Unsuitable as backfill material.

c. The definition and classification of soil materials shall be in accordance with ASTM D 2487.

Basement and foundation walls that have restricted horizontal movement at the top shall be designed for at-rest pressure. Retaining walls free to move and rotate at the top are permitted to be designed for active pressure. Where basement walls extend a maximum of 8 feet below grade and support a flexible floor system, they also may be designed for active pressure.

Code Text: *Special inspection is inspection as herein required of the materials, installation, fabrication, erection or placement of components and connections requiring special expertise to ensure compliance with approved construction documents and referenced standards. Structural observation is the visual observation of the structural system by a registered design professional for general conformance to the approved construction documents at significant construction stages and at completion of the structural system. Structural observation does not include or waive the responsibility for the inspection required by Section 109, Section 1704 or other sections of the IBC.*

Discussion and Commentary: In addition to the general inspections called for in Section 109 (footings, frame, final, etc.), it is often necessary to call for a more exacting review of the construction process. Through special inspections and structural observation, the work can be evaluated more closely for compliance with the approved construction documents.

Steel Construction	Section 1704.3
Concrete Construction	Section 1704.4
Masonry Construction	Section 1704.5
Wood Construction	Section 1704.6
Soils	Section 1704.7
Pile Foundations	Section 1704.8
Pier Foundations	Section 1704.9
Sprayed Fire-resistant Materials	Section 1704.10
Mastic and Intumescent Fire-resistant Coatings	Section 1704.11
Exterior Insulation and Finish Systems (EIFS)	Section 1704.12
Special Cases	Section 1704.13
Smoke Control	Section 1704.14

A statement of special inspections must be prepared by the registered design professional in responsible charge where special inspection or testing is required. The code specifies the content required to be provide in the statement. The code also sets forth unique provisions for special inspections and structural testing for seismic-resistance and wind-resistance considerations.

Code Text: *Where application is made for construction as described in Section 1704, the owner or the registered design professional in responsible charge acting as the owner's agent shall employ one or more special inspectors to provide inspections during construction on the types of work listed under Section 1704.* See three exceptions for work, components or occupancies where special inspection is not required.

Discussion and Commentary: Most building departments do not have the staff of inspectors to provide detailed inspections on large and complex projects. There are also projects where the nature of construction is such that extra care in quality control must be exercised to assure compliance. For these reasons, the code mandates continuous or periodic inspection by special inspectors for certain types of work.

TABLE 1704.4
REQUIRED VERIFICATION AND INSPECTION OF CONCRETE CONSTRUCTION

VERIFICATION AND INSPECTION	CONTINUOUS	PERIODIC	REFERENCED STANDARD[a]	IBC REFERENCE
1. Inspection of reinforcing steel, including prestressing tendons, and placement.	—	X	ACI 318: 3.5, 7.1-7.7	1913.4
2. Inspection of reinforcing steel welding in accordance with Table 1704.3, Item 5b.	—	—	AWS D1.4 ACI 318: 3.5.2	—
3. Inspect bolts to be installed in concrete prior to and during placement of concrete where allowable loads have been increased.	X	—	—	1911.5
4. Verifying use of required design mix.	—	X	ACI 318: Ch. 4, 5.2-5.4	1904.2.2, 1913.2, 1913.3
5. At the time fresh concrete is sampled to fabricate specimens for strength tests, perform slump and air content tests, and determine the temperature of the concrete.	X	—	ASTM C 172 ASTM C 31 ACI 318: 5.6, 5.8	1913.10
6. Inspection of concrete and shotcrete placement for proper application techniques.	X	—	ACI 318: 5.9, 5.10	1913.6, 1913.7, 1913.8
7. Inspection for maintenance of specified curing temperature and techniques.	—	X	ACI 318: 5.11-5.13	1913.9
8. Inspection of prestressed concrete: a. Application of prestressing forces. b. Grouting of bonded prestressing tendons in the seismic-force-resisting system.	X X	—	ACI 318: 18.20 ACI 318: 18.18.4	—
9. Erection of precast concrete members.	—	X	ACI 318: Ch. 16	—
10. Verification of in-situ concrete strength, prior to stressing of tendons in posttensioned concrete and prior to removal of shores and forms from beams and structural slabs.	—	X	ACI 318: 6.2	—
11. Inspect formwork for shape, location and dimensions of the concrete member being formed.	—	X	ACI 318: 6.1.1	—

For SI: 1 inch = 25.4 mm.
a. Where applicable, see also Section 1707.1, Special inspection for seismic resistance.

A statement of special inspections containing the work requiring special inspection, the specific inspections to be performed, and the individuals or firms to be retained for conducting special inspections must be submitted by the permit applicant prior to issuance of the building permit.

Topic: Inspector Qualifications   **Category:** Structural Tests and Special Inspections
Reference: IBC 1704.1   **Subject:** Special Inspections

Code Text: *The special inspector shall be a qualified person who shall demonstrate competence, to the satisfaction of the building official, for inspection of the particular type of construction or operation requiring special inspection. Special inspectors shall keep records of inspections. The special inspector shall furnish inspection reports to the building official, and to the registered design professional in responsible charge.*

Discussion and Commentary: It is the duty of the special inspector not only to observe the work, but also to furnish inspection reports indicating that the work inspected was done in accordance with the approved construction documents. If discrepancies are found in the work, the inspector should bring them to the immediate attention of the contractor for correction. If the discrepancies are not corrected, the building official and registered design professional in responsible charge should be notified.

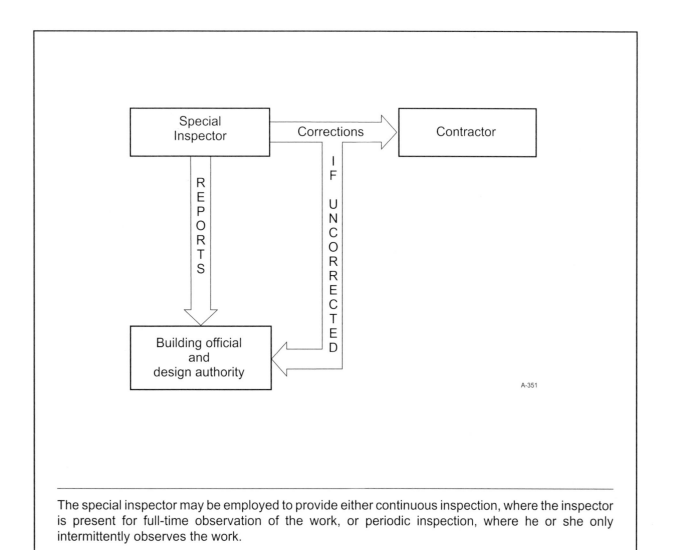

The special inspector may be employed to provide either continuous inspection, where the inspector is present for full-time observation of the work, or periodic inspection, where he or she only intermittently observes the work.

Code Text: *Structural observations shall be provided for those structures included in Seismic Design Category D, E or F, as determined in Section 1613, where one or more of the following conditions exist: 1) the structure is classified as Occupancy Category III or IV in accordance with Section 1604.5, 2) the height of the structure is greater than 75 feet above the base, 3) the structure is assigned to Seismic Design Category E, is classified as Occupancy Category I or II in accordance with Section 1604.5 and is greater than two stories in height, 4) when so designated by the registered design professional in responsible charge of the design, and 5) when such observation is specifically required by the building official.*

Discussion and Commentary: Structural observations not only are required for the higher seismic design categories but also are required in high-wind areas (basic wind speed > 110 mph) where the structure has a height greater than 75 feet or has an Occupancy Category of III or IV. The building official or registered design professional in responsible charge can also mandate such observation.

Structural observation is not to be confused with the mandated inspections specified in Section 109, or with the special inspections listed in Section 1704. This activity is intended to provide an additional level of expertise in the review of structures posing a very high level of complexity.

Code Text: *The provisions of Chapter 19 shall govern the materials, quality control, design and construction of concrete use in structures. Structural concrete shall be designed and constructed in accordance with the requirements of Chapter 19 and ACI 318 as amended in Section 1908 of the IBC. The format and subject matter of Sections 1902 through 1907 are patterned after, and in general conformity with, the provisions for structural concrete in ACI 318.*

Discussion and Commentary: Requirements for the design, testing, mixing, placing and protection of concrete construction are located in ACI 318, *Building Code Requirements for Structural Concrete*, a publication published and copyrighted by the American Concrete Institute. Section 1908 contains the primary modifications to the provisions of ACI 318.

Construction Requirements
for Concrete Work

Specifications for Tests and Materials (Chapter 3)	Section 1903
Durability Requirements (Chapter 4)	Section 1904
Concrete Quality, Mixing and Placing (Chapter 5)	Section 1905
Formwork, Embedded Pipes and Construction (Chapter 6)	Section 1906
Details of Reinforcement (Chapter 7)	Section 1907

Where modifications have been made to the text of ACI 318, the section designations of IBC Section 1908 are followed by those found in ACI 318. Italics are used to indicate where the IBC differs substantially from the ACI standard.

Code Text: *Concrete that will be subject to the following exposures shall conform to the corresponding maximum water-cementitious materials ratios and minimum specified concrete compressive strength requirements of ACI 318 4.2.2: 1) concrete intended to have low permeability where exposed to water; 2) concrete exposed to freezing and thawing in a moist condition or deicer chemicals; or 3) concrete with reinforcement where the concrete is exposed to chlorides from deicing chemicals, salt, salt water, brackish water, seawater or spray from these sources.* See exception for Group R occupancies less than four stories in height.

Discussion and Commentary: Concrete exposed to the conditions listed in this section must comply with the limitations on water-cementitious materials, or specified compressive strength, or both. The intent of the upper limit on the water-cementitious materials ratio and lower limit on the specified compressive strength is to achieve dense, impermeable or watertight concrete.

TABLE 1904.2.2
MINIMUM SPECIFIED COMPRESSIVE STRENGTH (f'_c)

TYPE OR LOCATION OF CONCRETE CONSTRUCTION	MINIMUM SPECIFIED COMPRESSIVE STRENGTH (f'_c at 28 days, psi)		
	Negligible exposure	Moderate exposure	Severe exposure
Basement walls[c] and foundations not exposed to the weather	2,500	2,500	2,500[a]
Basement slabs and interior slabs on grade, except garage floor slabs	2,500	2,500	2,500[a]
Basement walls[c], foundation walls, exterior walls and other vertical concrete surfaces exposed to the weather	2,500	3,000[b]	3,000[b]
Driveways, curbs, walks, patios, porches, carport slabs, steps and other flatwork exposed to the weather, and garage floor slabs	2,500	3,000[b, d]	3,500[b, d]

For SI: 1 pound per square inch = 0.00689 MPa.

a. Concrete in these locations that can be subjected to freezing and thawing during construction shall be of air-entrained concrete in accordance with Section 1904.2.1.

b. Concrete shall be air entrained in accordance with Section 1904.2.1.

c. Structural plain concrete basement walls are exempt from the requirements for exposure conditions of Section 1904.2.2 (see Section 1909.6.1).

d. For garage floor slabs where a steel trowel finish is used, the total air content required by Section 1904.2.1 is permitted to be reduced to not less than 3 percent, provided the minimum specified compressive strength of the concrete is increased to 4,000 psi.

In climatic areas where concrete construction is exposed to freeze-thaw cycles or deicer chemicals, concrete must be of the quality necessary to resist the harmful effects of such weather. In low-rise Group R occupancies, the compressive strength need only comply with Table 1904.2.2 which mandates a minimum compressive strength as a function of the specific concrete element under consideration, and the exposure of such element.

Topic: Weather Requirements

Category: Concrete

Reference: IBC 1905.12, 1905.13

Subject: Quality, Mixing and Placing

Code Text: *Concrete to be placed during freezing or near-freezing weather shall comply with the requirements of ACI 318, Section 5.12. Concrete to be placed during hot weather shall comply with the requirements of ACI 318, Section 5.13.*

Discussion and Commentary: Preparation prior to placing concrete is as critical to quality construction as is the actual concrete placement. Equipment for mixing and transporting concrete must be clean. Debris and ice shall be removed from any areas to be occupied by concrete. All laitance and other unsound material must be removed prior to the placement of additional concrete against hardened concrete. For the most part, IBC Section 1905 simply provides a reference to the appropriate provisions of ACI 318 for requirements regarding concrete quality, mixing and placing.

Possible Effects of Cold Weather

- Permanent damage to early freezing
- Slower setting and slower strength gain
- Freezing of fresh concrete prior to hardening
- Reduced durability . . . same as strength reduction
- Freezing of green concrete at edges and corners
- Dehydrated surface areas due to use of space heaters
- Cracking due to sudden temperature change if strength gain insufficient

Possible Effects of Hot Weather

- Increased water demand
- Difficult to control entrained air
- Rapid evaporation of mixing water if dry and windy
- Rapid slump loss
- Faster set
- Greater dimensional changes on cooling
- Increased tendency to crack or craze
- Reduced long term strength
- Possible "cold joints"
- Increased permeability

In hot weather, proper attention shall be given to any aspect that might impair the required strength or serviceability of the concrete. Adequate equipment must be provided for heating concrete materials and protecting concrete during freezing or near-freezing weather.

Code Text: *Unless otherwise required or indicated on the construction documents, head and bed joints shall be $^3/_8$ inch (9.5 mm) thick, except that the thickness of the bed joint of the starting course placed over foundations shall not be less than $^1/_4$ inch (6.4 mm) and not more than $^3/_4$ inch (19.1 mm). Units shall be placed while the mortar is soft and plastic. Any unit disturbed to the extent that the initial bond is broken after initial positioning shall be removed and relaid in fresh mortar.*

Discussion and Commentary: During masonry construction, it is important that the units be placed in a manner so as to form a solid bond. Hollow-masonry units must have all face shells of bed joints fully mortared. Solid masonry units must be placed in fully mortared bed and head joints, with the ends of the units completely buttered.

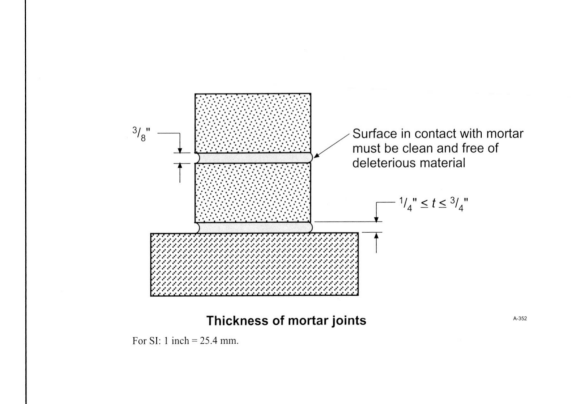

Thickness of mortar joints

A-352

For SI: 1 inch = 25.4 mm.

Masonry placed during periods of cold weather is regulated under Section 2104.3. In addition to general requirements for the protection of materials and partially completed masonry walls, provisions address procedures for construction spanning various temperature ranges.

Topic: Lateral Support   **Category:** Masonry

Reference: IBC 2109.4.1, 2109.4.3   **Subject:** Empirical Design

Code Text: *Masonry walls shall be laterally supported in either the horizontal or vertical direction at intervals not exceeding those given in Table 2109.4.1. Lateral support shall be provided by cross walls, pilasters, buttresses or structural frame members when limiting distance is taken horizontally, or by floors, roofs acting as diaphragms or structural frame members when the limiting distance is taken vertically.*

Discussion and Commentary: The empirical provisions of Section 2109 are design rules developed by experience rather than engineered analysis. This empirical design method is based on several premises: gravity loads are reasonably centered on bearing walls; effects of reinforcement are neglected; walls are laid in running bond; and buildings have limited height, seismic risk and wind loading. The requirements of Section 2109 do not apply for higher-risk structures or in others cases where an engineered masonry design method is utilized.

TABLE 2109.4.1
WALL LATERAL SUPPORT REQUIREMENTS

CONSTRUCTION	MAXIMUM WALL LENGTH TO THICKNESS OR WALL HEIGHT TO THICKNESS
Bearing walls	
Solid units or fully grouted	20
All others	18
Nonbearing walls	
Exterior	18
Interior	36

Minimum thicknesses have been established in the code for various types of masonry walls. Bearing walls shall be at least 6 inches in thickness for one-story buildings, with a minimum of 8 inches for multistory conditions. Masonry shear walls, foundation walls, parapet walls and foundation piers shall also be a minimum of 8 inches in thickness. Rough, random or coursed rubble stone walls shall be at least 16 inches thick.

Code Text: *Masonry fireboxes shall be constructed of solid masonry units, hollow masonry units grouted solid, stone, or concrete. When a lining of firebrick at least 2 inches (51 mm) in thickness or other approved lining is provided, the minimum thickness of back and sidewalls shall each be 8 inches (203 mm) of solid masonry, including the lining. The width of joints between firebricks shall not be greater than $^1/_4$ inch (6.4 mm). When no lining is provided, the total minimum thickness of back and sidewalls shall be 10 inches (254 mm) of solid masonry.*

Discussion and Commentary: Firebox thickness is regulated in order to insulate surrounding construction, both exposed and concealed, from excessive temperature levels. It is important that the walls be constructed in such a fashion that they are solid throughout.

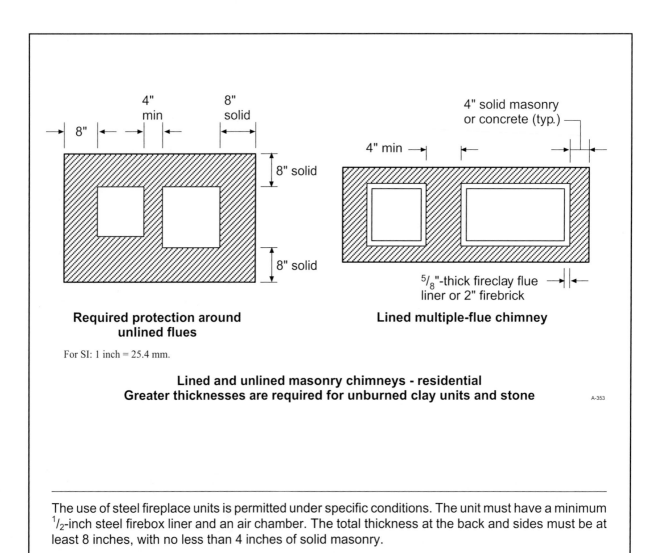

Required protection around unlined flues

For SI: 1 inch = 25.4 mm.

4" solid masonry or concrete (typ.)

Lined multiple-flue chimney

$^5/_8$"-thick fireclay flue liner or 2" firebrick

Lined and unlined masonry chimneys - residential
Greater thicknesses are required for unburned clay units and stone

A-353

The use of steel fireplace units is permitted under specific conditions. The unit must have a minimum $^1/_2$-inch steel firebox liner and an air chamber. The total thickness at the back and sides must be at least 8 inches, with no less than 4 inches of solid masonry.

Code Text: *Hearth extensions shall extend at least 16 inches (406 mm) in front of, and at least 8 inches (203 mm) beyond, each side of the fireplace opening. Where the fireplace opening is 6 square feet (0.557 m²) or larger, the hearth extension shall extend at least 20 inches(508 mm) in front of, and at least 12 inches (305 mm) beyond, each side of the fireplace opening. The minimum thickness of hearth extensions shall be 2 inches (51 mm).* See exception for raised firebox openings.

Discussion and Commentary: Hearth extensions are necessary to keep sparks and embers that fly from the firebox from igniting combustible material, such as carpet, on the floor. Radiated heat from the fireplace can also ignite combustible flooring materials located adjacent to the fireplace opening and adjacent fireplace walls.

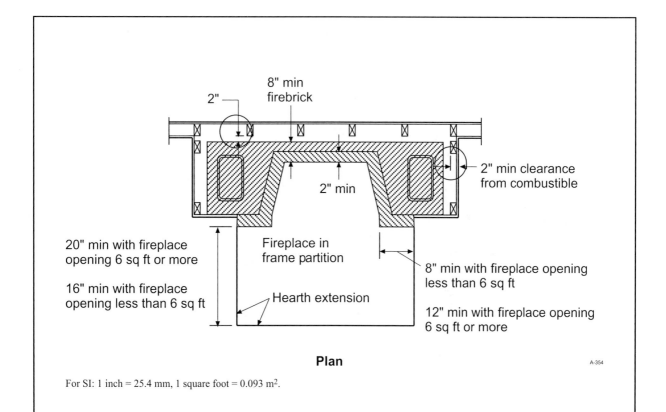

Plan

A-354

For SI: 1 inch = 25.4 mm, 1 square foot = 0.093 m².

For obvious reasons, the hearth and the hearth extension must be constructed of, and supported by, noncombustible materials. Hearths must be specifically constructed of either concrete or masonry. The minimum required thickness of fireplace hearths is 4 inches.

Code Text: *Any portion of a masonry fireplace located in the interior of a building or within the exterior wall of a building shall have a clearance to combustibles of not less than 2 inches (51 mm) from the front faces and sides of masonry fireplaces and not less than 4 inches (102 mm) from the back faces of masonry fireplaces. The airspace shall not be filled, except to provide fireblocking in accordance with Section 2111.12.* See four exceptions for alternate methods to the required clearances.

Discussion and Commentary: The radiant heat transfer through the materials used to construct a masonry fireplace and/or chimney necessitates a minimum separation between the masonry and combustible materials, such as wood floor, wall or ceiling framing. The depth of the noncombustible fireblocking, placed on metal strips or lath, is to be 1 inch.

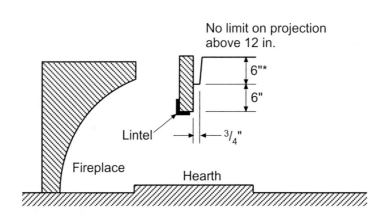

No limit on projection
above 12 in.

6"*

6"

Lintel

³/₄"

Fireplace

Hearth

* Combustible materials may project ¹/₈ in. for each 1 in. clearance. No combustible materials permitted within 6 in. of opening.

For SI: 1 inch = 25.4 mm.

Combustible materials projection from fireplace

A-355

No combustible materials, such as trim and ornamentation, are permitted within 6 inches directly above the opening at the face of the fireplace. Combustibles placed less than 12 inches from the opening, while permitted, are very limited in their projection from the fireplace opening.

Code Text: *Sawn lumber used for load-supporting purposes, including end-jointed or edge-glued lumber, machine stress-rated or machine-evaluated lumber, shall be identified by the grade mark of a lumber grading or inspection agency that has been approved by an accreditation body that complies with DOC PS 20 or equivalent. Wood structural panels, when used structurally (including those used for siding, roof and wall sheathing, subflooring, diaphragms and built-up members), shall conform to the requirements for its type in DOC PS 1 or PS 2.*

Discussion and Commentary: Obviously, the proper use of a wood structural member cannot be determined unless it has been identified. Grade marks, identification marks, certificates of inspection and quality marks are various methods of indicating the type and quality of wood members.

Visually Graded Lumber

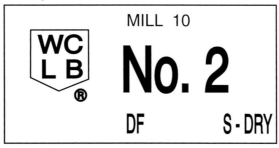

Machine Stress-rated lumber

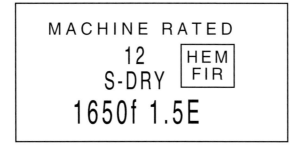

A-356

Lumber and, particularly, wood structural panels are highly variable in strengths and other mechanical properties; hence, such materials must conform to the applicable standards or grading rules specified in the code.

Topic: Protection Against Decay

Reference: IBC 2304.11.2

Category: Wood

Subject: General Construction Requirements

Code Text: *Wood installed above ground in the locations specified in Sections 2304.11.2.1 through 2304.11.2.7, 2304.11.3 and 2304.11.5 shall be naturally durable wood or preservative-treated wood using water-borne preservatives, in accordance with AWPA U1 (Commodity Specifications A or F) for above-ground use.*

Discussion and Commentary: To protect against decay and termite infestation, the code addresses those members for which care must be taken, including: joists, girders and subfloor adjacent to exposed ground in crawl spaces; framing members and wall sheathing that rest on exterior foundation walls; sleepers and sills on a concrete slab in direct contact with earth; girder ends in masonry or concrete walls; wood siding adjacent to the ground; and posts and columns supported by a concrete slab.

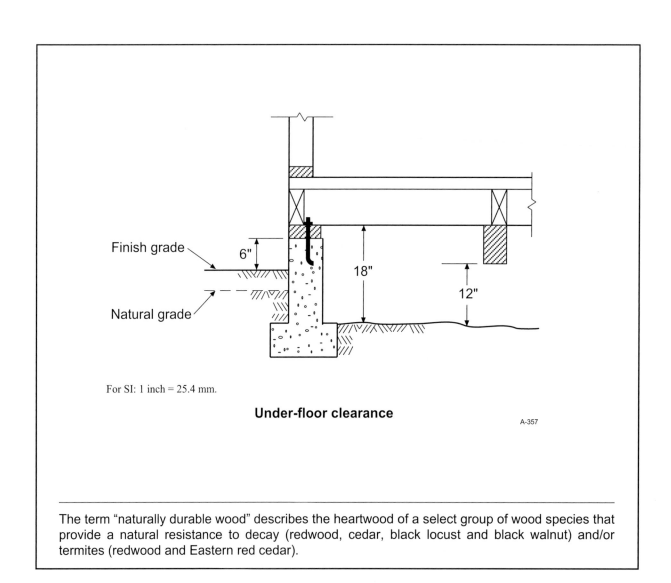

For SI: 1 inch = 25.4 mm.

Under-floor clearance

A-357

The term "naturally durable wood" describes the heartwood of a select group of wood species that provide a natural resistance to decay (redwood, cedar, black locust and black walnut) and/or termites (redwood and Eastern red cedar).

Code Text: *The requirements in Section 2308 are intended for conventional light-frame construction. Other methods are permitted to be used provided a satisfactory design is submitted showing compliance with other provisions of* the IBC. *Interior nonload-bearing partitions, ceilings and curtain walls of conventional light-frame construction are not subject to the limitations of Section 2308.*

Discussion and Commentary: Conventional light-frame wood construction is considered *a type of construction whose primary structural elements are formed by a system of repetitive wood-framing members.* The provisions of Section 2308 are based on experience gained over the last several decades. This experience has resulted in the prescriptive requirements contained in this section. Compliance with AF&PA's *Wood Frame Construction Manual* is considered equivalent to Section 2308.

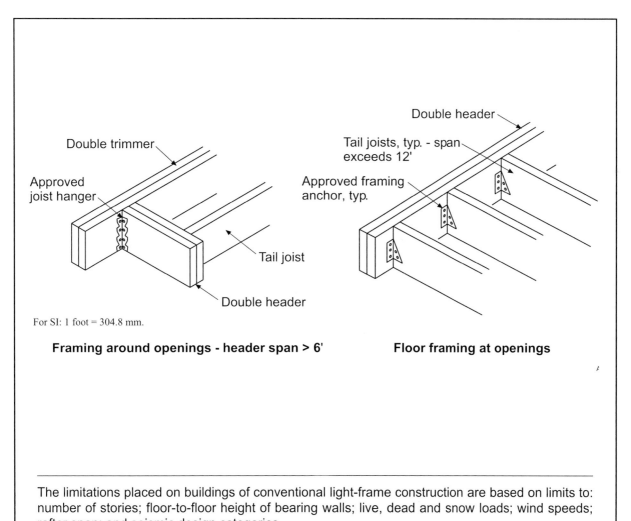

For SI: 1 foot = 304.8 mm.

Framing around openings - header span > 6'   **Floor framing at openings**

The limitations placed on buildings of conventional light-frame construction are based on limits to: number of stories; floor-to-floor height of bearing walls; live, dead and snow loads; wind speeds; rafter span; and seismic design categories.

Code Text: *Foundation plates or sills shall be bolted or anchored to the foundation with not less than $^{1}/_{2}$-inch (12.7 mm) diameter steel bolts or approved anchors. Bolts shall be embedded at least 7 inches (178 mm) into concrete or masonry, and spaced not more than 6 feet (1829 mm) apart. There shall be a minimum of two bolts or anchor straps per piece with one bolt or anchor strap located not more than 12 inches (305 mm) or less than 4 inches (102 mm) from each end of each piece. A properly sized nut and washer shall be tightened on each bolt to the plate.*

Discussion and Commentary: The prescriptive requirements for the installation of anchor bolts are applicable to all buildings of conventional light-frame wood construction. The provisions set forth the necessary criteria to adequately tie the framing system to the foundation.

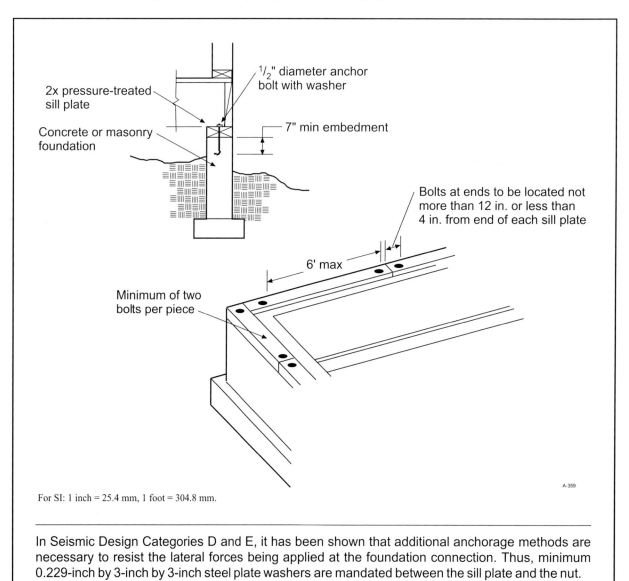

2x pressure-treated sill plate

Concrete or masonry foundation

$^{1}/_{2}$" diameter anchor bolt with washer

7" min embedment

Bolts at ends to be located not more than 12 in. or less than 4 in. from end of each sill plate

6' max

Minimum of two bolts per piece

For SI: 1 inch = 25.4 mm, 1 foot = 304.8 mm.

A-359

In Seismic Design Categories D and E, it has been shown that additional anchorage methods are necessary to resist the lateral forces being applied at the foundation connection. Thus, minimum 0.229-inch by 3-inch by 3-inch steel plate washers are mandated between the sill plate and the nut.

Quiz

Study Session 16
IBC Chapters 16, 17, 19, 21, 22 and 23

1. Live loads exceeding a minimum of _____ pounds per square foot shall be posted in applicable portions of commercial or industrial buildings.

    a. 50   b. 75

    c. 100   d. 125

    Reference_____

2. The minimum uniformly distributed live load used in the design of a fixed-seat arena is determined to be _____ pounds per square foot.

    a. 50   b. 60

    c. 100   d. 125

    Reference_____

3. In general, in a Group B office building, handrail assemblies and guards shall be designed to resist a minimum load of _____ pounds per lineal foot applied at any direction at the top.

    a. 20   b. 30

    c. 40   d. 50

    Reference_____

4. Roofs used for roof gardens shall be designed for a minimum live load of _____ pounds per square foot.

    a. 40   b. 60

    c. 75   d. 100

Reference_____

5. For wind design purposes, Exposure _____ describes flat open country and grasslands.

    a. A   b. B

    c. C   d. D

Reference_____

6. Special inspection for fill placement is not required during placement of controlled fill a maximum of _____ inches in depth.

    a. 4   b. 6

    c. 12   d. 24

Reference_____

7. Special inspection thickness testing for sprayed fire-resistant material applied to structural framing members shall be performed on a minimum of _____ percent of the structural members on each floor.

    a. 10   b. 25

    c. 50   d. 100

Reference_____

8. Structural observation shall be provided for structures assigned to Seismic Design Category E, classified as Occupancy Category I or II, and a minimum of _____ stories in height.

    a. two   b. three

    c. four   d. five

Reference_____

9. Strength tests of concrete are not required when the total volume of a given class of concrete is less than _____ cubic yards, provided evidence of satisfactory strength is submitted to the building official and approved.

　　a. 10　　　　　　　　　　b. 20

　　c. 25　　　　　　　　　　d. 50

Reference _____

10. As a modification to the text of ACI 318, a wall pier is defined as a wall segment with a horizontal length-to-thickness ratio between 2.5 and 6, with a clear height at least _____ times its horizontal length.

　　a. 2　　　　　　　　　　b. 3

　　c. 4　　　　　　　　　　d. 6

Reference _____

11. What is the minimum required thickness of structural plain concrete exterior basement walls?

　　a. 6 inches　　　　　　　b. $7^1/_2$ inches

　　c. 8 inches　　　　　　　d. $9^1/_2$ inches

Reference _____

12. The minimum thickness of a concrete floor slab supported directly on the ground shall be _____ inches.

　　a. $3^1/_2$　　　　　　　　b. 4

　　c. 5　　　　　　　　　　d. 6

Reference _____

13. What is the term used to describe masonry in which the tensile resistance of the masonry is taken into consideration and the effects of stresses in reinforcement are neglected?

　　a. ashlar masonry　　　　b. plain masonry

　　c. reinforced masonry　　d. solid masonry

Reference _____

14. Wire wall ties for masonry construction shall be embedded a minimum of _____ inch(es) into the mortar bed of solid-grouted hollow units.

    a. $^3/_4$ b. 1

    c. $1^1/_4$ d. $1^1/_2$

Reference_____

15. Weep holes provided in the outside wythe of masonry walls shall be located at a maximum spacing of _____ inches on center.

    a. 33   b. 48

    c. 64   d. 96

Reference_____

16. A hearth extension for a masonry fireplace with a 24-inch by 42-inch opening shall extend a minimum of _____ inches in front of, and a minimum of _____ inches beyond, each side of the fireplace opening.

    a. 16, 8   b. 16, 10

    c. 20, 12   d. 24, 12

Reference_____

17. Combustible materials located 10 inches directly above the opening of a masonry fireplace shall have a maximum projection of _____ inch(es).

    a. 0, no projection is permitted   b. $^3/_4$

    c. 1   d. $1^1/_4$

Reference_____

18. Masonry chimneys shall extend at least 2 feet higher than any portion of the building within 10 feet, with a minimum height of _____ feet above the point where the chimney passes through the roof.

    a. 2   b. $2^1/_2$

    c. 3   d. 4

Reference_____

19. Unless naturally durable or preservative-treated wood is used, what is the minimum air space required on the top, sides and end of wood girders entering exterior concrete walls?

    a. $^1/_2$ inch   b. 1 inch

    c. $1^1/_2$ inch   d. 2 inches

Reference_____

20. Which of the following types of naturally durable wood cannot qualify as decay resistant?

    a. redwood   b. Eastern red cedar

    c. black walnut   d. red oak

Reference_____

21. What is the minimum number and size of common nails used to connect a wood jack rafter to a hip?

    a. two 10d, face nail   b. three 10d, toenail

    c. two 16d, toenail   d. three 16d, face nail

Reference_____

22. In conventional light-frame construction, buildings shall be provided with exterior and interior braced wall lines at maximum intervals of _____ feet in both the longitudinal and transverse directions in each story.

    a. 20   b. 25

    c. 35   d. 50

Reference_____

23. In conventional light-frame construction, trimmer and header joists in framing around openings shall be doubled where the span of the header exceeds a minimum of _____ feet.

    a. 4   b. 6

    c. 10   d. 12

Reference_____

24. In conventional light-frame wall construction, the edge of a bored hole shall be a minimum of _____ inch(es) from the edge of the stud.

    a. $^5/_8$ b. $^3/_4$

    c. 1   d. $1^1/_4$

Reference_____

25. In conventional light-frame roof construction, the maximum span of a 2-inch by 6-inch purlin shall be _____ feet.

    a. 4   b. 6

    c. 8   d. 12

Reference_____

26. In the design of stairway treads, a minimum concentrated live load of _____ pounds must be used, determined on an area of 4 square inches.

    a. 40   b. 100

    c. 200   d. 300

Reference_____

27. In climates determined to be of moderate exposure, concrete foundation walls exposed to weather shall have a minimum specified compressive strength of _____ psi at 28 days.

    a. 2,000   b. 2,500

    c. 3,000   d. 3,500

Reference_____

28. Gypsum sheathing, where used to enclose wood-framed buildings, shall have a minimum thickness of _____ inch and the wall studs shall be spaced at a maximum of _____ inches on center.

    a. $^1/_2$, 16   b. $^1/_2$, 24

    c. $^5/_8$, 16   d. $^5/_8$, 24

Reference _____

29. Masonry nonstructural floor surfacing a maximum of _____ inches in thickness is permitted to be supported by wood members without checking for the effects of long-term loading.

    a. 2   b. 4

    c. 6   d. 8

    Reference_____

30. In conventional light-frame construction, cripple walls exceeding _____ in height shall be framed of wood studs having the size required for an additional story.

    a. 14 inches   b. 20 inches

    c. 30 inches   d. 48 inches

    Reference_____

31. Unless the specified live load exceeds 80 psf, the partition load in an office building shall be a minimum uniformly distributed live load of _____ pounds.

    a. 5   b. 10

    c. 15   d. 20

    Reference _____

32. Grab bars shall be designed to resist a single minimum concentrated load of _____ pounds applied in any direction at any point.

    a. 160   b. 225

    c. 250   d. 300

    Reference _____

33. Areas along the U.S. Gulf of Mexico coast are not considered hurricane-prone regions where the maximum basic wind speed is _____ miles per hour.

    a. 85   b. 90

    c. 100   d. 110

    Reference _____

34. Where special inspection is required for the welding of structural steel, which of the following welds does not require continuous inspection?

    a.   partial penetration groove welds

    b.   multipass fillet welds

    c.   single-pass fillet welds > $\frac{5}{16}$ inch

    d.   floor and roof deck welds

Reference _____

35. The statement of special inspections shall include wind requirements for structures constructed in areas of Exposure Category C where the minimum 3-second-gust basic wind speed is _____ miles per hour or greater.

    a.   90   b.   100

    c.   110   d.   120

Reference _____

2006 IBC Chapters 8, 12, 25 and 30

Interior Finishes, Interior Environment, Gypsum Board and Elevators

OBJECTIVE: To gain an understanding of the limitations on interior wall and ceiling finishes; the installation requirements for gypsum board, lath and plaster; the important issues concerning the interior environment, including light, ventilation and sound transmission; and the provisions for elevators and their hoistways.

REFERENCE: Chapters 8, 12, 25 and 30, 2006 *International Building Code*

KEY POINTS:
- Which building elements are considered to be interior wall and ceiling finishes?
- Which types of materials are not regulated as wall or ceiling finishes?
- What is the standard of quality that addresses interior wall and ceiling finishes?
- Based on flame-spread index, what are the various classes of finish materials?
- When tested in a manner consistent with their use, what is the maximum smoke density index permitted for finish materials?
- In general, what is the maximum thickness of an interior wall or ceiling finish that must be applied directly against a noncombustible backing?
- How is Table 803.5 used to regulate the finish materials on ceilings and walls?
- Under which conditions are textile wall coverings permitted as wall and ceiling finishes?
- In what portions of a building are the interior floor finishes regulated? In which specific occupancies?
- What are the limitations on combustible trim and decorative materials?
- How must the occupiable portions of buildings be illuminated? Ventilated?
- When yards and courts are adjacent to exterior openings providing natural light and ventilation, how shall the openings be located?
- In which type of occupancy is the transmission of sound regulated?
- How is air-borne sound to be controlled? Structure-borne sound?
- What are the minimum room widths of habitable spaces? Minimum ceiling heights of occupiable spaces?
- How are efficiency dwelling units regulated for interior environment?
- What minimum size access opening is required for a crawl space? An attic?
- How must walls and floors in toilet rooms and bathing rooms be surfaced?

KEY POINTS:
(Cont'd)

- What are the limitations for gypsum wallboard in regard to exterior installation and weather protection?
- How shall fasteners for gypsum wallboard be applied?
- What type of gypsum board assembly requires treated joints and fasteners?
- Where is water-resistant gypsum backing board required?
- Water-resistant gypsum backing board is not permitted for use in which three locations?
- Where used to provide a horizontal diaphragm, how shall gypsum board be installed in a ceiling application?
- What level of fire resistance is required for elevator shaft enclosures?
- What is the maximum number of elevator cars that may be located in a single elevator hoistway? At what point are two hoistways required?
- What types of doors are prohibited at the point of access to an elevator car?
- When must elevator hoistways be vented? Where are the vents to be located?

Code Text: *Interior wall and ceiling finish includes the exposed interior surfaces of buildings including, but not limited to: fixed or movable walls and partitions; toilet room privacy partitions; columns; ceilings; and interior wainscotting, paneling, or other finish applied structurally or for decoration, acoustical correction, surface insulation, structural fire resistance or similar purposes, but not including trim. These provisions shall limit the allowable flame spread and smoke development based on location and occupancy classification.* See exceptions for very thin materials (< 0.036 inches in thickness) and heavy-timber members.

Discussion and Commentary: It is the intent of the IBC to govern those materials applied to walls or ceilings that could contribute to the spread of flame or the development of smoke. Floor finishes are regulated in Section 804.

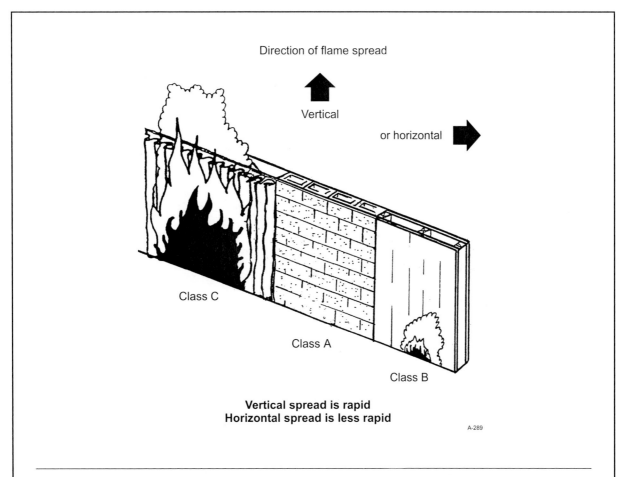

The classification of interior wall and ceiling finishes is based primarily on their flame spread index. Class A has an index of 0 to 25, Class B of 26 to 75, and Class C of 76 to 200. The smoke-developed index for all three classifications is limited to 450.

Code Text: *Interior wall and ceiling finish shall have a flame spread index not greater than that specified in Table 803.5 for the group and location designated.*

Discussion and Commentary: Based on fire statistics, the rapid spread of fire across an interior finish material has been second only to vertical fire spread through openings between floors as a cause of life loss during building fires. Therefore, limitations are placed on the materials that are used to cover the walls and ceilings of rooms and other enclosed spaces, corridors, and vertical exits and exit passageways. The rapid spread of fire and the increased contribution of fuel to the fire are the major reasons why finish materials must meet stringent criteria to gain acceptance.

TABLE 803.5
INTERIOR WALL AND CEILING FINISH REQUIREMENTS BY OCCUPANCY[k]

GROUP	SPRINKLERED[l]			NONSPRINKLERED		
	Exit enclosures and exit passageways[a,b]	Corridors	Rooms and enclosed spaces[c]	Exit enclosures and exit passageways[a,b]	Corridors	Rooms and enclosed spaces[c]
A-1 & A-2	B	B	C	A	A[d]	B[e]
A-3[f], A-4, A-5	B	B	C	A	A[d]	C
B, E, M, R-1, R-4	B	C	C	A	B	C
F	C	C	C	B	C	C
H	B	B	C[g]	A	A	B
I-1	B	C	C	A	B	B
I-2	B	B	B[h,i]	A	A	B
I-3	A	A[j]	C	A	A	B
I-4	B	B	B[h,i]	A	A	B
R-2	C	C	C	B	B	C
R-3	C	C	C	C	C	C
S	C	C	C	B	B	C
U	No restrictions			No restrictions		

For SI: 1 inch = 25.4 mm, 1 square foot = 0.0929 m².

a. Class C interior finish materials shall be permitted for wainscotting or paneling of not more than 1,000 square feet of applied surface area in the grade lobby where applied directly to a noncombustible base or over furring strips applied to a noncombustible base and fireblocked as required by Section 803.4.1.

b. In exit enclosures of buildings less than three stories in height of other than Group I-3, Class B interior finish for nonsprinklered buildings and Class C interior finish for sprinklered buildings shall be permitted.

c. Requirements for rooms and enclosed spaces shall be based upon spaces enclosed by partitions. Where a fire-resistance rating is required for structural elements, the enclosing partitions shall extend from the floor to the ceiling. Partitions that do not comply with this shall be considered enclosing spaces and the rooms or spaces on both sides shall be considered one. In determining the applicable requirements for rooms and enclosed spaces, the specific occupancy thereof shall be the governing factor regardless of the group classification of the building or structure.

d. Lobby areas in Group A-1, A-2 and A-3 occupancies shall not be less than Class B materials.

e. Class C interior finish materials shall be permitted in places of assembly with an occupant load of 300 persons or less.

f. For places of religious worship, wood used for ornamental purposes, trusses, paneling or chancel furnishing shall be permitted.

g. Class B material is required where the building exceeds two stories.

h. Class C interior finish materials shall be permitted in administrative spaces.

i. Class C interior finish materials shall be permitted in rooms with a capacity of four persons or less.

j. Class B materials shall be permitted as wainscotting extending not more than 48 inches above the finished floor in corridors.

k. Finish materials as provided for in other sections of this code.

l. Applies when the exit enclosures, exit passageways, corridors or rooms and enclosed spaces are protected by a sprinkler system installed in accordance with Section 903.3.1.1 or 903.3.1.2.

Textile materials, where applied to walls or ceilings, must meet additional criteria prior to approval. Finishes that have napped, tufted, looped, nonwoven, woven or similar surface characteristics present a unique hazard on account of their contribution to extremely rapid fire spread.

Topic: Floor Finish Requirements   **Category:** Interior Finishes

Reference: IBC 804.4.1   **Subject:** Interior Floor Finish

Code Text: *Interior floor finish and floor covering materials in exit enclosures, exit passageways and corridors shall not be less than Class I in Groups I-2 and I-3 and not less than Class II in Groups A, B, E, H, I-4, M, R-1, R-2 and S. In all areas, the floor covering materials shall comply with the DOC FF-1 "pill test."* See exception for the permitted reduction of classification in fully sprinklered buildings.

Discussion and Commentary: Although there are many different types of floor finishes and floor coverings, only those flooring materials composed of fibers are regulated by Section 804. Where required to be classified as Class I or Class II materials, the floor covering materials must be tested by an approved agency in accordance with NFPA 253. In order to verify compliance, the materials must be identified by a hang tag or other suitable method that identifies the manufacturer or supplier, style and finish classification.

Types of classifications: (in terms of heat flux, Sec. 804.2)

- Class I: Minimum 0.45 watts/cm^2 per NFPA 253
- Class II: Minimum 0.22 watts/cm^2 per NFPA 253
- DOC FF-1:   Minimum 0.04 watts/cm^2

Required classifications: (Sec. 804.5)

	Nonsprinklered[a]		Sprinklered (NFPA 13 only)	
	Exits/Corr.[b]	Other Areas	Exits/Corr.[b]	Other Areas
Groups I-2 and I-3	Class I	DOC FF-1	Class II	DOC FF-1
Groups F, I-1, R-3, R-4 and U	DOC FF-1	DOC FF-1	DOC FF-1	DOC FF-1
Other Groups	Class II	DOC FF-1	DOC FF-1	DOC FF-1

Note: [a]Section 903.2 requires sprinklers in various occupancies

[b]Includes exit enclosures, exit passageways, corridors and rooms or spaces not separated from corridors by full-height partitions.

DOC FF-1, often referred to as the Methenamine Pill Test, essentially evaluates the floor covering when subjected to a cigarette-type ignition by using a small methenamine tablet. All carpeting sold in the United States is required by federal law to pass this test procedure.

Code Text: *In occupancies of Groups A, E, I and R-1 and dormitories in Group R-2, curtains, draperies, hangings and other decorative materials suspended from walls or ceilings shall meet the flame propagation performance criteria of NDPA 701 in accordance with Section 806.2 or noncombustible. In Groups I-1 and I-2, combustible decorations shall meet the flame propagation criteria of NFPA 701 unless the decorative materials, such as photographs and paintings, are of such limited quantities that a hazard of fire spread or development is not present. In Group I-3, combustible decorative materials are prohibited.*

Discussion and In occupancies where large, concentrated occupant loads are expected, or where occupants
Commentary: have limited mobility due to physical limitations or restraint, the contribution of decorative materials to a fire condition is of concern. Therefore, the amount and characteristics of such materials are regulated to limit their impact on the fire severity.

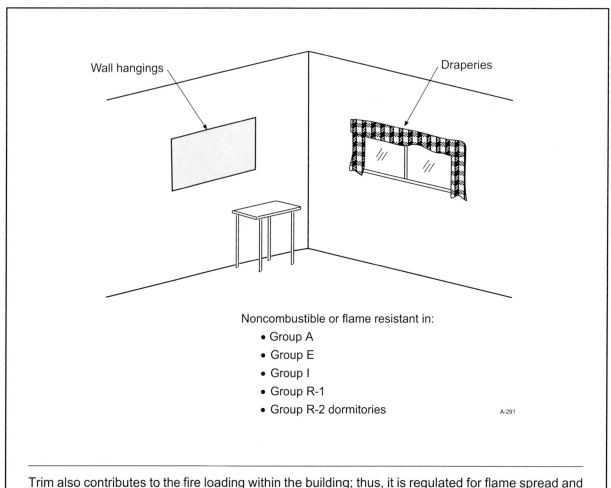

Wall hangings

Draperies

Noncombustible or flame resistant in:
- Group A
- Group E
- Group I
- Group R-1
- Group R-2 dormitories

A-291

Trim also contributes to the fire loading within the building; thus, it is regulated for flame spread and smoke contribution. Where trim consists of combustible materials, except for handrails and guardrails, its surface area is limited to 10 percent of the aggregate wall or ceiling area.

Code Text: *Buildings shall be provided with natural ventilation in accordance with Section 1203.4 or shall be provided with mechanical ventilation in accordance with the* International Mechanical Code. *Natural ventilation of an occupied space shall be through windows, doors, louvers or other openings to the outdoors. The minimum openable area to the outdoors shall be 4 percent of the floor area being ventilated.*

Discussion and Commentary: To obtain a minimum level of environmental comfort, as well as to maintain sanitary conditions, some form of ventilation must be provided to portions of a building that are normally occupied. The *International Mechanical Code* will usually be used to determine the minimum acceptable ventilation methods and quantities.

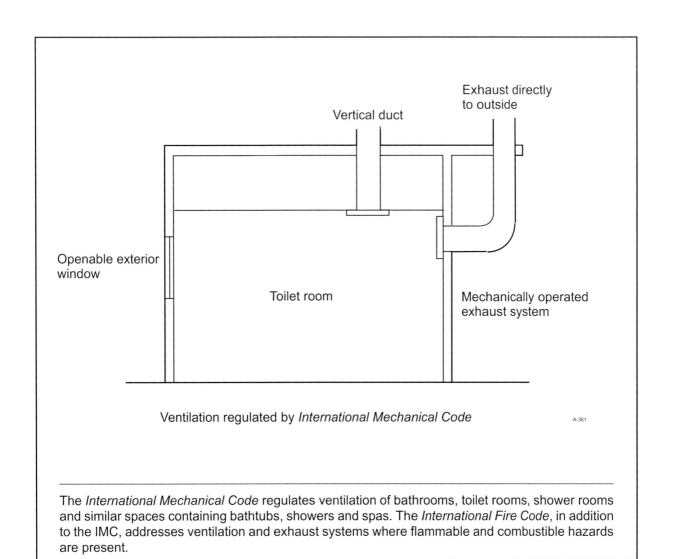

Ventilation regulated by *International Mechanical Code*

A-361

The *International Mechanical Code* regulates ventilation of bathrooms, toilet rooms, shower rooms and similar spaces containing bathtubs, showers and spas. The *International Fire Code*, in addition to the IMC, addresses ventilation and exhaust systems where flammable and combustible hazards are present.

Code Text: *Enclosed attics and enclosed rafter spaces formed where ceilings are applied directly to the underside of roof framing members shall have cross ventilation for each separate space by ventilating openings protected against the entrance of rain and snow. The space between the bottom of the floor joists and the earth under any building except spaces occupied by a basement or cellar shall be provided with ventilation openings through foundation walls or exterior walls. Such openings shall be placed so as to provide cross-ventilation of the under-floor space.* See exceptions for reductions and eliminations of required ventilation.

Discussion and Commentary: Ventilation of the attic and under-floor spaces prevents moisture condensation, which can have adverse effects on the materials of construction located in those spaces. Various exceptions are available that provide equivalent results.

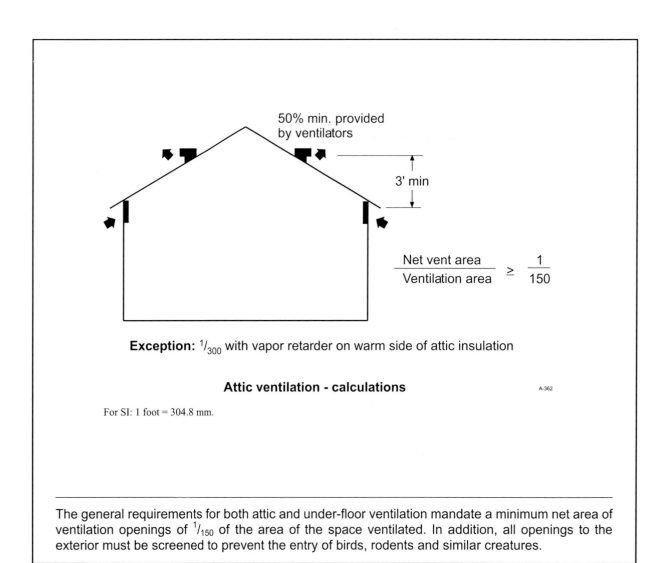

50% min. provided by ventilators

3' min

$$\frac{\text{Net vent area}}{\text{Ventilation area}} \geq \frac{1}{150}$$

Exception: $^1/_{300}$ with vapor retarder on warm side of attic insulation

Attic ventilation - calculations

A-362

For SI: 1 foot = 304.8 mm.

The general requirements for both attic and under-floor ventilation mandate a minimum net area of ventilation openings of $^1/_{150}$ of the area of the space ventilated. In addition, all openings to the exterior must be screened to prevent the entry of birds, rodents and similar creatures.

Code Text: *Every space intended for human occupancy shall be provided with natural light by means of exterior glazed openings in accordance with Section 1205.2 or shall be provided with artificial light in accordance with Section 1205.3. Exterior glazed openings shall open directly onto a public way or onto a yard or court in accordance with Section 1206. The minimum net glazed area shall not be less than 8 percent of the floor area of the room served. Artificial light shall be provided that is adequate to provide an average illumination of 10 foot-candles (107 lux) over the area of the room at a height of 30 inches (762 mm) above the floor level.*

Discussion and Commentary: It is fundamental that all occupiable areas of a building be provided with adequate illumination. The use of artificial light to satisfy the code is acceptable because it can produce the light necessary for occupancy at any time of the day or night.

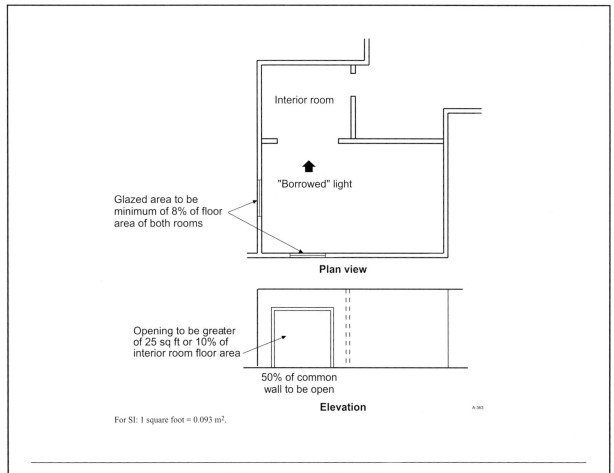

Interior room

"Borrowed" light

Glazed area to be minimum of 8% of floor area of both rooms

Plan view

Opening to be greater of 25 sq ft or 10% of interior room floor area

50% of common wall to be open

Elevation

A-363

For SI: 1 square foot = 0.093 m².

Where natural light is used to satisfy the provisions of the IBC, it can be shared by two rooms. The common wall between the rooms must be adequately open, and the total floor area of both rooms shall be used to calculate the minimum glazed area.

Code Text: *Walls, partitions and floor/ceiling assemblies separating dwelling units from each other or from public or service areas shall have a sound transmission class (STC) of not less than 50 (45 if field tested) for air-borne noise when tested in accordance with ASTM E 90. Floor/ceiling assemblies between dwelling units or between a dwelling unit and a public or service area within the structure shall have an impact insulation class (IIC) rating of not less than 50 (45 if field tested) when tested in accordance with ASTM E 492.*

Discussion and Commentary: To control sound transmission between areas of a residential building, insulated walls and floor/ceiling assemblies are necessary. The regulations address air-borne sound that may be carried throughout the structure, as well as impact noise created on the floor of a floor/ceiling assembly.

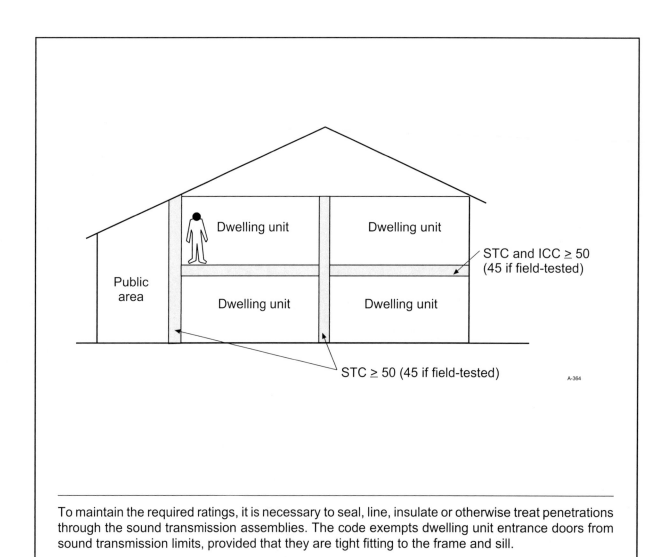

To maintain the required ratings, it is necessary to seal, line, insulate or otherwise treat penetrations through the sound transmission assemblies. The code exempts dwelling unit entrance doors from sound transmission limits, provided that they are tight fitting to the frame and sill.

Topic: Room Size and Ceiling Height

Category: Interior Environment

Reference: IBC 1208

Subject: Interior Space Dimensions

Code Text: *Habitable spaces, other than a kitchen, shall not be less than 7 feet (2134 mm) in any plan dimension. Kitchens shall have a clear passageway of not less than 3 feet (914 mm) between counter fronts and appliances or counter fronts and walls. Occupiable spaces, habitable spaces and corridors shall have a ceiling height of not less than 7 feet 6 inches (2286 mm). Bathrooms, toilet rooms, kitchens, storage rooms and laundry rooms shall be permitted to have a ceiling height of not less than 7 feet (2134 mm).* See exceptions for ceilings with exposed beams, sloped ceilings and mezzanines. *Every dwelling unit shall have at least one room that shall have not less than 120 square feet (13.9 m²) of net floor area. Other habitable rooms shall have a net floor area of not less than 70 square feet (6.5 m²).*

Discussion and Commentary: For fundamental usability and environmental purposes, it is necessary to mandate minimum requirements for the size and height of occupiable spaces.

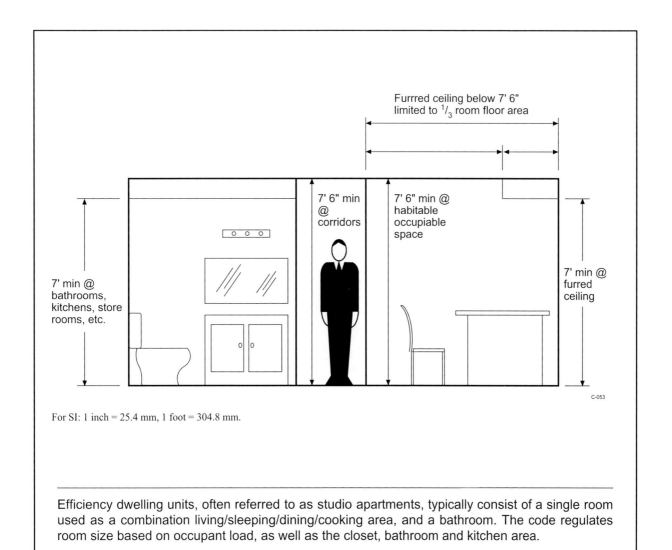

For SI: 1 inch = 25.4 mm, 1 foot = 304.8 mm.

Efficiency dwelling units, often referred to as studio apartments, typically consist of a single room used as a combination living/sleeping/dining/cooking area, and a bathroom. The code regulates room size based on occupant load, as well as the closet, bathroom and kitchen area.

Code Text: *Crawl spaces shall be provided with a minimum of one access opening not less than 18 inches by 24 inches (457 mm by 610 mm). An opening not less than 20 inches by 30 inches (559 mm by 762 mm) shall be provided to any attic area having a clear height of over 30 inches (762 mm). A 30-inch (762 mm) minimum clear headroom in the attic space shall be provided at or above the access opening.*

Discussion and Commentary: Items such as plumbing and wiring installations pass through crawl space at times. Required initial and periodic inspections and maintenance and repairs cannot be carried out without access to such crawl spaces. Attic access is also required for similar reasons. Although uncommon, access to the attic for fire department purposes can also be accomplished through such openings. The required openings are a convenient and nondestructive means for any user to access such concealed spaces.

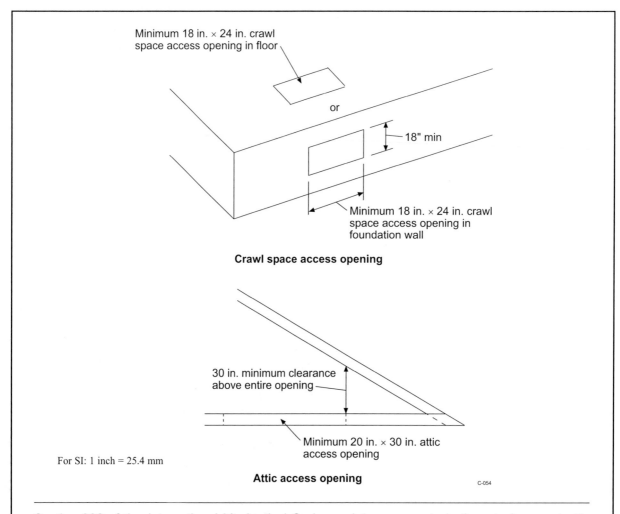

Minimum 18 in. × 24 in. crawl space access opening in floor

or

18" min

Minimum 18 in. × 24 in. crawl space access opening in foundation wall

Crawl space access opening

30 in. minimum clearance above entire opening

Minimum 20 in. × 30 in. attic access opening

For SI: 1 inch = 25.4 mm

Attic access opening

C-054

Section 306 of the *International Mechanical Code* regulates access to both underfloor and attic spaces for the inspection, service, repair or replacement of any mechanical equipment. In addition to the access opening, the passageway and service area sizes are also addressed.

Code Text: *In other than dwelling units, toilet and bathing room floors shall have a smooth, hard, nonabsorbent surface that extends upward onto the walls at least 6 inches (152 mm). Walls within 2 feet (610 mm) of urinals and water closets shall have a smooth, hard, nonabsorbent surface, to a height of 4 feet (1219 mm) above the floor, and except for structural elements, the materials used in such walls shall be of a type that is not adversely affected by moisture.* See exceptions for dwelling units and private toilet rooms. *Accessories such as grab bars, towel bars, paper dispensers and soap dishes, provided on or within walls, shall be installed and sealed to protect structural elements from moisture*

Discussion and Commentary: For obvious sanitary reasons, it is necessary to provide surfaces in bath and toilet areas that are easily cleaned and maintained.

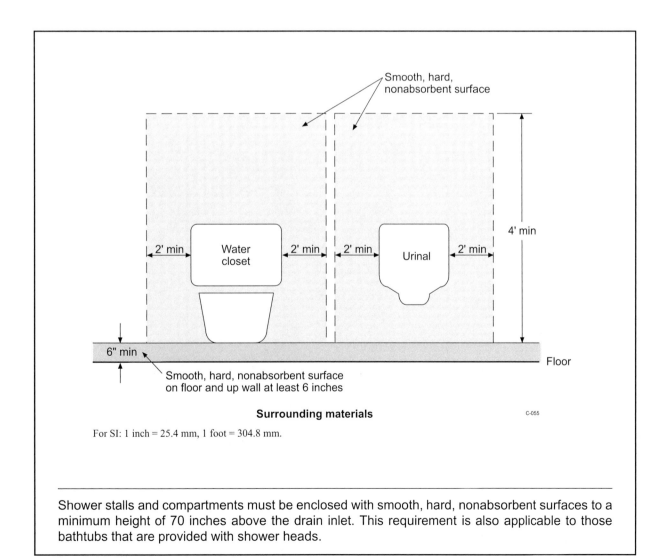

Surrounding materials

C-055

For SI: 1 inch = 25.4 mm, 1 foot = 304.8 mm.

Shower stalls and compartments must be enclosed with smooth, hard, nonabsorbent surfaces to a minimum height of 70 inches above the drain inlet. This requirement is also applicable to those bathtubs that are provided with shower heads.

Code Text: *Gypsum wallboard or gypsum plaster shall not be used in any exterior surface where such gypsum construction will be exposed directly to the weather. Gypsum wallboard, gypsum lath or gypsum plaster shall not be installed until weather protection for the installation is provided. Edges and ends of gypsum board shall occur on the framing members, except those edges and ends that are perpendicular to the framing members.*

Discussion and Commentary: Gypsum wallboard, like gypsum plaster, is subject to deterioration from moisture. Accordingly, the code does not permit such gypsum materials to be installed on weather-exposed surfaces, as defined in Section 2502. Gypsum materials shall not be installed on interior surfaces until adequate protection from the weather has been provided.

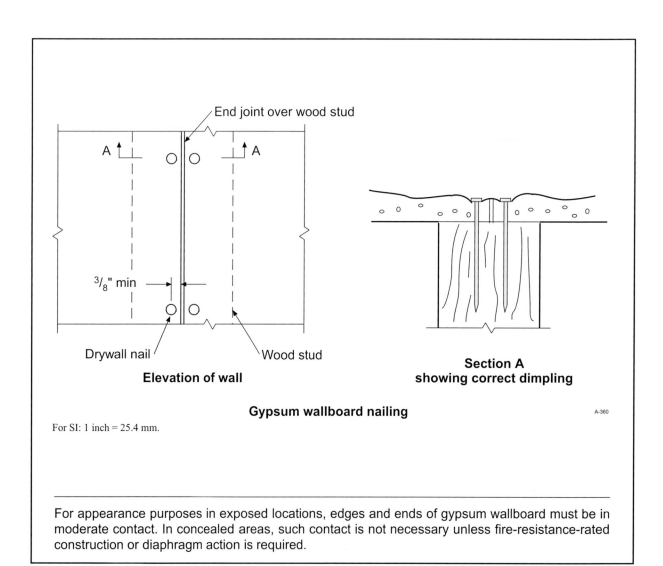

End joint over wood stud

$^3/_8$" min

Drywall nail / Wood stud

Elevation of wall

**Section A
showing correct dimpling**

Gypsum wallboard nailing

A-360

For SI: 1 inch = 25.4 mm.

For appearance purposes in exposed locations, edges and ends of gypsum wallboard must be in moderate contact. In concealed areas, such contact is not necessary unless fire-resistance-rated construction or diaphragm action is required.

Topic: Base for Tile

Reference: IBC 2509.2

Category: Gypsum Board and Plaster

Subject: Gypsum Board in Showers

Code Text: *Cement, fiber-cement or glass mat gypsum backers in compliance with ASTM C 1178, C 1288 or C 1325 and installed in accordance with manufacturer recommendations shall be used as a base for wall tile in tub and shower areas and wall and ceiling panels in shower areas. Water resistant gypsum backing board shall be used as a base for tile in water closet compartment walls when installed in accordance with GA-216 or ASTM C 840 and manufacturer recommendations. Regular gypsum wallboard is permitted under tile or wall panels in other wall and ceiling areas when installed in accordance with GA-216 or ASTM C 840.*

Discussion and Commentary: Because of the moisture-resistant qualities, cement, fiber-cement or glass mat gypsum backers are required when used as a backing material for tile in high-moisture areas. Although water-resistant gypsum backing board is required as a base for tile on public water closet compartment walls, such gypsum board is prohibited for use in three locations: 1) over a vapor retarder in tub or shower compartments, 2) in areas subject to continuous high humidity or where there will be direct exposure to water, and 3) on ceilings with excessive spacing between framing members.

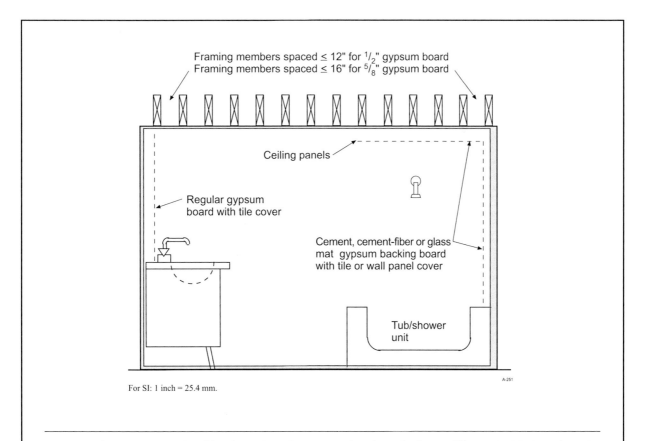

Framing members spaced ≤ 12" for $\frac{1}{2}$" gypsum board
Framing members spaced ≤ 16" for $\frac{5}{8}$" gypsum board

Ceiling panels

Regular gypsum board with tile cover

Cement, cement-fiber or glass mat gypsum backing board with tile or wall panel cover

Tub/shower unit

For SI: 1 inch = 25.4 mm.

A-251

Water-resistant gypsum backing board tends to sag when installed as ceiling material; therefore, its use is limited only to those applications where the ceiling framing is spaced at no more than 12 inches on center for $\frac{1}{2}$-inch board, or a maximum of 16 inches on center for $\frac{5}{8}$-inch board.

Code Text: *A minimum 0.019-inch (0.48 mm) (No. 26 galvanized sheet gage), corrosion-resistant weep screed with a minimum vertical attachment flange of 3½ inches (89 mm) shall be provided at or below the foundation plate line on exterior stud walls in accordance with ASTM C 926. the weep screed shall be placed a minimum of 4 inches (102 mm) above the earth or 2 inches (51 mm) above paved areas and be of a type that will allow trapped water to drain to the exterior of the building.*

Discussion and Commentary: Water can penetrate exterior plaster walls for a variety of reasons. Once it penetrates the plaster, the water will run down the exterior face of the weather-resistive barrier until it reaches the sill plate or mudsill. At this point, the water will seek exit from the wall and, if the exterior plaster is not applied to allow the water to escape, it will exit through the inside of the wall and into the interior of the building. Therefore, a weep screed, when properly installed, will permit the water's escape to the exterior of the building.

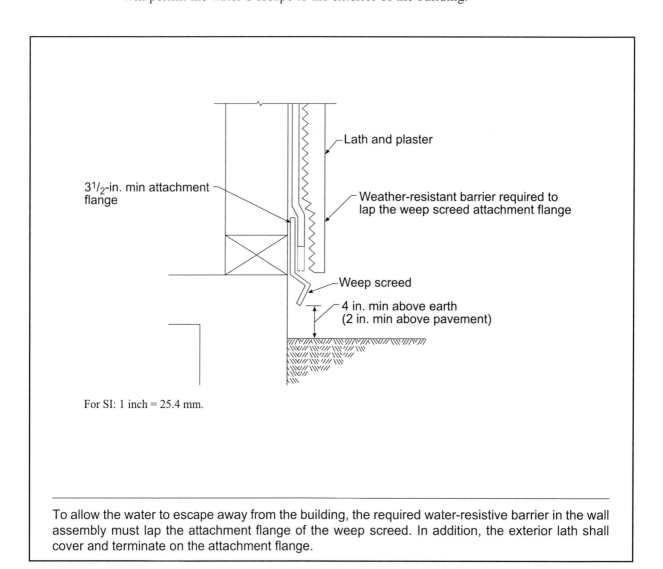

For SI: 1 inch = 25.4 mm.

To allow the water to escape away from the building, the required water-resistive barrier in the wall assembly must lap the attachment flange of the weep screed. In addition, the exterior lath shall cover and terminate on the attachment flange.

Code Text: *Elevator, dumbwaiter and other hoistway enclosures shall be shaft enclosures complying with Section 707. Openings in hoistway enclosures shall be protected as required in Chapter 7. Doors, other than hoistway doors and the elevator car door, shall be prohibited at the point of access to an elevator car unless such doors are readily openable from the car side without a key, tool, special knowledge or effort.*

Discussion and Commentary: An elevator shaft is regulated under the shaft enclosure provisions of Section 707. Generally, an elevator enclosure must be of 2-hour fire-resistance-rated construction in Type I buildings or where four or more stories are connected. A 1-hour rating is permitted where the shaft enclosure connects three stories or less.

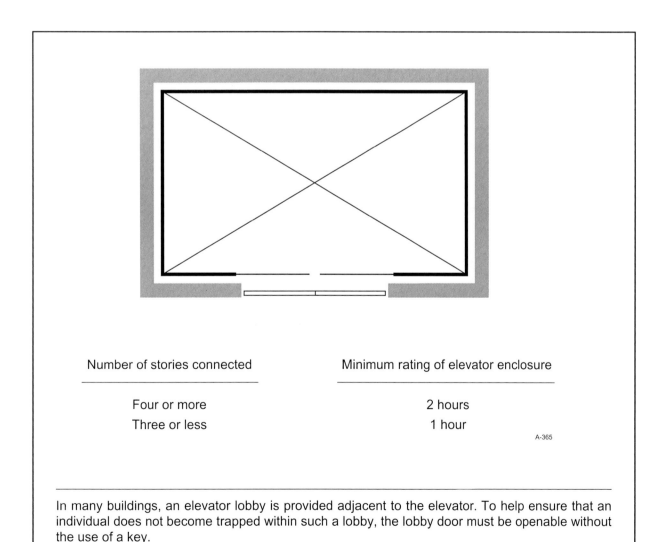

Number of stories connected	Minimum rating of elevator enclosure
Four or more	2 hours
Three or less	1 hour

A-365

In many buildings, an elevator lobby is provided adjacent to the elevator. To help ensure that an individual does not become trapped within such a lobby, the lobby door must be openable without the use of a key.

Code Text: *Where four or more elevator cars serve all or the same portion of a building, the elevators shall be located in at least two separate hoistways. Not more than four elevator cars shall be located in any single hoistway enclosure. Elevators shall not be in a common shaft enclosure with a stairway.*

Discussion and Commentary: The basis for limiting the number of elevator cars in a single hoistway is to provide a reasonable level of assurance that a multilevel building served by several elevators would not have all of its elevator cars disabled by a single fire incident. The provisions increase the chance that some of the elevators would remain operational during an emergency situation.

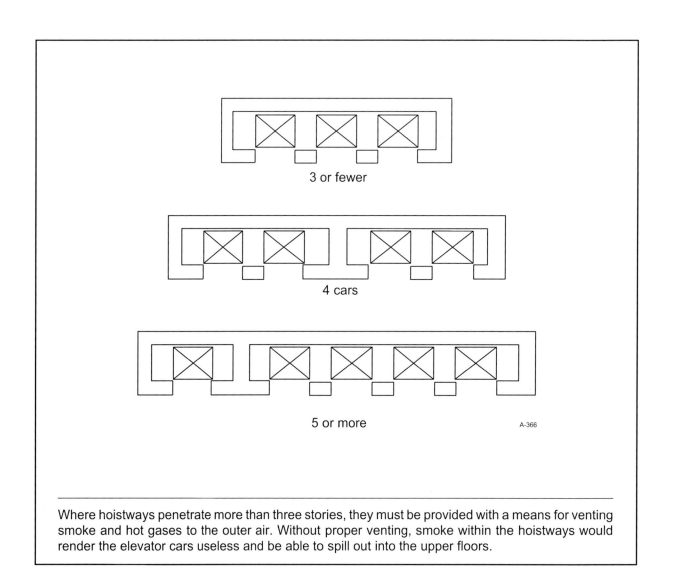

3 or fewer

4 cars

5 or more   A-366

Where hoistways penetrate more than three stories, they must be provided with a means for venting smoke and hot gases to the outer air. Without proper venting, smoke within the hoistways would render the elevator cars useless and be able to spill out into the upper floors.

Code Text: *Where elevators are provided in buildings four or more stories above grade plane or four or more stories below grade plane, at least one elevator shall be provided for fire department emergency access to all floors. The elevator car shall be of such a size and arrangement to accommodate a 24-inch by 84-inch (610 mm by 1930 mm) ambulance stretcher in the horizontal, open position.*

Discussion and Commentary: In those buildings over three stories in height with one or more elevators, which is essentially every building, it is necessary that a minimum of one elevator car be of sufficient size to hold an ambulance stretcher. The elevator car must access all floor levels within the building or additional complying cars must be provided to provide such access. The minimum size requirement is based on the stretchers now commonly in use by medical service personnel and emergency responders.

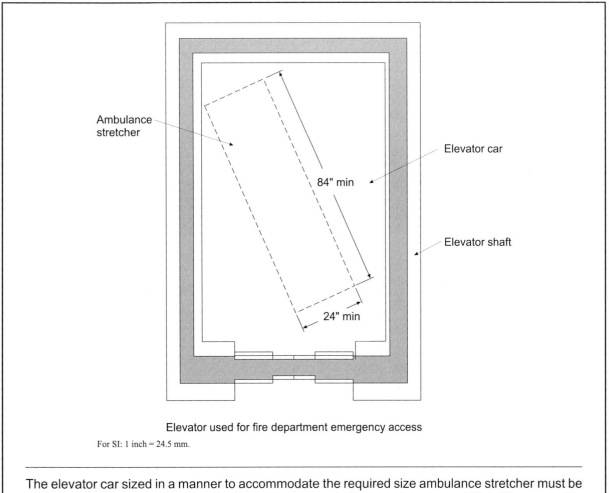

Elevator used for fire department emergency access

For SI: 1 inch = 24.5 mm.

The elevator car sized in a manner to accommodate the required size ambulance stretcher must be identified by the international symbol for emergency medical services (star of life). The symbol is required to be a minimum of 3 inches in height and is to be placed on both sides of the hoistway door frame.

Quiz

1. Where classified in accordance with ASTM E 84, Class B interior wall and ceiling finishes have a flame spread index of _____ and a smoke-developed index of 0-450.

    a. 0-25   b. 26-75

    c. 76-200   d. 201 and greater

    Reference_____

2. Unless a Class A material or qualified by tests, an interior wall or ceiling finish that is a maximum of _____-inch thick shall be applied directly against a noncombustible backing.

    a. 0.036   b. 0.125

    c. 0.25   d. 0.5

    Reference_____

3. The wall finish in a dining room classified as Group A-2 in a sprinklered building shall have a minimum flame spread index classification of Class _____.

    a. A   b. B

    c. C   d. no restrictions

    Reference_____

4. The wall finish in a vertical exit enclosure of a four-story nonsprinklered Group B office building shall have a minimum flame spread index classification of Class _____.

   a. A
   b. B
   c. C
   d. no restrictions

   Reference_____

5. In a patient room of a fully-sprinklered Group I-2 occupancy, the interior finish materials are permitted to be a maximum of Class C when the room has a maximum capacity of _____ persons.

   a. 1
   b. 2
   c. 3
   d. 4

   Reference_____

6. Unless in compliance with the Method B test protocol, textile wall coverings shall have a minimum Class _____ flame spread index where installed in a fully-sprinklered art gallery.

   a. A
   b. B
   c. C
   d. not permitted

   Reference_____

7. In a fully sprinklered Group I-2 occupancy, carpet installed as an interior floor finish in an exit passageway shall be a minimum _____.

   a. Class I
   b. Class II
   c. Class A
   d. DOC FF-1 "pill test"

   Reference_____

8. Combustible trim, excluding handrails and guardrails, shall be limited to a maximum of _____ percent of the aggregate wall or ceiling area in which it is located.

   a. 5
   b. 10
   c. 25
   d. 50

   Reference_____

9. In a fully sprinklered Group A auditorium, complying flame-resistant decorative material shall be limited to a maximum of _____ percent of the aggregate area of walls and ceilings.

    a. 5   b. 10

    c. 25   d. 50

Reference_____

10. What is the minimum permitted flame spread and smoke-developed index for materials, other than foam plastic, used as interior trim?

    a. Class A   b. Class B

    c. Class C   d. unlimited

Reference_____

11. Within a sleeping unit of a Group R-1 hotel, a space-heating system shall be provided that is capable of maintaining a minimum indoor temperature of _____ at a point 3 feet above the floor on the design heating day.

    a. 65 °F   b. 68 °F

    c. 70 °F   d. 72 °F

Reference_____

12. Where natural light by means of exterior glazed openings is utilized as the required lighting for an occupied space, the minimum net glazed area shall not be less than _____ percent of the floor area of the room served.

    a. 4   b. 5

    c. 8   d. 10

Reference_____

13. A stairway within a dwelling unit in a Group R-2 apartment building shall be provided with a minimum illumination level on tread runs of _____ foot-candle(s).

    a. 1   b. 2

    c. 5   d. 10

Reference_____

14. A floor/ceiling assembly between a dwelling unit and a public area within the same building shall have a minimum impact insulation class (IIC) rating of _____ when field tested.

    a. 40   b. 45

    c. 50   d. not regulated

Reference_____

15. Occupiable spaces in an office building shall have a minimum ceiling height of _____.

    a. 6 feet, 8 inches   b. 7 feet, 0 inches

    c. 7 feet, 6 inches   d. 8 feet, 0 inches

Reference_____

16. Every dwelling unit shall have at least one room with a minimum floor area of _____ square feet.

    a. 70   b. 120

    c. 150   d. 220

Reference_____

17. The minimum size of the required access opening to a crawl space shall be _____.

    a. 18 inches by 24 inches   b. 20 inches by 24 inches

    c. 22 inches by 30 inches   d. 22 inches by 30 inches

Reference_____

18. The minimum size of the required access opening to an attic area over 30 inches in clear height shall be _____.

    a. 18 inches by 24 inches   b. 20 inches by 24 inches

    c. 20 inches by 30 inches   d. 22 inches by 30 inches

Reference_____

19. Shower compartments shall be finished with a smooth, nonabsorbent surface to a minimum height of _____ inches above the drain inlet.

    a. 60   b. 66

    c. 70   d. 72

    Reference_____

20. Roof soffits are considered weather-exposed surfaces except for those portions located a minimum horizontal distance of _____ feet from the outer edges of the soffit.

    a. 3   b. 5

    c. 8   d. 10

    Reference_____

21. Water-resistant gypsum backing board that is $^5/_8$-inch-thick is permitted for use on ceilings having a maximum frame spacing of _____ inches on center.

    a. 12   b. 16

    c. 19.2   d. 24

    Reference_____

22. For exterior plastering, the second coat shall have a maximum variation of _____inch in any direction under a 5-foot straight edge.

    a. $^1/_8$ b. $^3/_{16}$

    c. $^1/_4$ d. $^3/_8$

    Reference_____

23. A maximum of _____ elevator car(s) shall be located in any single hoistway enclosure.

    a. 1   b. 2

    c. 3   d. 4

    Reference_____

24. Holes in the elevator machine room floor for the passage of cables or other moving elevator equipment shall be limited so as to provide a maximum clearance of _____ inch(es) on all sides.

a. $^1/_2$

b. 1

c. $1^1/_2$

d. 2

Reference_____

25. Where elevators are provided in buildings with four or more stories above grade plane, at least one elevator car shall be of such a size to accommodate a minimum _____ ambulance stretcher in the horizontal, open position.

a. 24-inch by 76-inch

b. 24-inch by 84-inch

c. 28-inch by 78-inch

d. 30-inch by 78-inch

Reference_____

26. A fibrous floor covering installed in the dining area of a Group A-2 sprinklered restaurant shall have a minimum classification of _____.

a. DOC FF-1 "pill test"

b. Class I

c. Class II

d. Class A

Reference_____

27. As a general rule, the minimum net area of ventilation openings for under-floor ventilation shall be based on 1 square foot for each _____ square feet of crawl-space area.

a. 100

b. 120

c. 150

d. 300

Reference_____

28. Where a court is adjacent to exterior openings on both sides of the court that provide for the required natural ventilation, the minimum court width shall be _____ feet if the building is five stories in height.

a. 3

b. 6

c. 9

d. 10

Reference_____

29. In the construction of a gypsum board fire-resistance-rated assembly, for which of the following applications is joint and fastener treatment required on single-layer systems?

    a.   walls that extend above a fire-rated ceiling

    b.   where joints occur over wood framing

    c.   assemblies tested without joint treatment

    d.   tongue-and-groove edge gypsum board

Reference_____

30. In a single-elevator building where standby power is provided to operate the elevator, the transfer to standby power shall occur automatically within a maximum of _____ seconds after failure of normal power.

    a. 10   b. 15

    c. 30   d. 60

Reference_____

31. Combustible decorative materials are prohibited in which one of the following occupancies?

    a. Group A-1   b. Group I-2

    c. Group I-3   d. Group R-4

Reference _____

32. Where ceilings are applied directly to the underside of roof framing members to form enclosed rafter spaces, such spaces shall be provided with cross ventilation with a minimum of _____ inch(es) of airspace provided between the insulation and the roof sheathing.

    a. 1   b. 2

    c. 3   d. 4

Reference _____

33. In other than dwelling units, sleeping units and toilet rooms not accessible to the public with a single water closet, walls within 2 feet of urinals and water closets shall have a smooth, hard, nonabsorbent surface to a minimum height of _____ feet above the floor.

    a. 3   b. 4

    c. 5   d. 6

Reference _____

34. Fasteners used to attach gypsum board to a horizontal diaphragm ceiling shall be spaced at a maximum of _____ inches on center at all supports and located a maximum of _____ inch from the edges and ends of the gypsum board.

    a.   7, $^3/_8$ b.   8, $^3/_8$

    c.   7, $^1/_2$ d.   8, $^1/_2$

Reference _____

35. Unless special provisions are made, exterior cement plaster shall not be applied when the ambient temperature is a maximum of _____ .

    a.   32°F   b.   35°F

    c.   40°F   d.   45°F

Reference _____

2006 IBC Chapters 24 and 26
Glazing, Skylights and Plastics

OBJECTIVE: To gain an understanding of the installation requirements for glass and glazing, glazing support and framing, safety glazing, skylights, foam plastics, light-transmitting plastics and plastic veneers.

REFERENCE: Chapters 24 and 26, 2006 *International Building Code*

KEY POINTS:
- How is the installation of replacement glass regulated?
- How must a pane of glass be identified? Tempered glass?
- How must glazing be supported?
- What are the limitations for louvered windows and jalousies?
- Sloped glazing provisions for skylights, roofs and sloped walls apply when the glazing material is installed at what minimum slope from the vertical plane?
- Which materials are permitted for sloped glazing? What limitations are placed on these materials?
- For which sloped glazing installations are screens mandated?
- Skylight frames must be constructed of noncombustible materials in which types of construction?
- When are curbs required for the mounting of skylights?
- How are unit skylights regulated?
- What is safety glazing? What are the standards that regulate safety glazing?
- How shall safety glazing be identified? What information must be included as a part of the identifying mark?
- Which types of doors are exempt from the glazing requirements for hazardous locations?
- When glazing is located adjacent to a door, how is it determined if safety glazing is required?
- Which areas of tub and shower enclosures are considered hazardous locations for glazing?
- Individual fixed or operable glazed panels exceeding nine square feet in area must be safety glazed unless which three conditions exist?
- How is glazing adjacent to stairways and landings to be addressed?

- What are the seven glazing products, materials and uses that are exempt from the requirements for hazardous locations?
- What are the three types of glass permitted to be used as structural balustrade panels in rails?
- What is the maximum flame spread index for foam plastic insulation used in building construction? What is the maximum smoke-developed index?
- When is a thermal barrier necessary to separate the interior of a building from foam plastic insulation?
- Under which conditions may foam plastic insulation be incorporated as a part of a roof covering assembly?
- What are the limitations for the use of plastic veneer within a building? On the exterior wall of a building?
- How are light-transmitting plastics used as wall or roof panels regulated?
- Which specific provisions apply to plastic used as exterior wall panels? As roof panels? In skylights? In light-diffusing systems?

Code Text: *Section 2405 applies to the installation of glass and other transparent, translucent or opaque glazing material installed at a slope more than 15 degrees from the vertical plane, including glazing materials in skylights, roofs and sloped walls. For monolithic glazing systems, the glazing material of the single light or layer shall be laminated glass with a minimum 30-mil (0.76 mm) polyvinyl butyral (or equivalent) interlayer, wired glass, light-transmitting plastic materials, heat-strengthened glass or fully tempered glass.*

Discussion and Commentary: The provisions for skylights are intended to protect such glazed openings from flying firebrands, to provide adequate strength to carry the load normally attributed to roofs, and to protect the occupants of a building from falling glazing materials.

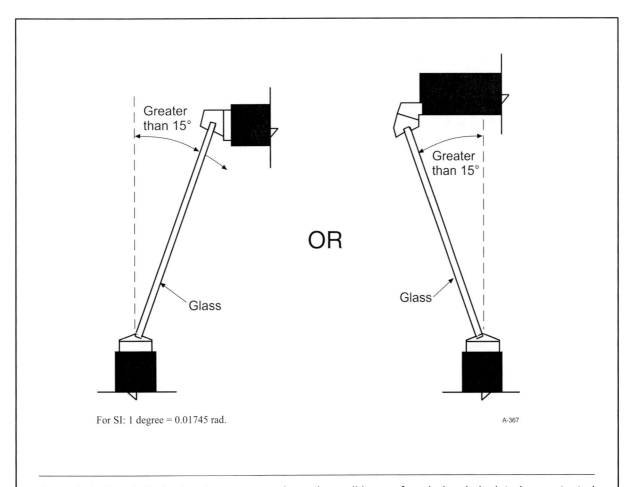

For SI: 1 degree = 0.01745 rad.

A-367

Annealed glass is limited to those areas where the walking surface below is isolated or protected from the risk of falling glass, or to specified greenhouses. Multiple-layer glazing systems must be glazed with only those materials permitted for single-layer glazing systems.

Topic: Screening

Reference: IBC 2405.3

Category: Glass and Glazing

Subject: Sloped Glazing and Skylights

Code Text: *Where used in monolithic glazing systems, heat-strengthened glass and fully tempered glass shall have screens installed below the glazing material.* See exceptions. *The screens and their fastenings shall: 1) be capable of supporting twice the weight of the glazing, 2) be firmly and substantially fastened to the framing members, and 3) be installed within 4 inches (102 mm) of the glass.*

Discussion and Commentary: Heat-strengthened glass has the undesirable characteristic of breaking into shards, whereas tempered glass has been shown to break spontaneously such that large chunks of glass may fall unexpectedly. Thus, these two types of glass require screen protection below the skylight to protect the occupants below.

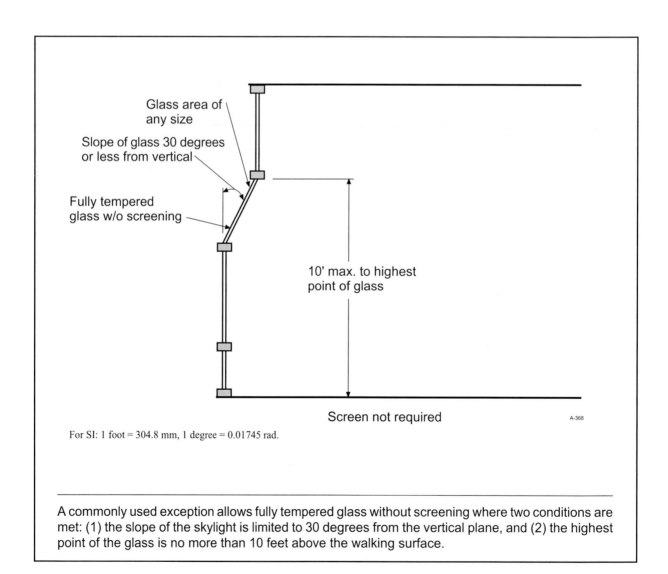

For SI: 1 foot = 304.8 mm, 1 degree = 0.01745 rad.

A commonly used exception allows fully tempered glass without screening where two conditions are met: (1) the slope of the skylight is limited to 30 degrees from the vertical plane, and (2) the highest point of the glass is no more than 10 feet above the walking surface.

Code Text: *Except as provided in Sections 2406.1.2 through 2406.1.4 (plastic glazing, glass block, louvered windows and jalousies), all glazing shall pass the test requirements of CPSC 16 CFR 1201, listed in Chapter 35. Glazing shall comply with the CPSC 16 CFR, Part 1201 criteria, for Category I or II as indicated in Table 2406.1.*

Discussion and Commentary: The only test standard recognized for the acceptance of safety glazing materials in the IBC is CPSC 16 CFR 1201. Developed by the Consumer Product Safety Commission in 1977, in cooperation with building officials and the glass industry, the standard sets forth the criteria for glazing that is required in areas subject to human impact.

TABLE 2406.1
MINIMUM CATEGORY CLASSIFICATION OF GLAZING

EXPOSED SURFACE AREA OF ONE SIDE OF ONE LITE	GLAZING IN STORM OR COMBINATION DOORS (Category class)	GLAZING IN DOORS (Category class)	GLAZED PANELS REGULATED BY ITEM 7 OF SECTION 2406.3 (Category class)	GLAZED PANELS REGULATED BY ITEM 6 OF SECTION 2406.3 (Category class)	DOORS AND ENCLOSURES REGULATED BY ITEM 5 OF SECTION 2406.3 (Category class)	SLIDING GLASS DOORS PATIO TYPE (Category class)
9 square feet or less	I	I	No requirement	I	II	II
More than 9 square feet	II	II	II	II	II	II

For SI: 1 square foot = 0.0929m².

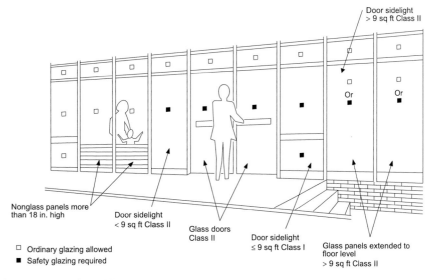

Door sidelight > 9 sq ft Class II

Nonglass panels more than 18 in. high

Door sidelight < 9 sq ft Class II

Glass doors Class II

Door sidelight ≤ 9 sq ft Class I

Glass panels extended to floor level > 9 sq ft Class II

□ Ordinary glazing allowed
■ Safety glazing required

For SI: 1 inch = 25.4 mm mm, 1 foot = 304.8 mm

The tests established in CPSC 16 CFR Part 1201 vary with the category classification. As a part of the test, Category I glazing is impacted from a drop height of 18 inches. Limitations are placed on the damage that can occur due to the impact. Category II glazing is impacted from a drop height of 48 inches, resulting in a more severe test of compliance. Category II glazing is permitted in all safety glazing locations, whereas glazing only recognized as Category I materials are limited to specific locations.

Code Text: *Except as indicated in Section 2406.2.1, each pane of safety glazing installed in hazardous locations shall be identified by a manufacturer's designation specifying who applied the designation, the manufacturer or installer and the safety glazing standard with which it complies, as well as the information specified in Section 2403.1 (general glazing identification). The designation shall be acid etched, sand blasted, ceramic fired, or an embossed mark, or shall be of a type that once applied cannot be removed without being destroyed.* See exceptions for certifications of compliance and tempered spandrel glass.

Discussion and Commentary: Improper glazing installed in areas subject to human impact can create a serious hazard. Accordingly, it is critical that glazing in such locations be appropriately identified to ensure that the proper glazing is in place.

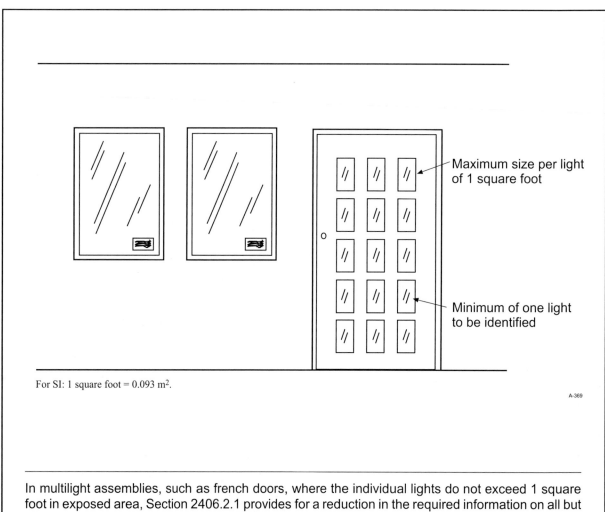

Maximum size per light of 1 square foot

Minimum of one light to be identified

For SI: 1 square foot = 0.093 m².

A-369

In multilight assemblies, such as french doors, where the individual lights do not exceed 1 square foot in exposed area, Section 2406.2.1 provides for a reduction in the required information on all but one light. At least one light must be fully identified.

Code Text: *The following shall be considered specific hazardous locations requiring safety glazing materials: 1) glazing in swinging doors except jalousies, 2) glazing in fixed and sliding panels of sliding door assemblies and panels in sliding and bifold closet door assemblies, 3) glazing in storm doors, and 4) glazing in unframed swinging doors.*

Discussion and Commentary: Collectively, Items 1 through 4 can be summarized by saying that any door containing glazing must be glazed with safety glass or other safety glazing material recognized by the code for that intended purpose. Glazing in doors is of particular concern due to the increased likelihood of accidental impact by individuals operating or opening the doors. In addition, a person may push against a glazed portion of the door to gain leverage in pushing it open. Therefore, it is important that only safety glazing materials be used for glazing in doors.

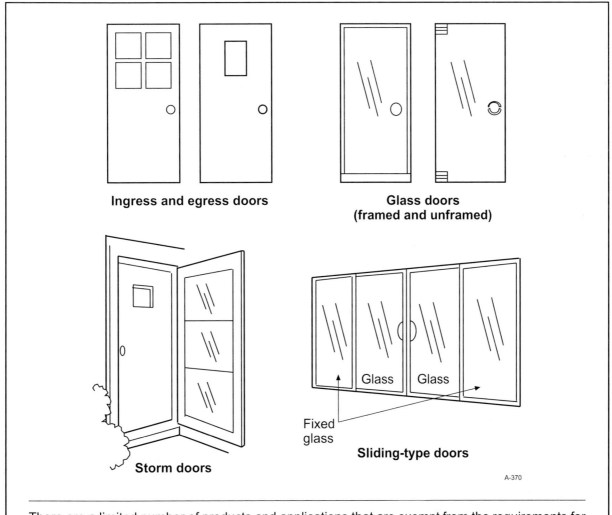

Ingress and egress doors

Glass doors (framed and unframed)

Storm doors

Fixed glass

Glass   Glass

Sliding-type doors

A-370

There are a limited number of products and applications that are exempt from the requirements for hazardous locations, including small openings in doors through which a 3-inch-diameter sphere will not pass, and specific decorative assemblies, such as leaded, faceted or carved glass.

Code Text: *The following shall be considered specific hazardous locations requiring safety glazing materials: (5) glazing in doors and enclosures for hot tubs, whirlpools, saunas, steam rooms, bathtubs and showers. Glazing in any portion of a building wall enclosing these compartments where the bottom exposed edge of the glazing is less than 60 inches (1524 mm) above a standing surface.*

Discussion and Commentary: Because the standing surfaces of bathtubs, showers, hot tubs and similar elements are wet and slippery, glazing adjacent to these elements must be regulated due to the potential for human impact. It is not uncommon for the user to slip while trying to enter or exit. Safety glazing is mandated where any of the glazing within the enclosed area extends to within 60 inches vertically of the standing surface.

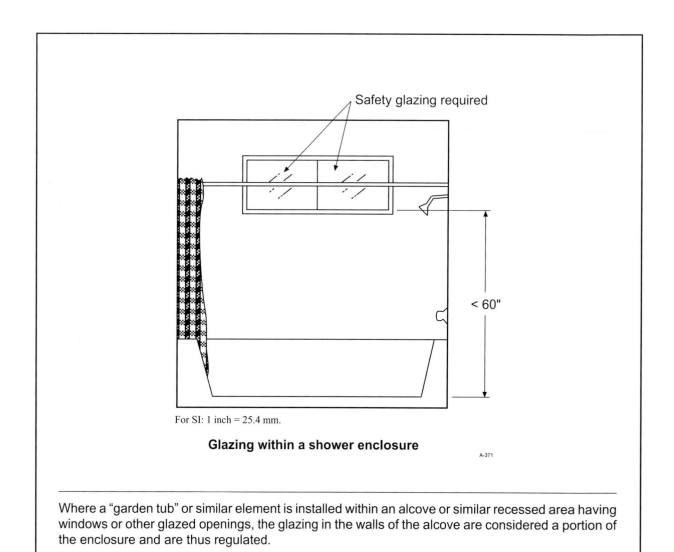

For SI: 1 inch = 25.4 mm.

Glazing within a shower enclosure

A-371

Where a "garden tub" or similar element is installed within an alcove or similar recessed area having windows or other glazed openings, the glazing in the walls of the alcove are considered a portion of the enclosure and are thus regulated.

Code Text: *The following shall be considered specific hazardous locations requiring safety glazing materials: (6) glazing in an individual fixed or operable panel adjacent to a door where the nearest exposed edge of the glazing is within a 24-inch (610 mm) arc of either vertical edge of the door in a closed position and where the bottom exposed edge of the glazing is less than 60 inches (1524 mm) above the walking surface.* See three exceptions where safety glazing is not mandated.

Discussion and Commentary: When an individual approaches a doorway, areas adjacent to the door pose a risk when glazing is within 60 inches vertically of the walking surface. A person may slip or mistake the glass panel adjacent to a door for a passageway and walk into the glass, or a person may push against the sidelight with one hand for support while opening the door with the other hand. Therefore, safety glazing is required for any glazed opening located within 24 inches horizontally of the vertical edge of the door.

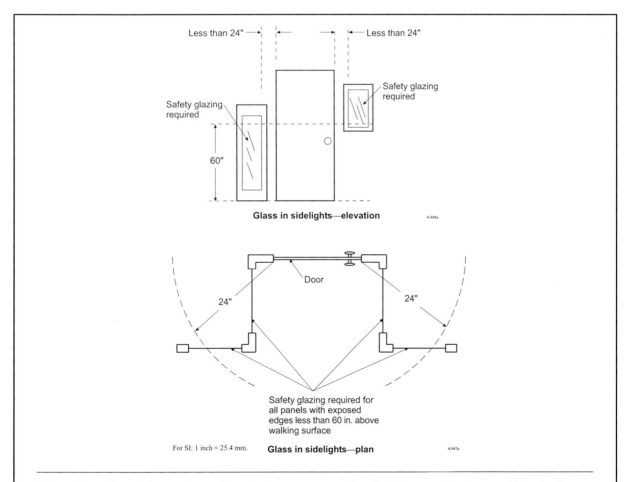

Glass in sidelights—elevation A-948a

Glass in sidelights—plan

For SI: 1 inch = 25.4 mm.

Where there is an intervening wall or similar permanent barrier between the door and the glazing, or where access through the door is to a closet or similar storage area of limited depth, safety glazing is not required, as the potential for contact is greatly reduced.

Code Text: *The following shall be considered specific hazardous locations requiring safety glazing materials: (7) glazing in a fixed or operable panel, other than in those locations described in preceding Items 5 and 6, which meets all of the following conditions: (7.1) exposed area of an individual pane greater than 9 square feet (0.84 m²), (7.2) exposed bottom edge less than 18 inches (457 mm) above the floor, (7.3) exposed top edge greater than 36 inches (914 mm) above the floor, and (7.4) one or more walking surface(s) within 36 inches (914 mm) horizontally of the plane of the glazing.*

Discussion and Commentary: Large pieces of glass create a hazard where located close to a travel path because it is possible to impact glazing where no obstacle or barrier is provided as an alternative impact area. Expansive glazing may also be mistaken for a clear opening in the wall.

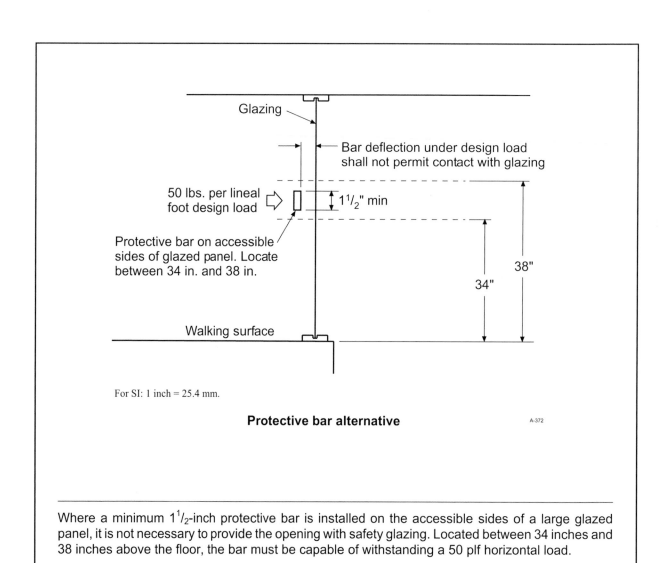

Glazing

Bar deflection under design load shall not permit contact with glazing

50 lbs. per lineal foot design load

1¹/₂" min

Protective bar on accessible sides of glazed panel. Locate between 34 in. and 38 in.

38"

34"

Walking surface

For SI: 1 inch = 25.4 mm.

Protective bar alternative   A-372

Where a minimum 1¹/₂-inch protective bar is installed on the accessible sides of a large glazed panel, it is not necessary to provide the opening with safety glazing. Located between 34 inches and 38 inches above the floor, the bar must be capable of withstanding a 50 plf horizontal load.

Topic: Glazing in Guards or Railings   **Category:** Glass and Glazing
Reference: IBC 2406.3, #8   **Subject:** Hazardous Locations

Code Text: *The following shall be considered specific hazardous locations requiring safety glazing materials: (8) glazing in guards and railings, including structural baluster panels and nonstructural in-fill panels, regardless of area or height above a walking surface.*

Discussion and Commentary: Both intentional and unintentional contact with guards and railings are expected to occur; therefore, the IBC mandates that glazing used in such applications always be safety glazing. Safety glazing is required when the glazed infill panel is nonstructural and is supported by a structural frame system. Occasionally, glazing is used as a structural guard rail system without any other means of support. In this case, the glazing is required to be safety glazing.

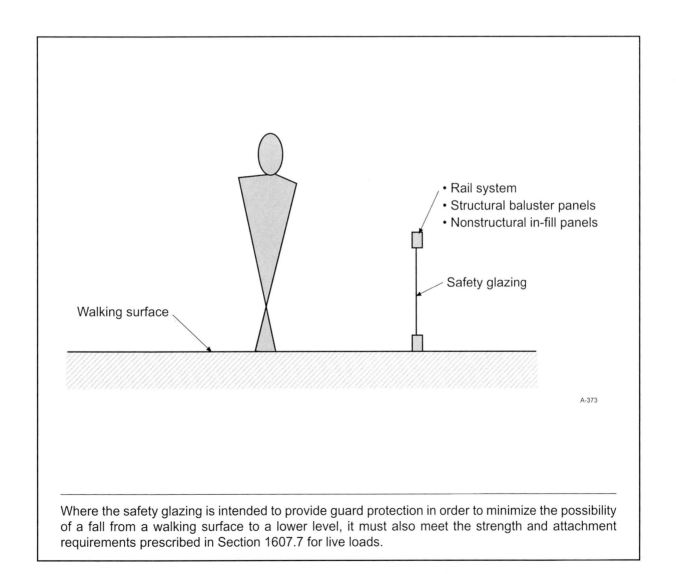

Where the safety glazing is intended to provide guard protection in order to minimize the possibility of a fall from a walking surface to a lower level, it must also meet the strength and attachment requirements prescribed in Section 1607.7 for live loads.

Code Text: *The following shall be considered specific hazardous locations requiring safety glazing materials: (9) glazing in walls and fences enclosing indoor and outdoor swimming pools, hot tubs and spas where all of the following conditions are present: (9.1) the bottom edge of the glazing on the pool or spa side is less than 60 inches (1524 mm) above a walking surface on the pool or spa side of the glazing, and (9.2) the glazing is within 60 inches (1524 mm) horizontally of the water's edge of a swimming pool or spa.*

Discussion and Commentary: The deck or similar area adjacent to a swimming pool or spa is often wet when the pool or spa is in use. The walking surface typically becomes slippery and causes a considerable number of slips and falls. It is important that any adjacent glazed area that is susceptible to human impact be provided with safety glazing.

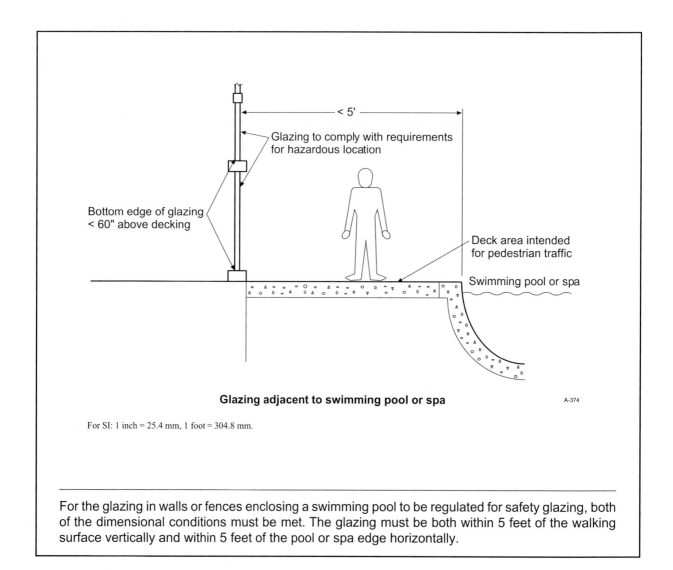

Glazing adjacent to swimming pool or spa

A-374

For SI: 1 inch = 25.4 mm, 1 foot = 304.8 mm.

For the glazing in walls or fences enclosing a swimming pool to be regulated for safety glazing, both of the dimensional conditions must be met. The glazing must be both within 5 feet of the walking surface vertically and within 5 feet of the pool or spa edge horizontally.

Topic: Glazing Adjacent to Stairways

Category: Glass and Glazing

Reference: IBC 2406.3, #10

Subject: Hazardous Locations

Code Text: *The following shall be considered specific hazardous locations requiring safety glazing materials: (10) glazing adjacent to stairways, landings and ramps within 36 inches (914 mm) horizontally of a walking surface; when the exposed surface of the glass is less than 60 inches (1524 mm) above the plane of the adjacent walking surface.* See exception where safety glazing is not required.

Discussion and Commentary: Stairways and ramps present users with a greater risk for injury caused by falling than does a flat surface. Not only is the risk of falling greater when using a stair, but the injuries are generally more severe. Unlike falling on a flat surface where the floor will break a person's fall, there is nothing to stop someone from continuing to fall until he or she reaches the bottom of the stair.

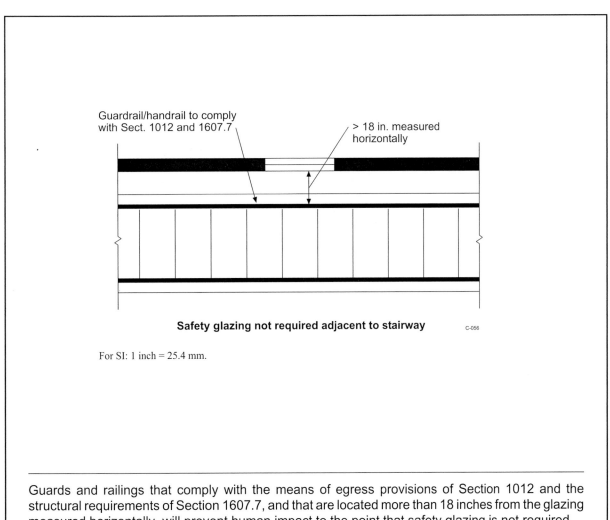

Guardrail/handrail to comply with Sect. 1012 and 1607.7

> 18 in. measured horizontally

Safety glazing not required adjacent to stairway

C-056

For SI: 1 inch = 25.4 mm.

Guards and railings that comply with the means of egress provisions of Section 1012 and the structural requirements of Section 1607.7, and that are located more than 18 inches from the glazing measured horizontally, will prevent human impact to the point that safety glazing is not required.

Code Text: *The following shall be considered specific hazardous locations requiring safety glazing materials: (11) glazing adjacent to stairways within 60 inches (1524 mm) horizontally of the bottom tread of a stairway in any direction when the exposed surface of the glass is less than 60 inches (1524 mm) above the nose of the tread.* See exception where safety glazing is not required.

Discussion and Commentary: Historically, stairways have been considered one of the most dangerous elements of a building. Missteps and falls on stairways are quite common; therefore, it is important that any glazing that may be impacted is made of safety glazing materials. Where a complying guard rail or handrail is provided and the glazing is located a sufficient distance from the railing, safety glazing is not mandated.

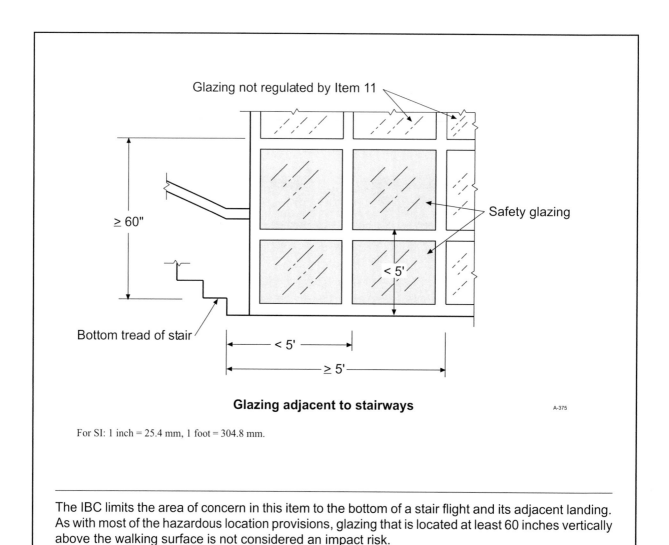

Glazing adjacent to stairways

A-375

For SI: 1 inch = 25.4 mm, 1 foot = 304.8 mm.

The IBC limits the area of concern in this item to the bottom of a stair flight and its adjacent landing. As with most of the hazardous location provisions, glazing that is located at least 60 inches vertically above the walking surface is not considered an impact risk.

Code Text: *The provisions of Section 2603 shall govern the requirements and uses of foam plastic insulation in buildings and structures. Foam plastic insulation is a plastic that is intentionally expanded by the use of a foaming agent to produce a reduced-density plastic containing voids consisting of open or closed cells distributed throughout the plastic for thermal insulating or acoustical purposes and that has a density less than 20 pounds per cubic foot.*

Discussion and Commentary: Two basic concepts address the hazards created when foam plastic is exposed to fire conditions: 1) limitation of flame spread and smoke development, and 2) separation from the interior of the building by an approved thermal barrier.

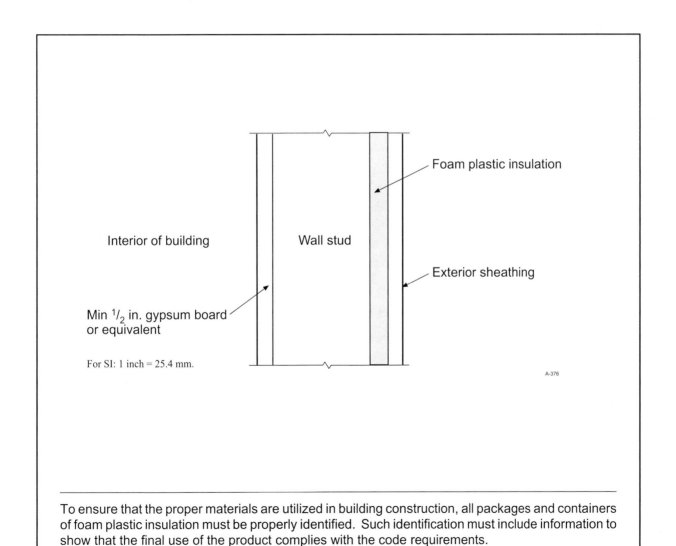

Interior of building

Wall stud

Foam plastic insulation

Exterior sheathing

Min $^1/_2$ in. gypsum board or equivalent

For SI: 1 inch = 25.4 mm.

A-376

To ensure that the proper materials are utilized in building construction, all packages and containers of foam plastic insulation must be properly identified. Such identification must include information to show that the final use of the product complies with the code requirements.

Code Text: *Except as provided for in Sections 2603.4.1 and 2603.9, foam plastic shall be separated from the interior of a building by an approved thermal barrier of 0.5 inch (12.7 mm) gypsum wallboard or equivalent thermal barrier material that will limit the average temperature rise of the unexposed surface to not more than 250 °F (120°C). after 15 minutes of fire exposure, complying with the standard time-temperature curve of ASTM E 119. The thermal barrier shall be installed in such a manner that it will remain in place for 15 minutes.*

Discussion and Commentary: A barrier is mandated to provide a minimum degree of protection between foam plastic materials and a building's occupants. Any type of separation equivalent to that provided by $^1/_2$-inch gypsum board is considered acceptable.

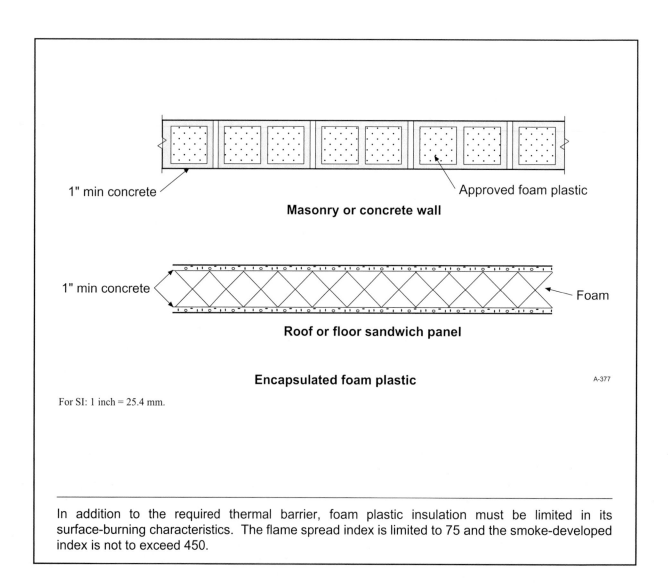

1" min concrete

Approved foam plastic

Masonry or concrete wall

1" min concrete

Foam

Roof or floor sandwich panel

Encapsulated foam plastic

A-377

For SI: 1 inch = 25.4 mm.

In addition to the required thermal barrier, foam plastic insulation must be limited in its surface-burning characteristics. The flame spread index is limited to 75 and the smoke-developed index is not to exceed 450.

Code Text: *Within an attic or crawl space where entry is made only for service of utilities, foam plastic insulation shall be protected against ignition by 1.5-inch-thick (38 mm) mineral fiber insulation; 0.25-inch-thick (6.4 mm) wood structural panel, particleboard, or hardboard; 0.375-inch (9.5 mm) gypsum wallboard, corrosion-resistant steel having a base metal thickness of 0.016 inch (0.4 mm) or other approved material installed in such a manner that the foam plastic insulation is not exposed.*

Discussion and Commentary: Although the general requirements of Section 2603.4 mandate a thermal barrier to isolated foam plastic from the interior of the building, Section 2603.4.1 reduces or eliminates the protective membrane requirements. One such reduction involves attic spaces and crawl spaces. Where access to the concealed attic or crawl space is only necessary to allow for the service of above-ceiling or under-floor utilities, such as plumbing, mechanical, electrical or communication elements, a reduced level of foam plastic protection is permitted.

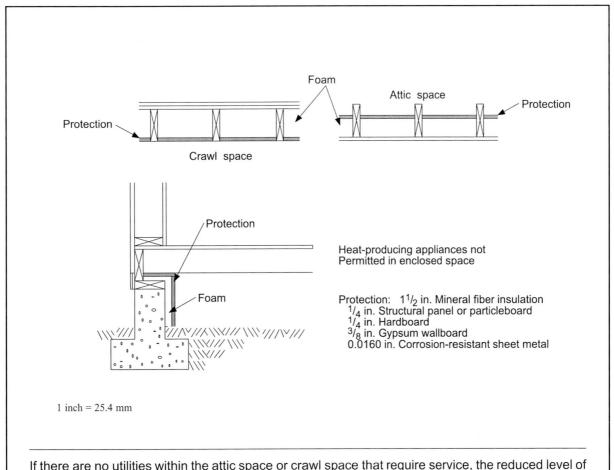

1 inch = 25.4 mm

If there are no utilities within the attic space or crawl space that require service, the reduced level of separation set forth in Section 2603.4.1.6 must still be provided. Where the attic or crawl space provides a usable area that exists for a purpose other than access to utilities, such as storage, a thermal barrier complying with Section 2603.4 is required.

Code Text: *For foam plastics used as interior trim: 1) the minimum density of the interior trim shall be 20 pcf (320 kg/m³), 2) the maximum thickness of the interior trim shall be 0.5 inch (12.7 mm) and the maximum width shall be 8 inches, 3) the interior trim shall not constitute more than 10 percent of the aggregate wall and ceiling area of any room or space, and 4) the flame spread index shall not exceed 75 where tested in accordance with ASTM E 84. The smoke-developed index shall not be limited.*

Discussion and Commentary: The general provisions regulating the use of decorations and trim in buildings are found in Section 806. They include plastics materials, other than foam plastics, used as interior trim. All foam plastic trim must comply with the provisions of Section 2604.2.

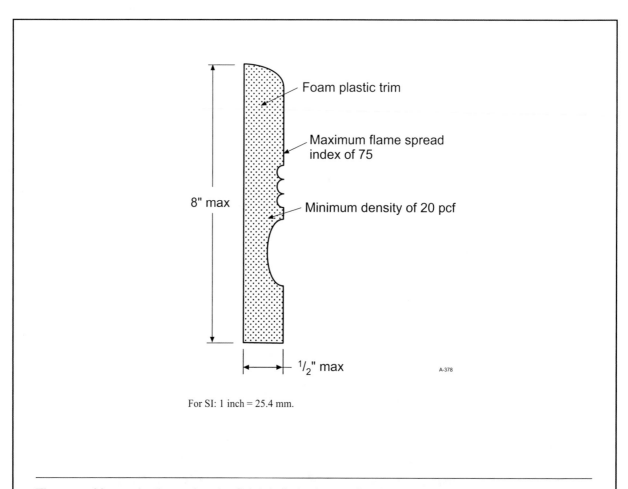

Foam plastic trim

Maximum flame spread index of 75

Minimum density of 20 pcf

8" max

$\frac{1}{2}$" max   A-378

For SI: 1 inch = 25.4 mm.

The use of foam plastics as interior finish is limited to applications that comply with Section 2603.9. This provision allows special approval based on the use of large-scale tests. Foam plastic finishes accepted by testing must also conform to Chapter 8 flame spread requirements.

Quiz

Study Session 18
IBC Chapters 24 and 26

1. Patterned glass in louvered windows and jalousies shall have a minimum thickness of
    _____ inch.

    a. $^3/_{16}$ b. $^1/_4$

    c. $^5/_{16}$ d. $^3/_8$

    Reference_____

2. Where screening is required below the glazing material in a skylight, the screen shall be
    installed a maximum of _____ inch(es) below the glass.

    a. 1   b. 3

    c. 4   d. 6

    Reference_____

3. When used as the bottom glass layer in a multiple-layer glazing system installed above a
    walking surface, which of the following glazing materials never needs to be equipped
    with complying screening?

    a. fully tempered glass   b. laminated glass

    c. heat-strengthened glass   d. wired glass

    Reference_____

4. When set at an angle less than 45 degrees from the horizontal plane, a skylight shall be mounted on a curb a minimum of _____ inch(es) above the plane of the roof of a Group B occupancy.

    a. 1   b. 2

    c. 3   d. 4

    Reference_____

5. Class I safety glazing with a maximum size of _____ per light is permitted in doors.

    a. 0 square inches (not permitted of any size)

    b. 100 square inches

    c. 144 square inches

    d. 9 square feet

    Reference_____

6. Which one of the following types of safety glazing materials shall not be identified by a certificate or affidavit?

    a. tempered glass   b. wired glass

    c. laminated glass   d. plastic glazing

    Reference_____

7. Glazing within a shower enclosure shall be safety glazing where the bottom exposed edge of the glazing is less than _____ inches above a standing surface.

    a. 60   b. 66

    c. 72   d. 78

    Reference_____

8. Where located less than 60 inches above the walking surface, glazing adjacent to a door within what maximum horizontal distance of either vertical door edge must be safety glazing?

    a. 12 inches   b. 18 inches

    c. 24 inches   d. 36 inches

    Reference_____

9. When a protective bar is permitted as protection in lieu of safety glazing, the bar must be capable of withstanding a minimum horizontal load of _____ pounds per linear foot.

    a. 15   b. 20

    c. 40   d. 50

    Reference_____

10. Glazing in walls enclosing a swimming pool need not be safety glazing where located a minimum of _____ feet horizontally from the water's edge.

    a. 5   b. 6

    c. 8   d. 10

    Reference_____

11. At other than the bottom tread, glazing adjacent to stairways does not need to be safety glazing where the plane of the glass is a minimum of _____ inches horizontally from the walking surface.

    a. 12   b. 18

    c. 24   d. 36

    Reference_____

12. Glazed openings in doors are not considered hazardous locations for glazing purposes, provided the openings will not allow passage of a _____-inch sphere.

    a. 3   b. 4

    c. 6   d. 9

    Reference_____

13. Fire department glass access panels shall be of _____ glass.

    a. heat-strengthened   b. tempered

    c. laminated   d. annealed

    Reference_____

14. Where glass is used as a handrail assembly or guard section, and unless otherwise adequately supported, each handrail or guard section shall be supported by a minimum of _____ glass balusters.

    a. 2   b. 3

    c. 4   d. 6

Reference_____

15. Foam plastic insulation shall have a density less than _____ pounds per cubic foot.

    a. 2   b. 4

    c. 10   d. 20

Reference_____

16. _____ is considered to be a plastic material that is capable of being changed into a substantially nonreformable product when cured.

    a. Approved plastic   b. Glass fiber reinforced plastic

    c. Thermoplastic material   d. Thermosetting material

Reference_____

17. Unless otherwise specified, foam plastic shall have a maximum flame spread index of _____ and a maximum smoke-developed index of _____ where tested in the maximum thickness intended for use.

    a. 25, 200   b. 25, 450

    c. 50, 200   d. 75, 450

Reference_____

18. What material is specifically identified as an approved thermal barrier for separating foam plastic from the interior of a building?

    a. 0.5-inch gypsum wallboard   b. 0.375-inch Type x gypsum wallboard

    c. 0.375-inch gypsum sheathing   d. 0.5-inch wood structural panel

Reference_____

19. Within a concrete floor system, all foam plastic insulation shall be covered on each face by a minimum of a _____ -inch thickness of concrete.

    a. $^1/_2$ b. 1

    c. $1^1/_2$ d. 2

    Reference_____

20. The maximum allowable thickness of foam plastic interior trim shall be _____ inch(es), with a maximum permitted width of _____ inches.

    a. 4, 12   b. 2, 12

    c. 1, 8   d. 0.5, 8

    Reference_____

21. Exterior plastic veneer may be installed on the exterior walls of buildings of any type of construction to a maximum height of _____ feet above grade.

    a. 25   b. 35

    c. 50   d. 75

    Reference_____

22. In a nonsprinklered Group B building, Class CC2 light-transmitting wall panels used in an exterior wall with a fire separation distance of 18 feet are limited to a maximum of _____ percent of the exterior wall.

    a. 10   b. 15

    c. 25   d. 50

    Reference_____

23. In a fully sprinklered building, what is the maximum height light-transmitting plastic glazing is permitted to be installed above grade level?

    a. 35 feet   b. 55 feet

    c. 75 feet   d. unlimited

    Reference_____

24. In which one of the following occupancies may light-transmitting plastic roof panels be installed?

    a.   Group A-1   b.   Group H-1

    c.   Group I-2   d.   Group I-3

Reference_____

25. An interior wall sign of light-transmitting plastic is limited in size to _____ square feet.

    a.   16   b.   24

    c.   48   d.   100

Reference_____

26. Glass in window walls sloped a maximum of _____ degrees from the vertical shall be designed to resist the required wind loads for components and cladding.

    a.   15   b.   30

    c.   45   d.   60

Reference_____

27. For which of the following applications is the use of a removable paper label specifically permitted for identifying tempered glass for safety glazing purposes?

    a.   storm doors   b.   spandrel panels

    c.   unframed swinging doors   d.   structural balustrade panels

Reference_____

28. Only one individual light in a multilight glazed door assembly requires safety glazing identification, provided each light is a maximum of _____ in size.

    a.   64 square inches   b.   100 square inches

    c.   1 square foot   d. $1^1/_2$ square feet

Reference_____

29. All glazing subject to human impact located in a(n) _____ shall comply with Category II of CPSC 16 CFR 1201.

   a. school classroom   b. detention facility

   c. multipurpose gymnasium   d. manufacturing building

   Reference_____

30. Foam plastic spray applied to a sill plate of Type V construction shall have a maximum thickness of _____.

   a. 1 inch   b. $1^1/_2$ inches

   c. 3 inches   d. $3^1/_4$ inches

   Reference_____

31. Any glazing material is permitted to be installed without screens in the sloped glazing system of commercial noncombustible greenhouses, provided the maximum ridge height is _____ feet above grade.

   a. 16   b. 20

   c. 25   d. 30

   Reference _____

32. Glazing installed in sliding glass patio doors shall have a minimum category classification of Class _____.

   a. A

   b. B

   c. I

   d. II

   Reference _____

33. For all glazing types, the minimum required thickness for glass used as a handrail assembly or guard section shall be _____ inch.

   a. $^1/_4$ b. $^3/_8$

   c. $^1/_2$ d. $^9/_{16}$

   Reference _____

34. Individual panels or units of light-transmitting plastic used in a light-diffusing system shall have a maximum size of _____ square feet and be a maximum of _____ feet in length.

    a. 10, 5   b. 15, 10

    c. 30, 10   d. 36, 12

Reference _____

35. In a nonsprinklered building with no smoke and heat vents, light-transmitting plastic skylight assemblies shall each have a maximum glazed area of _____ square feet within the curb.

    a. 16   b. 36

    c. 64   d. 100

Reference _____

Answer Keys

Study Session 1
2006 *International Building Code*

1.	c	Sec. 101.2, Exc. 1	31.	b	Sec. 105.2, #M7
2.	c	Sec. 101.2.1	32.	a	Sec. 106.1.2
3.	b	Sec. 102.1	33.	b	Sec. 108.3
4.	a	Sec. 103.1	34.	c	Sec. 109.1
5.	d	Sec. 104.1	35.	c	Sec. 114.2
6.	a	Sec. 104.9.1			
7.	b	Sec. 104.10			
8.	b	Sec. 104.11			
9.	c	Sec. 104.11.2			
10.	b	Sec. 105.2, #B1			
11.	c	Sec. 105.2, #B13			
12.	b	Sec. 105.5			
13.	a	Sec. 105.7			
14.	d	Sec. 106.3.1			
15.	b	Sec. 106.5			
16.	c	Sec. 109.3			
17.	a	Sec. 109.5			
18.	d	Sec. 110.2			
19.	d	Sec. 110.3			
20.	b	Sec. 112.2			
21.	b	Sec. 3403.2			
22.	c	Sec. 3404.3			
23.	a	Table 3409.8.5			
24.	d	Sec. 3410.5			
25.	a	Chapter 35			
26.	b	Sec. 105.2, #B9			
27.	d	Sec. 105.2, #B12			
28.	d	Sec. 105.6			
29.	b	Sec. 107.1			
30.	c	Sec. 3410.2			

Study Session 2

2006 *International Building Code*

1.	c	Sec. 302.1, #6	31.	c	Sec. 303.1, Exc. 1
2.	d	Sec. 302.1, #10	32.	b	Sec. 304.1
3.	a	Sec. 508.2.1	33.	c	Sec. 310.1
4.	c	Table 508.2	34.	b	Sec. 310.2
5.	d	Table 508.2	35.	a	Table 508.3.3
6.	d	Table 508.2			
7.	d	Sec. 508.2.2.1			
8.	c	Sec. 508.3.1.3, Exc.			
9.	a	Sec. 508.3.2.1			
10.	c	Table 508.3.3			
11.	c	Table 508.3.3			
12.	c	Sec. 508.3.1, Exc. 1			
13.	a	Sec. 508.3.1, Exc. 2			
14.	d	Sec. 303.1			
15.	a	Sec. 309.1			
16.	b	Sec. 306.3			
17.	c	Sec. 307.2			
18.	b	Sec. 307.4			
19.	c	Sec. 308.3.1			
20.	d	Table 307.1(1)			
21.	a	Sec. 308.2			
22.	c	Sec. 309.1, 304.1			
23.	b	Sec. 310.1			
24.	d	Sec. 310.2. 308.3			
25.	d	Sec. 311.3, 312.1			
26.	a	Sec. 508.3.1			
27.	b	Table 508.3.3			
28.	a	Sec. 303.1			
29.	d	Sec. 307.6			
30.	c	Sec. 307.1, Exc. 13			

Study Session 3
2006 *International Building Code*

1.	a	Sec. 602.2	31.	c	Table 601	
2.	c	Sec. 602.3	32.	c	Tables 601, 602	
3.	b	Sec. 602.4	33.	a	Table 602, Note d	
4.	a	Sec. 602.3, 602.4	34.	d	Table 602.4	
5.	d	Sec. 602.4.1	35.	a	Sec. 603.1, #2	
6.	c	Sec. 602.4.2				
7.	c	Sec. 602.4.5				
8.	d	Sec. 602.5				
9.	b	Table 601				
10.	d	Table 601, Note a				
11.	a	Table 601				
12.	b	Table 601				
13.	b	Table 601, Note b				
14.	d	Sec. 603.1, #1.3				
15.	b	Table 601, Note c				
16.	c	Tables 601, 602				
17.	c	Tables 601, 602				
18.	b	Tables 601, 602				
19.	d	Tables 601, 602				
20.	b	Table 602				
21.	d	Sec. 603.1, #1.1				
22.	c	Table 602, Note b; Sec. 406.1.2, #2				
23.	a	Sec. 603.1, #1.2				
24.	a	Sec. 603.1, #6				
25.	d	Sec. 603.1, #4, Table 1505.1				
26.	a	Sec. 602.4.6				
27.	a	Table 601				
28.	a	Sec. 603.1, #1.3, Exc.				
29.	c	Tables 601 and 602				
30.	d	Table 601, Note d				

Study Session 4

2006 *International Building Code*

1.	b	Sec. 501.2	31.	d	Sec. 505.2, Exc. 1
2.	a	Sec. 502.1	32.	d	Sec. 505.5.1
3.	c	Table 503	33.	b	Sec. 507.11
4.	c	Table 503	34.	c	Sec. 507.9, #3
5.	b	Sec. 506.1.1	35.	a	Sec. 506.3, Exc. 2
6.	d	Sec. 502.1			
7.	b	Sec. 506.4, #2			
8.	b	Sec. 504.2			
9.	c	Sec. 504.2, Exc. 1			
10.	b	Sec. 504.3			
11.	b	Sec. 505.1			
12.	b	Sec. 505.1			
13.	c	Sec. 505.2			
14.	a	Sec. 505.4, Exc. 3			
15.	a	Sec. 505.4, Exc. 1			
16.	b	Sec. 506.2			
17.	b	Sec. 506.2			
18.	b	Table 503			
19.	c	Sec. 504.2, Exc. 2			
20.	c	Sec. 506.3			
21.	d	Sec. 506.3			
22.	c	Sec. 507.2			
23.	d	Sec. 507.5, #2			
24.	d	Sec. 507.8			
25.	b	Sec. 509.2			
26.	d	Sec. 504.3			
27.	b	Sec. 507.4			
28.	a	Sec. 507.6, #3			
29.	b	Sec. 507.7, Table 503			
30.	c	Sec. 509.6			

Study Session 5
2006 *International Building Code*

1.	a	Sec. 702.1	31.	b	Sec. 703.4.1
2.	b	Sec. 702.1	32.	b	Table 704.8, Note d
3.	b	Sec. 702.1	33.	d	Sec. 704.11, Exc. 3
4.	d	Sec. 702.1	34.	b	Sec. 704.2.1, 1406.3, Exc. 1
5.	a	Sec. 702.1	35.	c	Sec. 704.8.2
6.	a	Sec. 702.1, 705.2			
7.	b	Sec. 702.1			
8.	b	Sec. 702.1			
9.	c	Sec. 703.2			
10.	d	Sec. 703.4			
11.	d	Sec. 703.4.2			
12.	d	Sec. 704.2			
13.	b	Sec. 704.2, #2			
14.	d	Sec. 704.2.2			
15.	c	Sec. 704.3			
16.	b	Sec. 704.5			
17.	a	Table 704.8			
18.	c	Table 704.8			
19.	b	Sec. 704.8.1, Table 704.8			
20.	c	Sec. 704.8.2			
21.	c	Sec. 704.9			
22.	c	Sec. 704.9			
23.	b	Sec. 704.11, Exc. 2			
24.	a	Sec. 704.11.1			
25.	b	Sec. 704.11.1			
26.	d	Sec. 703.2.1			
27.	d	Sec. 704.2.3			
28.	c	Sec. 704.8.1, Table 704.8			
29.	c	Sec. 704.11, #6; Table 704.8			
30.	d	Table 704.8, Note e			

Study Session 6
2006 *International Building Code*

1.	d	Sec. 705.3, Exc.	31.	a	Sec. 705.5.2
2.	c	Table 705.4	32.	b	Sec. 704.14.1
3.	b	Table 705.4, Note a	33.	c	Sec. 707.14.2.1
4.	a	Sec. 705.5	34.	b	Sec. 707.13.3
5.	b	Sec. 705.6	35.	b	Sec. 712.3.2, Exc. 2
6.	b	Sec. 705.6.1, Exc.			
7.	c	Sec. 705.7			
8.	b	Sec. 705.8			
9.	d	Sec. 706.3			
10.	c	Sec. 706.5			
11.	d	Sec. 706.7			
12.	a	Sec. 706.7, Exc. 2			
13.	c	Sec. 707.2, Exc. 2.1			
14.	c	Sec. 707.2, Exc. 7			
15.	c	Sec. 707.4			
16.	c	Sec. 707.14.1			
17.	c	Sec. 708.1			
18.	c	Sec. 708.3			
19.	b	Sec. 708.4, Exc. 3			
20.	d	Sec. 709.3			
21.	b	Sec. 709.4			
22.	b	Sec. 709.5, Table 715.4			
23.	a	Sec. 711.3.1			
24.	d	Sec. 711.3.3			
25.	c	Sec. 711.4			
26.	b	Table 706.3.9			
27.	c	Sec. 707.13.4			
28.	a	Sec. 710.3			
29.	a	Sec. 710.5			
30.	b	Sec. 708.3, Exc. 2; 711.3, Exc.			

Study Session 7
2006 *International Building Code*

1.	d	Sec. 712.4.1.1.2	31.	b	Sec. 714.4
2.	b	Sec. 712.4.2.1	32.	d	Sec. 714.6
3.	b	Sec. 713.1	33.	c	Sec. 716.3.2
4.	d	Sec. 714.2.1	34.	a	Sec. 717.2.6
5.	b	Sec. 714.2.4	35.	d	Sec. 718.2
6.	c	Sec. 714.2.5			
7.	d	Table 715.4			
8.	d	Table 715.4			
9.	b	Sec. 712.5			
10.	a	Sec. 715.4.6.1, Exc. 1			
11.	b	Sec. 715.4.7, Exc. 1			
12.	c	Sec. 715.4.7.3			
13.	b	Sec. 715.4.7.3			
14.	b	Table 715.5.3			
15.	d	Table 715.5			
16.	c	Sec. 715.5.7.2			
17.	d	Table 716.3.1			
18.	d	Sec. 716.3.2.1, #3			
19.	a	Sec. 716.5.1			
20.	c	Sec. 716.5.3			
21.	d	Sec. 717.2.1			
22.	b	Sec. 717.2.6			
23.	a	Sec. 717.3.1			
24.	b	Sec. 717.4.3			
25.	a	Sec. 719.2			
26.	a	Sec. 712.3.1.2			
27.	a	Sec. 712.3.2, Sec. 712.3.2, Exc. 1			
28.	b	Sec. 715.5, Exc. 2			
29.	c	Sec. 717.2.7			
30.	d	Sec. 718.4			

Study Session 8

2006 *International Building Code*

1.	b	Sec. 901.6.1, Exc. 2	31.	d	Sec. 903.3.1.2.1
2.	b	Sec. 902.1	32.	d	Sec. 907.2.4
3.	c	Sec. 902.1	33.	b	Sec. 909.20.1
4.	c	Sec. 903.2.1.2, #2	34.	c	Sec. 910.3.2.1
5.	b	Sec. 903.2.1.5	35.	a	Sec. 911.1
6.	d	Sec. 903.2.2, #1			
7.	b	Sec. 903.2.3.1			
8.	d	Table 903.2.4.2			
9.	c	Sec. 903.2.6, #2			
10.	d	Sec. 903.2.7			
11.	d	Sec. 903.2.8, #3			
12.	c	Sec. 903.2.8.1, #2			
13.	b	Sec. 903.2.9.1			
14.	b	Sec. 903.2.10.1.3			
15.	b	Sec. 903.3.1.3			
16.	d	Sec. 903.3.2			
17.	c	Sec. 903.3.3			
18.	b	Sec. 903.3.5.2			
19.	d	Sec. 904.11.1			
20.	b	Sec. 905.5			
21.	a	Sec. 907.2.2			
22.	a	Sec. 907.2.3, Exc. 1			
23.	a	Sec. 907.3.1			
24.	c	Table 907.9.1.3			
25.	a	Sec. 907.9.2			
26.	d	Sec. 904.11.1			
27.	d	Sec. 905.3.4			
28.	b	Sec. 909.20.3.3			
29.	d	Sec. 910.2.1			
30.	d	Table 910.3			

Study Session 9
2006 *International Building Code*

1.	b	Sec. 1002.1	31.	d	Sec. 1004.7
2.	d	Sec. 1002.1	32.	a	Sec. 1005.2
3.	c	Sec. 1002.1	33.	b	Sec. 1006.2
4.	b	Sec. 1002.1	34.	d	Sec. 1013.3
5.	c	Sec. 1002.1	35.	b	Sec. 1013.3, Exc. 5.
6.	b	Table 1004.1.1			
7.	b	Table 1004.1.1			
8.	b	Sec. 1004.7			
9.	c	Table 1005.1			
10.	c	Table 1005.1			
11.	d	Sec. 1005.1			
12.	d	Sec. 1005.2			
13.	b	Sec. 1003.3.1			
14.	b	Sec. 1003.3.1, Exc.			
15.	c	Sec. 1003.3.3			
16.	c	Sec. 1003.5, Exc. 1			
17.	c	Sec. 1011.1			
18.	b	Sec. 1011.5.2			
19.	b	Sec. 1006.4			
20.	b	Sec. 1013.1			
21.	c	Sec. 1013.2			
22.	d	Sec. 1013.3, Exc. 3			
23.	a	Sec. 1013.5			
24.	c	Sec. 1007.3			
25.	c	Sec. 1007.6.1			
26.	a	Sec. 1003.3.1			
27.	d	Sec. 1004.2			
28.	d	Sec. 1006.1			
29.	b	Sec. 1007.8			
30.	c	Sec. 1011.3			

Study Session 10
2006 *International Building Code*

1.	b	Sec. 1008.1.1	31.	c	Sec. 1008.1.9
2.	c	Sec. 1008.1.1	32.	c	Sec. 1009.1
3.	a	Sec. 1008.1.1, Exc. 3	33.	b	Sec. 1010.6.3, Exc. 2
4.	d	Sec. 1008.1.1.1	34.	c	Sec. 1012.5
5.	d	Sec. 1008.1.2, Exc. 1	35.	a	Sec. 1012.8
6.	c	Sec. 1008.1.2			
7.	b	Sec. 1008.1.2			
8.	b	Sec. 1008.1.3.1, #4			
9.	c	Sec. 1008.1.3.4, #3			
10.	b	Sec. 1008.1.4			
11.	c	Sec. 1008.1.6			
12.	a	Sec. 1008.1.8.6			
13.	c	Sec. 1008.1.8.2			
14.	c	Sec. 1008.1.9			
15.	c	Sec. 1009.2			
16.	b	Sec. 1009.3, Exc. 4			
17.	c	Sec. 1009.4			
18.	d	Sec. 1009.3			
19.	b	Sec. 1009.8			
20.	d	Sec. 1012.2			
21.	a	Sec. 1012.5			
22.	d	Sec. 1012.7			
23.	a	Sec. 1010.4			
24.	a	Sec. 1010.8			
25.	d	Sec. 1008.3.1			
26.	d	Sec. 1008.1.3.2			
27.	c	Sec. 1008.1.7			
28.	b	Sec. 1009.1, Exc. 1			
29.	a	Sec. 1012.3			
30.	a	Sec. 1010.9.1			

Study Session 11
2006 *International Building Code*

1.	d	Table 1015.1	31.	c	Sec. 1014.2, #2, Exc. 2.4
2.	d	Table 1015.1	32.	a	Sec. 1014.4.2
3.	c	Sec. 1015.1.1, 1019.1	33.	c	Sec. 1015.3
4.	b	Sec. 1015.2.1, Exc. 2	34.	c	Sec. 1015.6.1
5.	c	Sec. 1014.2, #1	35.	d	Sec. 1017.2, Exc. 6
6.	a	Sec. 1014.2, #2			
7.	c	Sec. 1014.2.2, Exc. 2			
8.	b	Sec. 1014.2.2			
9.	c	Table 1016.1			
10.	b	Table 1016.1			
11.	c	Sec. 1016.3			
12.	a	Sec. 1014.3			
13.	c	Sec. 1014.3, Exc. 2			
14.	b	Sec. 1014.4.1			
15.	b	Sec. 1014.4.1, Exc.			
16.	c	Sec. 1014.4.3.3			
17.	d	Sec. 1017.2, Exc. 4			
18.	a	Sec. 1017.2, Exc. 2			
19.	c	Table 1017.1			
20.	d	Table 1017.1			
21.	b	Sec. 1017.3			
22.	b	Sec. 1017.3, Exc. 2			
23.	b	Sec. 1017.4, Exc. 3			
24.	c	Sec. 1014.5, 1017.2			
25.	c	Sec. 1014.5.2			
26.	b	Sec. 1014.3			
27.	d	Sec. 1015.4			
28.	a	Sec. 1015.6.1, Exc. 4			
29.	d	Sec. 1016.2			
30.	d	Sec. 1017.3, Exc. 3			

Study Session 12

2006 *International Building Code*

1.	b	Table 1019.1	31.	b	Sec. 1022.2
2.	d	Sec. 1019.1.1	32.	a	Sec. 1025.10
3.	b	Table 1019.2	33.	c	Sec. 1025.13.1
4.	b	Sec. 1020.1	34.	b	Sec. 1026.1, Exc. 7
5.	a	Sec. 1020.1, Exc. 1	35.	a	Sec. 1026.2.1
6.	c	Sec. 1020.1.7.1			
7.	c	Sec. 1020.1.6			
8.	c	Sec. 1020.1.7			
9.	b	Sec. 1021.3			
10.	a	Sec. 1022.4			
11.	d	Sec. 1023.2			
12.	d	Sec. 1023.3			
13.	c	Sec. 1024.3			
14.	b	Sec. 1024.5.1			
15.	d	Sec. 1024.5.2			
16.	b	Sec. 1025.2			
17.	c	Sec. 1025.5			
18.	d	Sec. 1025.6.1, #1, #2			
19.	d	Table 1025.6.2			
20.	a	Sec. 1025.6.2.1			
21.	d	Sec. 1025.7, Exc. 2			
22.	b	Sec. 1025.9.1, #2			
23.	b	Sec. 1025.11.1, Exc.			
24.	c	Sec. 1025.14.3			
25.	c	Sec. 1026.2, Exc.			
26.	d	Sec. 1022.1, Exc. 1			
27.	c	Sec. 1024.1, Exc. 2			
28.	b	Sec. 1024.6, Exc.			
29.	c	Sec. 1026.3			
30.	c	Sec. 1026.5.1			

Study Session 13
2006 *International Building Code*

1.	a	Sec. 1102.1	31.	a	Sec. 1106.7.1
2.	a	Sec. 1102.1	32.	c	Sec. 1108.2.3, Exc. 1
3.	d	Sec. 1103.2.5, #1	33.	d	Sec. 1109.2, Exc. 3
4.	a	Sec. 1103.2.11	34.	c	Sec. 1109.5.2
5.	b	Sec. 1104.5	35.	b	Sec. 1109.14.4.1
6.	c	Sec. 1105.1			
7.	b	Table 1106.1			
8.	b	Sec. 1106.2			
9.	c	Sec. 1106.5			
10.	c	Table 1108.2.2.1			
11.	a	Sec. 1108.4.3.1			
12.	b	Sec. 1109.2.2			
13.	a	Sec. 1107.6.2.2.2			
14.	c	Table 1108.2.6.1			
15.	c	Sec. 1107.5.2.1			
16.	d	Table 1107.6.1.1			
17.	a	Table 1107.6.1.1			
18.	a	Sec. 1107.6.2.1.1			
19.	d	Table 1108.3			
20.	b	Sec. 1109.2.1			
21.	c	Sec. 1109.2.1.4			
22.	d	Sec. 1109.5.2			
23.	a	Sec. 1109.11			
24.	b	Table 1109.12.2			
25.	c	Sec. 1110.1, #1			
26.	a.	Sec. 1104.3.1, Exc. 1			
27.	a	Sec. 1104.3.2, Exc. 1			
28.	d	Sec. 1006.4			
29.	b	Sec. 1108.2.8.1			
30.	c	Sec. 1110.2			

Study Session 14
2006 *International Building Code*

1.	b	Sec. 402.4.4		31.	b	Sec. 406.1.2, #2
2.	c	Sec. 402.10, #3		32.	b	Sec. 410.2
3.	b	Sec. 403.1		33.	b	Sec. 412.2.1
4.	d	Sec. 403.10.1		34.	a	Sec. 419.2, 708
5.	b	Sec. 404.7		35.	d	Sec. 420.6.1
6.	d	Sec. 404.8				
7.	b	Sec. 405.4.1				
8.	a	Sec. 406.2.4				
9.	b	Sec. 406.5.2				
10.	b	Table 406.3.5				
11.	d	Sec. 407.4				
12.	d	Sec. 407.4.1				
13.	c	Sec. 408.3.1				
14.	d	Sec. 408.6.1				
15.	b	Sec. 409.2				
16.	b	Sec. 410.3.4, 410.3.5				
17.	d	Sec. 410.3.7				
18.	a	Sec. 411.8				
19.	b	Sec. 412.4.1				
20.	c	Table 414.2.2				
21.	c	Sec. 415.3, Exc. 1				
22.	d	Sec. 415.3.1, #3				
23.	a	Sec. 415.8.4.4				
24.	b	Sec. 417.2				
25.	b	Sec. 418.5				
26.	c	Sec. 410.5.1				
27.	d	Sec. 413.2				
28.	c	Sec. 414.2.4				
29.	b	Sec. 414.6.1.3				
30.	c	Sec. 416.2				

Study Session 15
2006 *International Building Code*

1.	b	Sec. 1402.1	31.	b	Sec. 1405.12.2
2.	a	Sec. 1402.1	32.	c	Sec. 1405.17.2
3.	a	Sec. 1404.2	33.	a	Sec. 1406.3, Exc. 1
4.	b	Table 1405.2	34.	d	Table 1504.8
5.	a	Sec. 1405.9.1	35.	d	Sec. 1810.8.2
6.	a	Sec. 1405.10			
7.	d	Sec. 1406.2.2			
8.	b	Sec. 1502.1			
9.	b	Table 1505.1			
10.	b	Sec. 1505.3, Exc.			
11.	c	Sec. 1507.2.8			
12.	d	Sec. 1507.2.8.1			
13.	c	Sec. 1507.3.6			
14.	b	Sec. 1507.5.2			
15.	d	Sec. 1507.9.7			
16.	d	Table 1507.8.6			
17.	b	Sec. 1507.14.1			
18.	c	Sec. 1509.2			
19.	c	Sec. 1805.1			
20.	b	Sec. 1805.2			
21.	d	Table 1804.2			
22.	c	Sec. 1805.4.2.1			
23.	d	Sec. 1805.4.2.5			
24.	b	Table 1805.4.2			
25.	b	Sec. 1805.5.4			
26.	d	Sec. 1405.13			
27.	c	Sec. 1505.7			
28.	d	Sec. 1507.7.2			
29.	d	Sec. 1509.5			
30.	b	Sec. 1803.3			

Study Session 16
2006 *International Building Code*

1.	a	Sec. 1603.3		29.	b	Sec. 2304.12, Exc.
2.	b	Table 1607.1, #4, #35		30.	d	Sec. 2308.9.4
3.	d	Sec. 1607.7.1		31.	c	Sec. 1607.5
4.	d	Sec. 1607.11.2.2, Table 1607.1, #30		32.	c	Sec. 1607.7.2
				33.	b	Sec. 1609.2
5.	c	Sec. 1609.4.2, 1609.4.3		34.	d	Table 1704.3, #5
				35.	c	Sec. 1705.4, #2
6.	c	Sec. 1704.7, Exc.				
7.	b	Sec. 1704.10.3.2				
8.	b	Sec. 1709.2, #3				
9.	d	Sec. 1905.6.2				
10.	a	Sec. 1908.1.3 ACI 318, Sec. 21.1				
11.	b	Sec. 1909.6.1				
12.	a	Sec. 1910.1				
13.	b	Sec. 2102.1				
14.	d	Sec. 2104.1.3				
15.	a	Sec. 2104.1.8				
16.	c	Sec. 2111.10				
17.	d	Sec. 2111.11, Exc. 4				
18.	c	Sec. 2113.9				
19.	a	Sec. 2304.11.2.5				
20.	d	Sec. 2302.1				
21.	b	Table 2304.9.1, #27				
22.	c	Sec. 2308.3.1				
23.	a	Sec. 2308.8.3				
24.	a	Sec. 2308.9.11				
25.	b	Sec. 2308.10.5				
26.	d	Table 1607.1, #36, Note f				
27.	c	Table 1904.2.2				
28.	a	Table 2304.6				

Study Session 17
2006 *International Building Code*

1.	b	Sec. 803.1		31.	c	Sec. 806.1
2.	c	Sec. 803.4.4		32.	a	Sec. 1203.2
3.	c	Table 803.5		33.	b	Sec. 1210.2
4.	a	Table 803.5		34.	a	Sec. 2508.5.4
5.	d	Table 803.5, Note i		35.	c	Sec. 2512.4
6.	a	Sec. 803.6.1				
7.	b	Sec. 804.4.1, Exc.				
8.	b	Sec. 806.5				
9.	d	Sec. 806.1.2, Exc.				
10.	c	Sec. 806.5				
11.	b	Sec. 1204.1				
12.	c	Sec. 1205.2				
13.	a	Sec. 1205.4				
14.	b	Sec. 1207.3				
15.	c	Sec. 1208.2				
16.	b	Sec. 1208.3				
17.	a	Sec. 1209.1				
18.	c	Sec. 1209.2				
19.	c	Sec. 1210.3				
20.	d	Sec. 2502.1				
21.	b	Sec. 2509.3, #3				
22.	c	Sec. 2512.5				
23.	d	Sec. 3002.2				
24.	d	Sec. 3004.2				
25.	b	Sec. 3002.4				
26.	a	Sec. 804.4.1				
27.	c	Sec. 1203.3.1				
28.	c	Sec. 1206.3				
29.	a	Sec. 2508.4				
30.	d	Sec. 3003.1.2				

Study Session 18

2006 *International Building Code*

1.	a	Sec. 2403.5	31.	d	Sec. 2405.3, Exc. 3
2.	c	Sec. 2405.3	32.	d	Table 2406.1
3.	b	Sec. 2405.3	33.	a	Sec. 2407.1
4.	d	Sec. 2405.4	34.	c	Sec. 2606.7.3
5.	d	Table 2406.1	35.	d	Sec. 2610.4
6.	a	Sec. 2406.2, Exc. 1			
7.	a	Sec. 2406.3, #5			
8.	c	Sec. 2406.3, #6			
9.	d	Sec. 2406.3, #7, Exc.1			
10.	a	Sec. 2406.3, #9.2			
11.	d	Sec. 2406.3, #10			
12.	a	Sec. 2406.3.1, #1			
13.	b	Sec. 2406.4			
14.	b	Sec. 2407.1.2			
15.	d	Sec. 2602.1			
16.	d	Sec. 2602.1			
17.	d	Sec. 2603.3			
18.	a	Sec. 2603.4			
19.	b	Sec. 2603.4.1.1			
20.	d	Sec. 2604.2.2			
21.	c	Sec. 2605.2, #2			
22.	b	Table 2607.4			
23.	d	Sec. 2608.2, #3, Exc.			
24.	a	Sec. 2609.1			
25.	b	Sec. 2611.3			
26.	a	Sec. 2404.1			
27.	b	Sec. 2406.2, Exc. 2			
28.	c	Sec. 2406.2.1			
29.	c	Sec. 2408.3			
30.	d	Sec. 2603.4.1.13			